W0105721

New Vistas in Drug Research

Vol. 1

P. Dostert, P. Riederer,
M. Strolin Benedetti, and R. Roncucci (eds.)

Early Markers in Parkinson's and Alzheimer's Diseases

Springer-Verlag Wien New York

Dr. Philippe Dostert
Farmitalia Carlo Erba, Milan, Italy

Prof. Dr. Peter Riederer
Department of Psychiatry, University of Würzburg, Federal Republic of Germany

Dr. Margherita Strolin Benedetti
Farmitalia Carlo Erba, Milan, Italy

Dr. Romeo Roncucci
Farmitalia Carlo Erba, Milan, Italy

With 38 Figures (1 in color)

Library of Congress Cataloging-in-Publication Data. Early markers in Parkinson's and Alzheimer's diseases/P. Dostert (et al.) (eds.)   p.,   cm. (New vistas in drug research; vol. 1.) Based on a conference which was organized by Erbamont and Carlo Erba Foundation. Includes index. ISBN-13:978-3-7091-9100-2 1. Parkinsonism—Diagnosis—Congresses. 2. Alzheimer's disease—Diagnosis—Congresses. 3. Biochemical markers—Congresses. I. Dostert, P. II. Erbamont Inc. III. Fondazione Carlo Erba. IV. Series. [DNLM: 1. Alzheimer's Disease—diagnosis—congresses. 2. Alzheimer's Disease—physiopathology—congresses. 3. Biological Markers—congresses. 4. Parkinson's Disease—diagnosis—congresses. 5. Parkinson's Disease—physiopathology—congresses. WL 359 E12] RC382.E38 1990. 616.8'31075—dc20. 90-10419

ISSN 0938-9245
ISBN-13:978-3-7091-9100-2   e-ISBN-13:978-3-7091-9098-2
DOI: 10.1007/978-3-7091-9098-2

Welcome address

I am really honored to open this first meeting of "New Vistas in Drug Research—Dialogues for the future" and to heartily welcome all the scientists, coming from 14 countries, who have accepted to offer here their contribution to exchange different experiences and to create new grounds for further scientific advances.

The fact that Erbamont, along with the Carlo Erba Foundation, has organized this meeting is only a detail. We believe, as all the other pharmaceutical companies based on a strong research and development, that we need to be more and more open to the scientific world, whose frontiers are getting broader day by day at an increasing speed. We have to run fast in this world and moments like these can give us the intellectual energy to run even faster.

I am sure you will have a very productive and stimulating meeting. Thank you again for coming.

P. Morrione
Erbamont, President and Chief Executive Officer

New Vistas today and tomorrow

Dr. Montalcini, Dr. Guillemin, Dear Colleagues, the title of this series of meetings, "New Vistas in Drug Research" reflects the magic moment the pharmaceutical sciences are presently living, due to the tremendous progress of basic research, and of molecular biology in particular. Thanks to this progress, we are now sitting on the top of a hill from which we can look at the huge horizon with the various disciplines growing fast and cross-reacting with each other. Basic research is no longer isolated in an ivory tower but it's becoming the foundation on top of which new intelligent drugs can be built up.

Based on the above, aim of these meetings is to stimulate dialog and constructive interaction between scientists with different background and of different affiliations. We are confident of the results of this cross-fertilization which could lead to new approaches for the topics that will be discussed in our New Vistas meetings. Today we are inaugurating the first of these meetings, whose title is "Early Markers in Parkinson's and Alzheimer's Diseases". Already nowadays, but even more in the future, neurodegenerative disorders will constitute a major health problem, especially in developed countries. Despite advances in research, the management of these diseases is still hampered by the lack of a full understanding of their etiology and physiopathology. It is, however, our belief that with the increased knowledge of the real mechanisms involved in these pathological conditions, it seems now possible, within a reasonable period of time, to find more dedicated drug treatments for these terrible diseases. Parallely with this goal, it appears now possible to identify also reliable biological markers which could predict the onset of the disease at very early stages, possibly before lesions appear, giving doctors efficacious tools to check and monitor the presence of the disease or its evolution in association with new treatments.

I sincerely hope that the discussions which will take place during this two-day meeting will contribute to the progress in this direction. Before leaving the floor to Dr. Montalcini, I wish to express my sincere thanks to all of you for having accepted to participate in this first New Vistas meeting. Thank you.

R. Roncucci
Erbamont, Research and Development, Vice President

Acknowledgments

The Editors are grateful to Patrizia Di Rosa for her dedicated assistance in the preparation of this volume.

The Organizing Committee is deeply indebted to Patrizia Di Rosa, Giovanna Garattini, Nadia La Spada, Luisa Stea and Rosanna Romaniello for their skilful contribution towards the success of the meeting.

Contents

Contributors XIII
List of abbreviations XIX
Levi Montalcini, R.: Introduction 1

Early markers in Parkinson's disease

Przuntek, H.: Early markers in Parkinson's disease 5
Agnoli, A., Stocchi, F., Carta, A.: Differential diagnosis of Parkinson's disease 13
Poewe, W.: The premorbid personality of patients with Parkinson's disease 23
Carvey, P. M., Klawans, H. L., Kao, L. C., Dahlström, A., McRae, A.: An antibody in the CSF of Parkinson's disease patients: summary of data and potential role as a diagnostic marker 31
Kraus, P. H., Klotz, P., Steinberg, R., Przuntek, H.: Contribution of motor performance tests to the early diagnosis of Parkinson's disease 41
Lücking, C. H., Hufschmidt, A., Deuschl, G.: Electrophysiological methods in the early diagnosis of Parkinson's disease 49
Leenders, K. L.: Characterization of Parkinson's disease using positron emission tomography 59
Nagatsu, T., Mogi, M., Harada, M., Kojima, K.: Dopamine beta-hydroxylase and beta 2-microglobulin in cerebrospinal fluid: early markers in Parkinson's disease? 71
Kienzl, E., Eichinger, K., Jellinger, K., Kuhn, W., Fuchs, G., Danielczyk, W., Wesemann, W., Riederer, P.: Urinary dopamine sulfate conjugates in Parkinson's disease 79
Dostert, P., Strolin Benedetti, M., Dordain, G.: Salsolinol and the early detection of Parkinson's disease 93
Olanow, C. W., Alberts, M., Djang, W., Stajich, J.: MR imaging of putamenal iron predicts response to dopaminergic therapy in parkinsonian patients 99
Youdim, M. B. H., Ben-Shachar, D.: The neurotoxic component in Parkinson's disease may involve iron-melanin interaction and lipid peroxidation in the substantia nigra 111
Round table on Parkinson's disease 123

Early markers in Alzheimer's disease

Murphy, D. L., Sunderland, T.: The need for early markers in Alzheimer's disease 137
Amaducci, L., Lippi, A.: Descriptive and analytic epidemiology of Alzheimer's disease 147

XII   Contents

Gottfries, C. G.: Differential diagnosis of early Alzheimer's disease 155

Delabar, J. M., Blouin, J. L., Rahmani, Z., Créau-Goldberg, N., Chettouh,
Z., Nicole, A., Bruel, A., de Blois, M. C., Sinet, P. M.: Down syndrome:
a model for the study of Alzheimer's disease and aging 165

Heiss, W.-D., Szelies, B., Adams, R., Kessler, J., Pawlik, G., Herholz, K.:
PET scanning for the detection of Alzheimer's disease 181

Dierks, T., Maurer, K.: Reference-free evaluation of auditory evoked potentials—P300 in aging and dementia 197

McBean, G. J., Horner, E. B., Couée, I., Phillips, J. P., O'Brien, M., Lee,
T. C., Tipton, K. F.: Enzymes and glial cells in brain damage and neurodegenerative diseases 209

Riederer, P., Sofic, E., Moll, G., Freyberger, A., Wichart, I., Gsell, W.,
Jellinger, K., Hebenstreit, G., Youdim, M. B. H.: Senile dementia of
Alzheimer's type and Parkinson's disease: neurochemical overlaps and specific
differences 221

Hoyer, S.: Changes in brain energy metabolism and the early detection of
Alzheimer's disease 233

Nordberg, A.: Choline metabolism in Alzheimer's disease: hints as to possible
markers 245

McRae, A., Blennow, K., Wallin, A., Fredman, P., Gottfries, C. G., Dahlström, A.: The presence of antibrain antibodies in the CSF of some Alzheimer
disease patients: correlation with CSF parameters 257

Oreland, L., Hiraga, Y., Jossan, S. S., Regland, B., Gottfries, C. G.: Increased
monoamine oxidase activity and vitamin B-12 deficiency in dementia disorders 267

Round table on Alzheimer's disease 287

Guillemin, R.: Closing remarks 299

Subject index 303

Contributors *

Adams, R. (181) Universitätsklinik für Neurologie, Joseph-Stelzmann-Strasse 9, D-W-5000 Köln 41 (Lindenthal), Federal Republic of Germany.

Agnoli, A. (13) I Clinic of Neurology, Department of Neurosciences, University "La Sapienza", Viale dell'Università 30, I-00185 Rome, Italy.

Alberts, M. (99) Duke University, Department of Medicine, Service of Neurology, Durham, NC, U.S.A.

Amaducci, L. (147, 287) Istituto delle Malattie Nervose e Mentali dell'Università di Firenze, Viale Morgagni 85, I-50134 Florence, Italy.

Ben-Shachar, D. (111) Rappaport Research Institute, Department of Pharmacology, Faculty of Medicine, Technion, Haifa, Israel.

Blennow, K. (257) Department of Psychiatry and Neurochemistry, St. Jörgens Hospital, University of Göteborg, S-422 03 Hisings Backa, Sweden.

de Blois, M. C. (165) Hôpital Necker, Service de Cytogénétique, 149 rue de Sèvres, F-75743 Paris, France.

Blouin, J. L. (165) Hôpital Necker, Laboratoire de Biochimie Génétique, 149 rue de Sèvres, F-75743 Paris, France.

Bruel, A. (165) Hôpital Necker, Laboratoire de Biochimie Génétique, 149 rue de Sèvres, F-75743 Paris, France.

Brun, A. (287) Department of Pathology, Division of Neuropathology, University Hospital, S-221 85 Lund, Sweden.

Calne, D. B. (123) U.B.C. Health Sciences Centre Hospital, Faculty of Medicine, 2211 Wesbrook Mall, Vancouver B.C., Canada V6T 1W5.

Caraceni, T. (123) Istituto Neurologico C. Besta, Via Celoria 11, I-20133 Milan, Italy.

Carta, A. (13) I Clinic of Neurology, Department of Neurosciences, University "La Sapienza", Viale dell'Università 30, I-00185 Rome, Italy.

Carvey, P. M. (31) Department of Neurological Sciences, Rush-Presbyterian St. Lukes Medical Center, 1753 West Harrison St., Chicago, IL 60616, U.S.A.

Chettouh, Z. (165) Hôpital Necker, Laboratoire de Biochimie Génétique, 149 rue de Sèvres, F-75743 Paris, France.

Couée, I. (209) Department of Biochemistry, Trinity College, Dublin 2, Ireland.

Créau-Goldberg, N. (165) Hôpital Necker, INSERM U173, 149 rue de Sèvres, F-75743 Paris, France.

Dahlström, A. (31, 257) University of Göteborg, Institute of Neurobiology, P.O.B. 33031, S-40033 Göteborg, Sweden.

Danielczyk, W. (79) LBI für Altersforschung, Neurologische Abteilung des Pflegeheimes Lainz, Versorgungsheimplatz 1, A-1130 Wien, Austria.

---

* Numbers in parentheses indicate the pages on which the authors' contributions begin.

Da Prada, M. (287) Hoffmann La Roche, Pharmaceutical Research Department, CH-4002 Basel, Switzerland.

Delabar, J. M. (165) Hôpital Necker, Laboratoire de Biochimie Génétique, 149 rue de Sèvres, F-75743 Paris, France.

Deuschl, G. (49) Neurologische Klinik und Poliklinik der Universität, Hansastrasse 9, D-W-7800 Freiburg i. Br., Federal Republic of Germany.

Dierks, T. (197) Department of Psychiatry, University of Würzburg, Füchsleinstrasse 15, D-W-8700 Würzburg, Federal Republic of Germany.

Djang, W. (99) Duke University, Department of Medicine, Service of Neurology, Durham, NC, U.S.A.

Dordain, G. (93) Hôpital Nord, Service de Neurologie, BP 145, F-63020 Clermont Ferrand, France.

Dostert, P. (93) Farmitalia Carlo Erba, R & D, Via Imbonati 24, I-20159 Milan, Italy.

Eichinger, K. (79) Division of Chromatography and Spectroscopy, Institute of Organic Chemistry, A-1040 Vienna, Austria.

Fredman, P. (257) Department of Psychiatry and Neurochemistry, St. Jörgens Hospital, University of Göteborg, S-422 03 Hisings Backa, Sweden.

Freyberger, A. (221) Clinical Neurochemistry, Department of Psychiatry, University of Würzburg, Füchsleinstrasse 15, D-W-8700 Würzburg, Federal Republic of Germany.

Fuchs, G. (79) Klinik Dr. med. E. Wohlauf, Postfach 12 80, D-W-7620 Wohlfach, Federal Republic of Germany.

Garattini, S. (123) Mario Negri Institute, Via Eritrea, 62, I-20157 Milan, Italy.

Gonce, M. (123) Service Universitaire de Neurologie, Hôpital de la Citadelle, B-4000 Liège, Belgium.

Gottfries, C. G. (155, 257, 267, 287) Department of Psychiatry and Neurochemistry, St. Jörgens Hospital, University of Göteborg, S-422 03 Hisings Backa, Sweden.

Gsell, W. (221) Clinical Neurochemistry, Department of Psychiatry, University of Würzburg, Füchsleinstrasse 15, D-W-8700 Würzburg, Federal Republic of Germany.

Guillemin, R. (299) The Whittier Institute for Diabetes and Endocrinology, 9894 Genesee Avenue, La Jolla, CA 92037, U.S.A.

Harada, M. (71) Department of Oral Biochemistry, Matsumoto Dental College, Shiojiri, Japan.

Hebenstreit, G. (221) Department of Psychiatry, Landeskrankenhaus, A-3300 Amstetten/Mauer, Austria.

Heiss, W. D. (181) Universitätsklinik für Neurologie, Joseph-Stelzmann-Strasse 9, D-W-5000 Köln 41 (Lindenthal), Federal Republic of Germany.

Herholz, K. (181) Universitätsklinik für Neurologie, Joseph-Stelzmann-Strasse 9, D-W-5000 Köln 41 (Lindenthal), Federal Republic of Germany.

Hiraga, Y. (267) Department of Medical Pharmacology, University of Uppsala, S-751 24 Uppsala, Sweden.

Horner, E. B. (209) Department of Biochemistry, Trinity College, Dublin 2, Ireland.

Hoyer, S. (233) Department of Pathochemistry and General Neurochemistry, University of Heidelberg, Im Neuenheimer Feld 220-221, D-W-6900 Heidelberg 1, Federal Republic of Germany.

Hufschmidt, A. (49) Neurologische Klinik und Poliklinik der Universität, Hansastrasse 9, D-W-7800 Freiburg i. Br., Federal Republic of Germany.

Jellinger, K. (79, 221) Ludwig-Boltzmann-Institute of Clinical Neurobiology, Lainz-Hospital, Wolkersbergenstrasse 1, A-1130 Vienna, Austria.

Jossan, S. S. (267) Department of Medical Pharmacology, University of Uppsala, S-751 24 Uppsala, Sweden.

Kao, L. C. (31) Department of Neurological Sciences, Rush-Presbyterian St. Lukes Medical Center, 1753 West Harrison St., Chicago, IL 60616, U.S.A.

Kay, D. W. K. (287) MRC Neurochemical Pathology Unit, Newcastle General Hospital, Westgate Road, Newcastle upon Tyne NE4 68E, United Kingdom.

Kessler, J. (181) Universitätsklinik für Neurologie, Joseph-Stelzmann-Strasse 9, D-W-5000 Köln 41 (Lindenthal), Federal Republic of Germany.

Kienzl, E. (79) Ludwig-Boltzmann-Institute of Clinical Neurobiology, Lainz-Hospital, Wolkersbergenstrasse 1, A-1130 Vienna, Austria.

Klawans, H. L. (31) Department of Neurological Sciences, Rush-Presbyterian St. Lukes Medical Center, 1753 West Harrison St., Chicago, IL 60616, U.S.A.

Klotz, P. (41) Neurologische Universitätsklinik der Ruhr-Universität Bochum, Gudrunstrasse 56, D-W-4630 Bochum, Federal Republic of Germany.

Kojima, K. (71) Hatano Research Institute, Food and Drug Safety Center, Hatano, Japan.

Kraus, P. H. (41) Department of Neurology, St. Josef Hospital, Gudrunstrasse 56, D-W-4630 Bochum, Federal Republic of Germany.

Kuhn, W. (79) Neurologische Universitätsklinik, Josef-Schneider-Strasse, D-W-8700 Würzburg, Federal Republic of Germany.

Lee, T. C. (209) Department of Anatomy, Trinity College, Dublin 2, Ireland.

Leenders, K. L. (59) PET group, Paul Scherrer Institute, CH-5232 Villigen, Switzerland.

Levi Montalcini, R. (1) National Research Council (CNR), Institute of Neurobiology, Viale Marx 15, I-00156 Rome, Italy.

Lippi, A. (147) S.M.I.D. Center, Via Il Prato, 59/62, I-50123 Florence, Italy.

Lücking, C. H. (49) Neurologische Klinik und Poliklinik der Universität, Hansastrasse 9, D-W-7800 Freiburg i. Br., Federal Republic of Germany.

Maurer, K. (197) Department of Psychiatry, University of Würzburg, Füchsleinstrasse 15, D-W-8700 Würzburg, Federal Republic of Germany.

McBean, G. J. (209) Department of Biochemistry, Trinity College, Dublin 2, Ireland.

McRae, A. (31, 257) University of Göteborg, Institute of Neurobiology, P.O. 33031, S-40033 Göteborg, Sweden.

Mogi, M. (71) Department of Oral Biochemistry, Matsumoto Dental College, Shiojiri, Japan.

Moll, G. (221) Clinical Neurochemistry, Department of Psychiatry, University of Würzburg, Füchsleinstrasse 15, D-W-8700 Würzburg, Federal Republic of Germany.

Murphy, D. L. (137, 287) Laboratory of Clinical Sciences, National Institute of Mental Health, NIH Clinical Center, 9000 Rockville Pike, Bethesda, MD 20892, U.S.A.

Nagatsu, T. (71) Department of Biochemistry, Nagoya University School of Medicine, Nagoya 466, Japan.

Nicole, A. (165) Hôpital Necker, Laboratoire de Biochimie Génétique, 149 rue de Sèvres, F-75743 Paris, France.

Nordberg, A. (245) Biomedical Center, Department of Pharmacology, Husargatan 3, S-751 24 Uppsala, Sweden.

O'Brien, M. (209) Department of Anatomy, Trinity College, Dublin 2, Ireland.

Olanow, C. W. (99) University of South Florida, Department of Neurology, 4 Columbia Drive #410, Tampa, FL 33606, U.S.A.

Oreland, L. (267) Department of Medical Pharmacology, University of Uppsala, S-751 24 Uppsala, Sweden.

Pawlik, G. (181) Universitätsklinik für Neurologie, Joseph-Stelzmann-Strasse 9, D-W-5000 Köln 41 (Lindenthal), Federal Republic of Germany.

Phillips, J. P. (209) Department of Neurosurgery, Beaumont Hospital, Dublin, Ireland.

Poewe, W. (23) Neurologische Klinik der Freien Universität, Klinikum Rudolf Virchow, Spandauer Damm 130, D-W-1000 Berlin 19, Federal Republic of Germany.

Przuntek, H. (5, 41, 123) Department of Neurology, St. Josef Hospital, Gudrunstrasse 56, D-W-4630 Bochum, Federal Republic of Germany.

Rahmany, Z. (165) Hôpital Necker, Laboratoire de Biochimie Génétique, 149 rue de Sèvres, F-75743 Paris, France.

Regland, B. (267) Department of Psychiatry and Neurochemistry, St. Jörgens Hospital, University of Göteborg, S-422 03 Hisings Backa, Sweden.

Riederer, P. (79, 123, 221) Clinical Neurochemistry, Department of Psychiatry, University of Würzburg, Füchsleinstrasse 15, D-W-8700 Würzburg, Federal Republic of Germany.

Rinne, U. K. (123) Department of Neurology, University of Turku, SF-20520 Turku 52, Finland.

Sinet, P. M. (165) Hôpital Necker, Laboratoire de Biochimie Génétique, 149 rue de Sèvres, F-75743 Paris, France.

Sofic, E. (221) Clinical Neurochemistry, Department of Psychiatry, University of Würzburg, Füchsleinstrasse 15, D-W-8700 Würzburg, Federal Republic of Germany.

Sorbi, S. (287) Istituto delle Malattie Nervose e Mentali dell'Università di Firenze, Viale Morgagni 85, I-50134 Florence, Italy.

Stajich, J. (99) Duke University, Department of Medicine, Service of Neurology, Durham, NC, U.S.A.

Steinberg, R. (41) Department of Neurology, St. Josef Hospital, Gudrunstrasse 56, D-W-4630 Bochum, Federal Republic of Germany.

Stocchi, F. (13) I Clinic of Neurology, Department of Neurosciences, University "La Sapienza", Viale dell'Università, 30, I-00185 Rome, Italy.

Strolin Benedetti, M. (93) Farmitalia Carlo Erba, R & D, Via Imbonati 24, I-20159 Milan, Italy.

Sunderland, T. (137) National Institute of Mental Health, NIH Clinical Center, 9000 Rockville Pike, Bethesda, MD 20892, U.S.A.

Szelies, B. (181) Universitätsklinik für Neurologie, Joseph-Stelzmann-Strasse 9, D-W-5000 Köln 41 (Lindenthal), Federal Republic of Germany.

Tipton, K. F. (209) Department of Biochemistry, Trinity College, Dublin 2, Ireland.

Wallin, A. (257) Department of Psychiatry and Neurochemistry, St. Jörgens Hospital, University of Göteborg, S-422 03 Hisings Backa, Sweden.

Wesemann, W. (79) Department of Neurochemistry, Institute of Physiology II, University of Marburg, D-W-3550 Marburg, Federal Republic of Germany.

Wichart, I. (221) Ludwig-Boltzmann-Institute of Clinical Neurobiology, Lainz-Hospital, Wolkersbergenstrasse 1, A-1130 Vienna, Austria.

Youdim, M. B. H. (111, 221) Rappaport Research Institute, Department of Pharmacology, Faculty of Medicine, Technion, Haifa, Israel.

List of abbreviations

AChE   acetylcholine esterase
AD   Alzheimer's disease
AEP   auditory evoked potentials
APV   D,L-aminophosphonovalerate

BBB   blood-brain barrier

CAT   choline acetyltransferase
CMRGl   cerebral metabolic rate of glucose
COMT   catechol-O-methyltransferase
CSF   cerebrospinal fluid
CT   computer tomography

DA   dopamine
DBH   dopamine beta-hydroxylase

DL-AA   D,L-α-aminoadipate

EEG   electro-encephalography

FDG   2(^{18}F)-fluoro-2-deoxy-D-glucose

GABA   γ-aminobutyric acid
GDH   glutamate dehydrogenase
GFP   global field power

5-HIAA 5-hydroxyindolacetic acid
5-HT   serotonin
HVA   homovanillic acid

L-AP4   L-2-amino-4-phosphonobutyrate

MAO   monoamine oxidase
MID   multi-infarct dementia
MPP+   1-methyl-4-phenyl pyridinium ion
MPTP   1-methyl-4-phenyl-1,2,3,6-tetrahydropyridine
MRI   magnetic resonance imaging
MSA   multiple system atrophy

NA   noradrenaline
NGF   nerve growth factor
NMDA   N-methyl-D-aspartate
NMR   nuclear magnetic resonance

OPCA   olivopontocerebellar atrophy

PD   Parkinson's disease
PDH   pyruvate dehydrogenase
PET   positron emission tomography
PHF   paired helical filaments
PMF   platelet membrane fluidity
PNMT   phenylethanolamine-N-methyltransferase
PSP   progressive supranuclear palsy

PST	phenolsulfotransferase	SDAT	senile dementia of the Alzheimer type
QNB	3-quinuclidinyl-4-iodo-benzilate	SN	substantia nigra
		SOD	superoxide dismutase
rCBF	regional cerebral blood flow	SPECT	single photon emission computed tomography
rCMRGl	regional cerebral metabolic rate of glucose	TH	tyrosine hydroxylase
		THA	tetrahydroaminoacridine

Introduction

R. Levi Montalcini

Distinguished Scientist

National Research Council (CNR), Institute of Neurobiology, Rome, Italy

I am very grateful for having been invited to this meeting. Even though I have not been personally working on Parkinson's disease, I am obviously very interested in the study of degenerative diseases, in particular of Alzheimer's disease, in the treatment of which there is some hope that the Nerve Growth Factor (NGF) may be of use.

After two decades entirely devoted to the study of the action of NGF on the primary sensory and sympathetic nerve cells, my opinion today is that NGF should not be regarded only as a nerve growth factor but rather as a specific growth factor acting on those particular nerve cells, sensory, sympathetic, cognitive and others, which are involved in homeostatic function. A new working hypothesis is being developed, according to which NGF acts as a modulator of the nervous, endocrine and immune systems.

The message I would like to convey today, is that degenerative diseases should be looked at in a broad perspective. Degenerative diseases, e.g. Parkinson's disease, are far more than diseases of one system, just like NGF is far more than a factor acting on particular cells.

Early markers in Parkinson's disease

Early markers in Parkinson's disease

H. Przuntek

Department of Neurology, St. Josef Hospital, University of Bochum,
Federal Republic of Germany

Summary

None of the instrumental methods used for the detection of Parkinson's disease, including biochemical tests and imaging techniques, have sufficiently been proved to be more reliable than clinical methods in terms of accuracy and specificity.

Nonetheless, the integration of multiple instrumental methods in field study designs could establish their diagnostic value as early markers in the preclinical stages of Parkinson's disease.

Introduction

Considering that Parkinson's disease manifests subclinically years before clinical symptoms become evident, it should ideally be possible to detect it by a well-targeted early diagnostic methodology.

If it is true that substances, such as the MAO-B inhibitor l-deprenyl or the diphenylpiperidine budipine are prophylactically effective in Parkinson's disease, it might be possible to investigate such drugs for their effectiveness in preventing progression of the disease. Considering this hypothesis it appears logical to start protective therapy as early as possible. For this particular reason, a diagnosis of Parkinson's disease has to be made as soon as possible. In principle, two different methods are available for the early diagnosis of Parkinson's disease: clinical methods and instrumental methods.

Rigidity, tremor, akinesia, bradyphrenia, depression and autonomic phenomena represent the main clinical signs and symptoms. Their expression and presence vary a great deal from patient to patient, making a thorough physical examination and follow-up investigations necessary. Additionally, it must be taken into account that instrumental methods may reveal findings not clinically detectable.

Clinical signs and symptoms in the early stages of Parkinson's disease

Consistently with the complex clinical picture, patients initially present the following unspecific general symptoms: reduced general fitness, reduced stress tolerance at work or endurance of physical exertion, generalized stiffness, shoulder pain, pain in the upper limb or calf muscles, toe cramps (occasionally with downward motion). A patient may complain of his voice having lost its volume and modulation when singing, of shortness of breath when whistling or of less emotional expression in his face and body gestures. He reports frequently stumbling on uneven ground. When trying to swim straight ahead he may find himself going round in a semi-circle towards the affected side. Fine finger movement loses smoothness, e.g. impaired performance playing the piano occurs. A patient may complain about the difficulty in making decisions, about depressed mood, agoraphobia, and social isolation.

Other frequent complaints are loss of appetite, loss of weight, constipation, sleeplessness, loss of libido, cold sensations, drooling of saliva, temperature dysregulation, i.e. excessive heat accumulation in hot and humid weather conditions, as well as pale and mask-like face with greasy skin and dry mouth. Emotional stress may lead to restlessness which manifests itself in trembling and, later, in shaking.

As to the *neurological examination*, early parkinsonian symptoms are: unilateral loss of facial expression, widened palpebral fissure on the side where the symptoms of parkinsonism first appear, occasionally reduced spontaneous eye movements, and paralysis of convergence. Furthermore, monotonous speech melody together with insufficient modulation of vowels and indistinct articulation of words with many consonants can be observed. Turning of the head is slow. Body gestures appear to be unilaterally diminished and there is less swinging of one arm when walking. Sometimes, the upper limbs are kept in adduction, flexion of the fingers in the metacarpophalangeal joints and extension in the interphalangeal joints show a different pattern on either side. The examiner will find adduction of the thumb and middle phalanx of the index finger on the more severely affected side of the body. Additionally, the angle of opposition between thumb and little finger appears to be more acute on the more severely affected side. When pressing both hands on an even surface, the angle between thumb and index finger seems to be reduced on the affected side.

When drawing a spiral, the distance between two lines decreases from the centre to the periphery. Typical micrographia occurs when writing numbers from 1 to 10, within the first 3. There is increased rigidity on rotation of the thumb in its carpometacarpal articulation.

Particularly, hypertonia in the carpal joints can be demonstrated with the forearms in upright position.

In the lower limbs, alternating movements on imitation of bicycle pedalling are slower, again unilaterally; similarly, steps are shorter on the affected side when mimicking forward movement. Sitting on the side of the bed with swinging legs, one notices less movement in the affected leg. Occasionally a patient limps with one leg, moving it in an extended way rather than with circumduction.

Instrumental methods

Considering the difficulty in interpreting the clinical signs in the early stages of Parkinson's disease, further investigations using instrumental methods may lead earlier to a correct diagnosis of the disease.

The following methods will be discussed: kinesiologic methods, including video recording; electrophysiological methods; psychometric tests; biochemical methods; imaging techniques. Among them, the most widely used methods are: computer-assisted electronic motion analysis; accelerometric examination; Motor Performance Test.

Kinesiologic methods

Kinesiologic investigations try to evaluate voluntary and involuntary movements both qualitatively and quantitatively. Johnels et al. (1987) use computer-assisted electronic motion analysis, which proves particularly useful in quantifying postural changes, locomotion and target-directed manual movements.

Further tests comprise examination of fast alternating movements. Pathologic slowing of the fastest alternating finger movements can be demonstrated by an accelerometric testing. Some patients show the so-called hastening phenomenon. Normally these patients are unable to achieve alternating movements of a frequency higher than $4-5$ Hz. Occasionally, however, when following a target, these patients without tremor show alternating movements of $7-8$ Hz, although the frequency of the pacing signal lies between $3-4$ Hz. On examination of complex movements, their duration of performance rises with amplitude. Corresponding data can be gathered using an accelerometer or motor tableau.

Additional testings involve measurement of oscillation in isometric contractions. The actual specificity of this "kinetic tremor" has been further investigated (Fitts, 1954; Freund and Büdingen, 1978; Hefter et al., 1987, 1989; Hömberg et al., 1987).

The Motor Performance Test represents another kinesiologic method, originally designed by Schoppe (1974) and later modified by Kraus et al. (1987). The Motor Performance Test consists of several subtests, namely tapping, aiming, tracing, steadiness and plugging of pins, through which changes in the speed of voluntary movements (plugging, tapping, etc.) can be recorded. Furthermore, changes in unsteadiness consisting of tremor and various other involuntary movement disturbances can be examined. Tracing and aiming, two further subtests concerning motor skills, are more appropriate for evaluating atactic disturbances than parkinsonian symptoms. The great advantage of motor performance testing lies in the low cost of the device used and its easy handling, which can be carried out by paramedical staff. Therefore, in our opinion, this test represents the most suitable method for the early diagnosis of Parkinson's disease, even more so because broad field studies must be undertaken and patients with abnormal findings may subsequently undergo further more specific, but also more time-consuming tests.

Electromyographical methods

In the early stages of Parkinson's disease, tremor is only encountered in stress situations, e.g. under mental or emotional strain. EMG-activity of tremor can be measured by surface electrodes from extensor carpi radialis and flexor carpi ulnaris during stress provocation. 24 h-recording has shown to be particularly useful. However, these elegant methods may be too sophisticated to be used in routine field studies (Scholz et al., 1989).

Evoked potentials

Normal somatosensory evoked potential can be elicited in most parkinsonian patients. Interestingly, $30-40\%$ of patients demonstrate abnormal visual evoked potential (P 2). However, this is of little diagnostic value, being a rather unspecific finding (Jörg and Gerhard, 1987).

Psychometric investigations

In principle, psychometric tests are designed to evaluate 3 aspects: personality changes, emotional disturbances, and disorders of cognitive function.

Personality changes can be studied by the use of personality inventory tests, such as the Freiburg Personality Inventory. Patients with Parkinson's disease have shown a consistent premorbid personality pattern characterized by signs of depression, introversion, rigidity and inflexibility (Poewe et al., 1983).

Measurement of bradyphrenia or impairment of cognitive functions can best be achieved by using a whole set of subtests of the Wechsler Adult Intelligence Scale: KAI (a short IQ test), d_2-test (attention-stress test), Aachener Aphasia Test (AAT) naming subtest, and the Beck Depression Inventory (BDI). At the same time, measurements of simple and complex reaction times are made using the Wiener reaction and determination device (El-Awar et al., 1987; Lees and Smith, 1983; Lehrl et al., 1980; Mildworf et al., 1986; Steinberg and Przuntek, 1989). All these tests revealed that parkinsonian patients perform more poorly in speed-dependent cognitive function tasks (Steinberg and Przuntek, 1989). The Beck Depression Inventory is applied to exclude depression, which by itself could impair cognitive performance.

Evaluation of emotional changes

Emotional changes or changes of mood mainly consist of various grades of depression. Today, the Hamilton Depression Scale, the Zung-Scale, and the Beck Depression Scale are mainly used in the clinical context. The Zung-Scale seems to provide particularly abnormal scores in early Parkinson's disease, because it includes questions about motor impairment.

Biochemical methods

Potential, early biochemical markers of Parkinson's disease are the ^{3}H-spiperone-binding of lymphocytes, platelet MAO-B activity and aspartate/glutamate and glutamine concentration in platelets (Bondy et al., 1989; Rolf et al., 1989).

Imaging techniques

With the advent of positron-emission tomography, it has become possible to quantitatively describe energy metabolism in certain areas of the brain and to give measurements of dopaminergic neurotransmission in humans. Striatal influx of labelled L-dopa correlates, somehow, to the severity of the disease.

Using recepetor markers PET may be of value in differential diagnosis, considering that the number of receptors remains normal in idiopathic Parkinson's disease, whereas it is reduced in certain groups of patients with Parkinson's syndrome (Leenders et al., 1986).

Discussion and conclusion

The question about the availability of subclinical markers with diagnostic value in non-manifest parkinsonism remains open. The instrumental methods mentioned above refer to studies of patients with advanced pathogenetic changes. It has to be assumed that dopamine concentrations in the nigrostriatal system of these patients is only 20% of the normal value. Additional studies are needed to find out whether instrumental methods are more sensitive and/or accurate than clinical methods.

Field studies should be carried out in a population preferably 5 − 10 years prior to the normal age at which clinical manifestation of Parkinson's disease occurs. In order to obtain an acceptable number of individuals developing Parkinson's disease later in life, a field study in a well-defined sample of people aged between 45 and 55 years is deemed appropriate to test the validity of the different methods described above. As large numbers of participants are needed, all time-consuming methods, such as 24-h tremor recordings or positron-emission tomography, are not practical for an early diagnosis of Parkinson's disease as methods of first choice. First of all the following methods should be included in a field study: Personality Inventory

Tests; measurement of simple and complex reaction times; a depression score, the Zung-Score being the most suitable rating scale as it correlates closely with the severity of parkinsonian symptoms; Motor Performance Test. Patients with abnormal results in psychometric tests or Motor Performance Test should then undergo more detailed examinations, e.g. positron-emission tomography.

Potential early biochemical markers of Parkinson's disease can only be developed from urine or blood samples. However, none of the currently available biochemical tests has been sufficiently validated in a large patient population.

References

Bondy B, Dengler FX, Oertel WH, Ackenheil M (1989) ^{3}H-spiperone binding to lymphocytes is increased in schizophrenic patients and decreased in Parkinson patients. In: Przuntek H, Riederer P (eds) Early diagnosis and preventive therapy in Parkinson's disease. Springer, Wien New York, pp 205–212

El-Awar M, Becker JT, Hammond KM, Nebes RD, Boller F (1987) Learning deficits in Parkinson's disease. Comparison with Alzheimer's disease and normal aging. Arch Neurol 44: 180–184

Fitts PM (1954) The information capacity of the human motor system in controlling the amplitude of movement. J Exp Psychol 47: 381–391

Freund H-J, Büdingen HJ (1978) The relationship between speed and amplitude of the fastest voluntary contractions of human arm muscles. Exp Brain Res 31: 1–12

Hefter H, Hömberg V, Lange H, Freund H-J (1987) Impairment of rapid movements in Huntington's disease. Brain 110: 585–612

Hefter H, Hömberg V, Freund H-J (1989) Quantitative analysis of voluntary and involuntary motor phenomena in Parkinson's disease. In: Przuntek H, Riederer P (eds) Early diagnosis and preventive therapy in Parkinson's disease. Springer, Wien New York, pp 65–73

Hömberg V, Hefter H, Reiners K, Freund H-J (1987) Differential effects of changes in mechanical limb properties on physiological and pathological tremor. J Neurol Neurosurg Psychiatry 50: 568–579

Jörg J, Gerhard H (1987) Somatosensory motor and special visual evoked potentials to single and double stimulation in "Parkinson's disease"–an early diagnostic test? J Neural Transm [Suppl 25]: 81–88

Johnels B, Ingvarsson PE, Rydgren U, Thorselius M, Steg G (1987) Measuring motor function in Parkinson's disease. In: Marsden CD, Conrad B, Benecke R (eds) Motor disturbances I. Academic Press, London, pp 131–144

Kraus PH, Klotz P, Fischer A, Przuntek H (1987) Assessment of symptoms of Parkinson's disease by apparative methods. In: Riederer P, Przuntek H (eds) MAO-B inhibitor selegiline (R-(–)-deprenyl), a new therapeutic concept in the treatment of Parkinson's disease. J Neural Transm [Suppl 25]: 89–96

Leenders KL, Palmer AJ, Quinn N, Clark JC, Firnau G, Garnett ES, Nahmias C, Jones T, Marsden CD (1986) Brain dopamine metabolism in patients with Parkinson's disease measured with positron emission tomography. J Neurol Neurosurg Psychiatry 49: 853–856

Lees AJ, Smith E (1983) Cognitive deficits in the early stages of Parkinson's disease. Brain 106: 257–270

Lehrl S, Gallwitz A, Blaha L (1980) KAI; Kurztest für allgemeine Intelligenz. VLESS Verlagsgesellschaft mbH, Vaterstetten

Mildworf B, Globus M, Melamed E (1986) Patterns of cognitive impairment in patients with DAT and Parkinson's disease. Adv Behav Biol 29: 135–140

Poewe W, Gerstenbrand F, Ransmayr G, Plörer S (1983) Premorbid personality of Parkinson patients. J Neural Transm [Suppl 19]: 215–224

Rolf LH, Klauke Th, Fünfgeld EW, Brune GG (1989) Aspartate, glutamate, and glutamine in platelets of patients with Parkinson's disease. In: Przuntek H, Riederer P (eds) Early diagnosis and preventive therapy in Parkinson's disease. Springer, Wien New York, pp 221–227

Scholz E, Bacher M, Bellenberg A, Hart S, Diener HC, Dichgans J (1989) Long-term measurement of tremor: early diagnostic possibilities. In: Przuntek H, Riederer P (eds) Early diagnosis and preventive therapy in Parkinson's disease. Springer, Wien New York, pp 93–102

Schoppe KJ (1974) Das MLS-Gerät: Ein neuer Testapparat zur Messung feinmotorischer Leistungen. Diagnostica 20: 43–46

Steinberg R, Przuntek H (1989) Psychometric assessment of early signs of dementia in special consideration of Parkinson's and Alzheimer's disease — an update. In: Przuntek H, Riederer P (eds) Early diagnosis and preventive therapy in Parkinson's disease. Springer, Wien New York, pp 19–32

Differential diagnosis of Parkinson's disease

A. Agnoli, F. Stocchi, and A. Carta

I Clinic of Neurology, Department of Neurosciences, University "La Sapienza",
Rome, Italy

Summary

Diagnosis of idiopathic Parkinson's disease (PD) is essentially clinical and is reached by exclusion. An akinetic rigid syndrome frequently means PD, although a number of other neurodegenerative diseases can share bradykinesia, rigidity, postural instability, and sometimes tremor. Parkinsonism can be classified as follows: pure parkinsonism; parkinsonism associated with other neurological and clinical signs; parkinsonism plus dementia; post-intoxication parkinsonism and parkinsonism due to other causes. The diagnostic approach will be discussed.

Introduction

A number of diseases may share the signs of idiopathic Parkinson's disease (PD) and are thus called parkinsonisms (Denny-Brown, 1968). Among others, a precise differential diagnosis of the akinetic rigid syndrome is of peculiar complexity (Marsden, 1984). At present, careful clinical examination, neuroradiological and neuropsychological assessment, and response to oral and intravenous levodopa are to be considered the best diagnostic criteria. However, for a correct diagnosis of parkinsonism, at least two of the following signs have to be observed: tremor, rigidity, akinesia, and posture instability. A number of classifications of parkinsonism have been suggested by different authors. This comprehensive clinical entity might be divided, for practical pruposes, into five groups (Table 1). Of all this impressive list of syndromes, we will discuss here only the most important ones from a clinical viewpoint.

Table 1. Classification of parkinsonism

Pure parkinsonism

Post-encephalitic parkinsonism
Drug-induced parkinsonism

Parkinsonism + other neurological or clinical signs

Progressive Supranuclear Palsy
Multiple System Atrophy
Creuzfeldt-Jacob disease
Wilson's disease
Hallervorden-Spatz disease
Fahr's disease

Parkinsonism associated with dementia

Alzheimer's disease
Pick's disease
Westphal variant of Huntington's chorea
Cerebrovascular diseases
Parkinson plus
Dementia pugilistica

Post-intoxication parkinsonism

Manganese
Carbon monoxide
Carbon disulfide
Cyanide
Methanol
MPTP

Other causes of parkinsonism

Normal pressure hydrocephalus
Stroke
Tumor
Trauma
Subdural haematoma
Syringomesencephalia

Pure parkinsonisms

Postencephalitic parkinsonism

Encephalitis lethargica, or Von Economo's encephalitis, is at present the only viral infection recognized as a cause of persistent parkinsonism (Miyasaki and Fujita, 1977). Today such patients are extremely uncommon, but a great number of them has been described in the thirties after the viral epidemic that spread over Europe and the U.S.A. between 1914 and 1918 (Von Economo, 1931). Although the presence of oculogyric crises and early age of onset are highly characteristic of postencephalitic parkinsonism, it has to be reminded that some cases do not present these signs (post-mortem diagnosis) and that other patients with idiopathic PD may have a history of encephalitis and oculogyric crises (Duvoisin and Yahr, 1965). Diagnosis may be helped by the presence of ocular movement disturbances, dystonias, and stabilization of the disease after an initial rapid worsening. Post-mortem examination of subjects who had been affected by postencephalitic parkinsonism shows bilateral diffuse degeneration and gliosis of the substantia nigra and locus coeruleus, without Lewy bodies but with neurofibrillary filaments in the brainstem, substantia nigra, locus coeruleus, mesencephalic tegmentum, hypothalamus, and hippocampus (Alvord, 1965). These patients do not respond well to levodopa, while positive results are obtained with anticholinergic drugs.

Drug-induced parkinsonism

This is a widespread affection and its incidence increases with aging, as is the case with PD. A large number of neuroleptics and major tranquillizers may induce a parkinsonian syndrome (Table 2). The onset is generally rapid and patients show rigidity, tremor, and a slight bradykinesia. Akathisia and neurodysleptic crises may complete the clinical picture. Patients usually recover within 6 to 24 months after drug withdrawal. Some cases of parkinsonism are also due to calcium-entry blocker compounds (flunarizine and cinnarizine), as has recently been reported especially in elderly patients given long-term high doses of these drugs.

Table 2. Pharmacologic causes of parkinsonism

— Neuroleptics	phenotiazines
	butyrophenones
	thioxanthines
	benzamides
— Reserpine, tetrabenazine	
— Calcium-entry blockers	flunarizine
	cinnarizine
— Miscellaneous agents	alpha-methyldopa
	lithium

Parkinsonism associated with other neurological and clinical signs

Progressive Supranuclear Palsy (PSP) (Steele-Richardson-Olszewski syndrome)

This entity was first described by Richardson and his colleagues in 1963. According to the authors, this is a progressive, non-familial disease, with onset during the 5th — 6th decade of life, marked by supranuclear ocular paralysis and by at least two of the following primary symptoms: axial dystonia and rigidity; pseudobulbar palsy; bradykinesia and rigidity; signs of frontal lobe lesion; postural instability with retropulsion (Lees, 1987). The PSP syndrome differs from PD in that it has a bilateral symmetric onset with frequent gait disorders; tremor is extremely rare while extensive stiffness may be present. As the disease progresses, the pseudobulbar signs become more evident with difficulties in swallowing and phonation. There are no signs of autonomic nervous system impairment and little or no response to L-dopa therapy. Life expectancy is 4 — 6 years from the onset (Jackson et al., 1983). The pathology is characterized by neurofibrillary degeneration, neuronal loss and gliosis of brainstem structures, subthalamic nucleus of Luys, pallidum, substantia nigra, periaqueductal gray, superior colliculus, locus coeruleus, dentate and raphe nuclei (Alvord, 1965). A substantial damage in the projections arising from the brainstem to the frontal cortex seems to be quite common in these subjects (Whitehouse et al., 1983).

Multiple System Atrophy

Multiple System Atrophy (MSA) includes a number of disorders often associated with PD, as well as with other neurological and disautonomic signs. Classically, three entities are grouped under the MSA label: the striato-nigral degeneration (Takei and Mirra, 1973); the Shy-Drager syndrome (Shy and Drager, 1960) and the different forms of olivopontocerebellar atrophy (Duvoisin, 1987). Other inherited maladies are considered as part of MSA, such as the Machado-Joseph disease and the nigral-subthalamic-pallidal atrophy (Duvoisin, 1987). Mainly, all MSA patients show pyramidal and cerebellar signs associated with stiffness and akinesia, and sometimes ocular voluntary motion palsy. However, typical signs are the autonomic nervous system failures (Oppenheimer, 1983). In a considerable number of cases, MSA patients are thought to be truly affected by PD, MSA reproducing the exact clinical picture of idiopathic PD, especially during the early years of the disease. A mild of poor response to L-dopa can arise suspicions as to the nature of the disease, in the absence of gross autonomic failure. CT scan or NMR examination show sometimes a marked atrophy at the level of the brainstem or cerebellum, but a negative neuroradiological picture does not necessarily exclude MSA. Therefore, one should always consider the possibility of MSA whenever a patient shows akinesia and rigidity with poor response to oral or, better, intravenous L-dopa administration. Post-mortem examination shows focal degeneration of the substantia nigra, locus coeruleus and putamen, associated in some cases with a true olivopontocerebellar atrophy or with the diffuse presence of Lewy bodies. The degeneration may be even of larger proportions in the case of additional clinical complications (Forno et al., 1986).

Wilson's disease (hepatolenticular degeneration)

The classic description of "progressive lenticular degeneration: a familial neurologic disease associated with cirrhosis of the liver" was made by Wilson in 1912. The disease is an autosomal recessive disorder and its incidence is of about 1/200,000. The first clinical signs, consisting of hepatosplenomegaly, thrombocytopenia and consequent haemorrhages, usually become evident during the second or third decade of life. Early neurological signs may be tremor, bradykinesia, dysarthria, dysphagia, choreic-athetosic movements, dystonic postures. Occasionally, psychiatric disturbances, such as behavioural disorders, depression and intellectual impairment, can be present from the beginning of the illness. Usually, cerebellar ataxia and action tremor are also evident and these signs, together with those previously described, are helpful for a differential diagnosis vs PD. The Kayser-Fleischer ring is another important element as regards differential diagnosis, although not evident in about 25% of patients already in hepatic stage. Serum ceruloplasmin values are quite low (less than 20 mg/dl) in about 95% of the patients, with consequently low bound copper serum values (below 60 pg/dl). Urinary copper excretion is significantly increased (more than 100 pg/in 24 h). Total copper content may be normal due to the large quantity of free circulating copper. In several cases, only a liver biopsy can be reliable to diagnose with certainty. Hepatic copper concentration is generally in excess (more than 250 µg/g of dry tissue), but this may not always be the case even for patients with neurological symptoms. Notably, in some cases, hepatic copper content may be reduced to as little as 100 µg/g, since copper tends to be accumulated in the brain tissue. Consequently, values below 250 µg/g do not allow to exclude definitely Wilson's disease. On the other hand, it has to be underlined that high copper content in liver tissue, with normal serum ceruloplasmin, is not indicative of Wilson's disease even in the presence of a Kayser-Fleischer ring. In this respect, a test with labelled Cu might be helpful. CT scan and NMR examination show alterations in the basal ganglia, cerebellar nuclei, trunk, and white matter. It is quite rare to observe the loss of matter in the basal ganglia as described by Wilson, while it is common to find atrophy and a light brown pigmentation of these nuclei. There may also be a loss of neurons and white matter in the putamen, substantia nigra, and caudate. Typically, a diffuse gliosis (with type I and II Alzheimer cells) in the cortex, basal ganglia, brainstem nuclei and cerebellum is observed (Walshe, 1976). Treatment mainly consists in removing the excess copper from brain and liver. This can be done by administering 1 g/day of penicillamine as first choice, dimercaprol (BAL®) (3 g/day), or triethylenetetramine (1 g/day). Recent reports show that the daily administration of zinc (100 − 200 mg) inhibits the absorption of copper from the gastrointestinal tract.

Idiopathic calcification of the basal ganglia (Fahr's disease)

These patients present a calcification of the basal ganglia with a clinical evidence of a parkinsonian syndrome with choreic-athetosic movements preceding or following it. The age of onset is generally the fifth decade. Serum calcium levels are usually normal. A familial form with onset during adolescence is also known, mainly characterized by ataxia and dementia. Serum calcium levels are low in this form. Serum calcium levels and CT scan findings are mandatory for diagnosis.

Parkinsonism associated with dementias

Pick's disease

This is a cerebral degenerative syndrome in which atrophy is confined mainly to the frontal and temporal lobes and involves both the gray and the white matter. The caudate nucleus, thalamus, subthalamic nucleus, substantia nigra, and globus pallidus may be affected, too. Clinically, the symptoms of frontal deficit prevail: apathy, abulia, gait impairment, urinary incontinence, extrapyramidal signs, and reflexes of liberation (Brown and Marsden, 1984). The histologic features consist of a loss of neurons, mainly in the first three cortical layers, and of neuronal swelling in the surviving cells with argentophilic bodies in the cytoplasm (Alvord, 1965). Problems of differential diagnosis with Pick's disease may occur in the early stages of this affection; a cerebral CT scan may prove to be helpful.

Cerebrovascular diseases (Pseudobulbar palsy; Binswanger's disease; arteriopathic parkinsonism)

Pseudobulbar palsy: this syndrome is caused by multiple lacunar infarcts of both corticospinal tracts. Parkinsonism is associated with pyramidal signs and partial impairment of dystal cranial nerves. Patients present also progressive dementia, spasmodic crying and laughing, and a mixed rigid-spastic hypertonia.

Binswanger's subcortical encephalopathy: an affection caused by multiple infarcts in the periventricular white matter. Cortex and basal ganglia are unaffected. Symptoms are both of pyramidal and extra-pyramidal origin: gait apraxia, urinary incontinence, and dementia are the most notable (Boller et al., 1980; Parkes et al., 1974). CT scan is crucial for an accurate diagnosis. NMR shows typical periventricular luminescence.

Arteriopathic parkinsonism: this entity is not entirely accepted, though histological studies on brains from presumed idiopathic PD patients claim that about 6% of them presented serious vascular damage with a mild degeneration of the substantia nigra (Alvord, 1965). However, clinical elements to be considered are the age of onset, the presence of vascular risk factors, together with additional pyramidal signs or intellectual impairment. Obviously, NMR can be quite useful.

Huntington's chorea

This affection may cause an akinetic-rigid syndrome, known as the Westphal rigid form. It usually occurs during the second decade of life, but also, rarely, in adults (Marsden, 1984). Differential diagnosis is possible considering behavioural disorders, dementia, and inheritance.

Post-intoxication parkinsonism (Langston and Ballard, 1983)
(see Table 1)

Parkinsonism due to other causes
(see Table 1)

Essential tremor

Although quite different as concerns clinical symptomatology, the most common condition misdiagnosed as PD is essential tremor, in a proportion varying from 4 to 6% of patients. However, the positive family history, the type of tremor (postural-kinetic, 5 – 9 Hz), the extent to which head and voice are affected, the absence of bradykinesia, rigidity and postural instability, the slowly progressive course of the disease, the benefit from alcohol, provide sufficient elements to make a correct diagnosis of essential tremor.

References

Alvord EC (1965) The pathology of parkinsonism: etiologic, pathogenetic and prognostic implications. Trans Am Neurol Assoc 90: 167–168

Boller F, Mizutani R, Roessman U, Gambetti F (1980) Parkinson's disease, dementia and Alzheimer's disease: clinicopathological correlations. Ann Neurol 1: 329–335

Brown RG, Marsden CD (1984) How common is dementia in Parkinson's disease? Lancet i: 1262–1265

Denny-Brown D (1968) Clinical symptomatology of diseases of the basal ganglia. In: Vinken PJ, Bruyn GW (eds) Handbook of clinical neurology, vol 6. Diseases of the basal ganglia. North-Holland, Amsterdam, pp 133–211

Duvoisin RC (1987) The olivopontocerebellar atrophies. In: Marsden CD, Fahn S (eds) Movement disorders, 2nd edn. Butterworths Scientific, London, pp 249–271

Duvoisin RC, Yahr MD (1965) Encephalitis and parkinsonism. Arch Neurol 12: 227–239

Forno LS, Langston JW, Delanney LE, Irwin I, Ricaurte GA (1986) Locus ceruleus lesions and eosinophilic inclusions in MPTP-treated monkeys. Ann Neurol 20: 449–455

Jackson JA, Jankovic J, Ford J (1983) Progressive supranuclear palsy: clinical features and response to treatment in 16 patients. Ann Neurol 13: 273–278

Langston JW, Ballard PA (1983) Parkinson's disease in a chemist working with 1-methyl-4-phenyl-1,2,5,6 tetrahydropyridine (MPTP). N Engl J Med 309: 310

Lees AJ (1987) The Steele-Richardson-Olszewski syndrome (progressive supranuclear palsy). In: Marsden CD, Fahn S (eds) Movement disorders, 2nd edn. Butterworths Scientific, London, pp 272–287

Marsden CD (1984) Motor disorders in basal ganglia disease. Hum Neurobiol 2: 245–250

Miyasaki K, Fujita T (1977) Parkinsonism following encephalitis of unknown etiology. J Neuropathol Exp Neurol 34: 1–8

Oppenheimer DR (1983) Neuropathology of progressive autonomic failure. In: Bannister R (ed) Autonomic failure. Oxford University Press, New York, pp 267–283

Parkes JD, Marsden CD, Rees JE, Carjon M (1974) Parkinson's disease, cerebral arteriosclerosis and senile dementia. Q J Med 43: 49–61

Richardson JC, Steele J, Olszewski J (1963) Supranuclear ophthalmoplegia, pseudobulbar palsy, dystonia and dementia. Trans Am Neurol Assoc 88: 25–27

Shy GM, Drager GA (1960) A neurological syndrome associated with orthostatic hypotension. Arch Neurol 2: 511–527

Takei Y, Mirra SS (1973) Striato-nigral degeneration: a form of multiple system atrophy with clinical parkinsonism. Prog Neuropathol 21: 26–32

Von Economo C (1931) Encephalitis lethargica. Its sequelae and treatment. Newman KO (transl), Oxford University, London

Walshe JM (1976) Wilson's disease (Hepatolenticular degeneration). In: Vinken PJ, Bruyn GW, Klawans HL (eds) Handbook of clinical neurology, vol 27. Metabolic and deficiency disease of the nervous system, part 1. Elsevier, New York, pp 379–414

Whitehouse P, Hedreen JC, White C, DeLong M, Price DL (1983) Basal forebrain neurons in dementia of Parkinson's disease. Ann Neurol 13: 243–248

Wilson SAK (1912) Progressive lenticular degeneration. A familial nervous disease associated with cirrhosis of the liver. Brain 34: 295–309

The premorbid personality of patients with Parkinson's disease

W. Poewe

Universitätsklinik für Neurologie, Innsbruck, Austria

Summary

Psychoanalysts in the first half of this century have speculated on the pathogenetic role of premorbid suppression of impulse in an introverted personality for later development of Parkinson's disease. More recent studies seem to support introversion, a predisposition towards depression and mental inflexibility as frequent premorbid personality features of parkinsonians. Neurobehavioural abnormalities found in the early stages of the disease include set-shifting difficulties, cognitive slowing and depression. Premorbid behavioural traits might represent the earliest sign of brain neurochemical dysfunction in Parkinson's disease.

Introduction

Clinical symptoms of Parkinson's disease are believed to become fully apparent after approximately 60% of nigrostriatal projection neurons have degenerated (Hornykiewicz, 1982; Gibb, 1989). The rate of nigral cell loss in Parkinson's disease is not known but there is evidence suggesting that pathology in the substantia nigra may have started decades before the clinical manifestation of the disease (Gibb, 1989) and subclinical damage of the nigrostriatal system can be demonstrated in vivo using positron emission tomography (Calne et al., 1985). Detection of preclinical Parkinson's disease could lead to major advances in preventive treatment and, ultimately, in the cure of this malady. Against this background, the old concept of a possibly distinctive premorbid parkinsonian personality has received new attention in recent years.

The present paper will review findings related to premorbid personality characteristics of Parkinson patients and will examine their possible relation to neurobehavioural abnormalities found in the early stages of the disease.

The concept of a premorbid parkinsonian personality

Ever since the beginning of this century, there has been a steady appearance of occasional reports on some peculiar features in the premorbid personality of patients with Parkinson's disease (Table 1). Traits of introversion, moral rigidity, and inflexibility with a predisposition to depression were consistently mentioned in descriptions by neurologists and psychiatrists alike (Todes and Lees, 1985). Particularly in the first half of the century, much of what had been written on the subject was based on a psychoanalytical approach, and lifelong suppression of aggressive tendencies and sexual urge were seen as possible causative mechanisms of the disease (Sands, 1942; Mitscherlich, 1960). Others, however, could not find convincing evidence for the existence of a particular premorbid parkinsonian personality and suggested that the psychological alterations observed in patients were the result of living with a severely disabling disease (Riklan et al., 1959; de Ajuriaguerra, 1970).

In recent years, new support for characteristic premorbid parkinsonian personality features has emerged. In our own studies, we initially tested 28 patients with idiopathic Parkinson's disease and some of their closest relatives, using the Giessen-Test personality inventory (Poewe et al., 1983). This test is specifically designed to allow both self- and foreign-assessment of personality traits; patients and relatives were asked to fill in the test retrospectively so as to describe the patients' character, as had been perceived before the onset of the disease. The most marked deviations of scores from those of a normal control population were found in the subscales for "control" and "basic mood" highlighting an "overcontrolled" and "depressive" premorbid personality type defined by such features as "overorderly", "talented to deal with money", "overambitious", "unable to be outgoing", "timid", "self-reflective", "swallowing anger", or "frequently depressed".

Table 1. Premorbid personality in Parkinson's disease

Camp (1913) — "Industriousness", "Moralistic"

Cohen-Booth (1935) — "Suppression of aggressive tendencies"

Sands (1942) — "Masked personality"

Mitscherlich (1960) — "Neurotic-compulsive character"

Korten and Kettering (1972) — "Reliable, sense of duty, moral rigidity"

Ward et al. (1984) — "Self-controlled, quiet, unhappy"

In a more recent study of 33 patients and 17 controls, patients appeared more socially alert, apprehensive, worrying, tense, and introverted in their actual personality profile as assessed by Cattell's 16 PF (Poewe et al., 1989). In this study semistandardized interviews of probands and spouses had been employed as a means of assessing premorbid habits and personality features. Evaluation of the material, including ratings made by a "blinded" psychologist, again showed traits of rigidity, introversion, depression, and obsession, more often in patients than in controls.

In these studies no attempt had been made to define different epochs in the patients' premorbid life so as to find when personality changes might have first become apparent. In the twin study of Parkinson's disease conducted in North America, several personality differences between affected and unaffected monozygotic twins have been detected dating back to different premorbid periods. From as early as childhood age the affected sibling appeared less dominant and more self-controlled than his healthy co-twin; by the age of 16 he was also more nervous, while within 10 years prior to the onset of the disease he had become less aggressive, more quiet, less confident, and lighthearted. At the time of the test, the parkinsonian twin showed the most marked personality differences from his non-affected co-twin, being less outgoing, unhappy, depressed, reluctant at making decisions, and shier (Ward et al., 1984). This development was taken as evidence that in Parkinson's disease neurochemical abnormalities might be present in the brain as early as in childhood leading to certain behavioural changes which might contribute to what has been perceived as the parkinsonian premorbid personality.

A different string of evidence for a premorbid behavioural pattern peculiar to patients with Parkinson's disease comes from epidemiological studies, where the number of habitual non-smokers among parkinsonians has been found to be significantly greater than would correspond to the smoking behaviour of the general population (Kessler and Diamond, 1967; Godwin-Austen et al., 1982). In the American twin study the unaffected twin had also smoked significantly more cigarettes than the twin with Parkinson's disease prior to the onset of his illness (Ward et al., 1984). This has occasionally been interpreted as a protective effect of some component of cigarette smoke against the development of Parkinson's disease, particularly since environmental toxins are being suggested to be causative agents of nigral cell death (Snyder and D'Amato, 1986). According to a report by the British Royal College of Physicians (1977), however, smokers tend to be more arousal-seeking, danger-loving, belligerent, and extroverted personalities compared to non-smokers. This and the absence of a dose-response relationship between the number of cigarettes smoked and severity of disease in those parkinsonians who did smoke (Golbe et al., 1986) make it more plausible to view the premorbid smoking behaviour of patients as a reflection of their character. Premorbid non-smoking is also associated with ulcerative colitis, another disease characterized by an introverted and rigid personality type (Bihari and Lees, 1987).

Neurobehavioural abnormalities in the early stages of Parkinson's disease

The majority of neuropsychological studies of Parkinson's disease have been performed in patients with varying duration of their illness receiving various drug regimens and have frequently included cases in whom overt cognitive decline had already set in. Only few of them have examined untreated patients in the early stages of their disease, pointing out slight neuropsychological abnormalities (Lees, 1989; see Table 2).

A number of studies have tested patients for their ability to initiate, execute or shift cognitive sets. Using the Wisconsin Card Sorting Test several authors have found abnormalities in patients with Parkinson's disease compared to controls indicating difficulties in initiating or switching cognitive sets with an increased number of perseverative errors (Lees and Smith, 1983; Taylor et al., 1986; Canavan et al., 1989). Taken together, the results of this and other tests of set shifting ability in Parksinon's disease are similar to those seen in patients with frontal lobe damage and it has been suggested that the difficulties of parkinsonian patients might be due to disrupted function of the complex loop system connecting the frontal association cortex with the basal ganglia via the caudate nucleus (Taylor et al., 1986).

Another neuropsychological abnormality found already in early Parkinson's disease is a peculiar slowing of thinking processes or

Table 2. Neurobehavioural abnormalities in Parkinson's disease

---

1. Difficulties executing "cognitive sets"
   - initiation of sets
   - completion of sets
   - shifting of sets
2. Cognitive slowing ("Bradyphrenia")
3. Obsessional slowness (?)
4. Depression
5. Dementia

"psychic akinesia", which was originally described as "bradyphrenia" by Naville (1922) in postencephalitic patients. It has been difficult to dissect from motor slowness but several reaction time paradigms have been employed to show increased central processing time in Parkinson's disease (Wilson et al., 1980; Hansch et al., 1982). Rogers et al. (1987) studied 30 newly diagnosed and untreated patients with Parkinson's disease with two computerized tests requiring the same motor response but differing in the complexity of the cognitive task. The results were compared to those obtained in 30 patients with primary depressive illness and 30 healthy controls and revealed cognitive slowing in patients who had evidence either of structural brain disorder on CT scanning or of affective impairment. In the group with primary depressive illness cognitive slowing was related to concomitant motor impairment. The authors concluded that bradyphrenia in Parkinson's disease and psychomotor retardation in depression might be related phenomena.

Depression is found in up to 50% of patients with Parkinson's disease and it has been estimated that this may be twice the percentage found in most other chronically disabling diseases (Lees, 1989). Episodes of major depression often antedate the onset of motor symptoms by more than a decade and 15 of the 34 newly diagnosed patients of Santamaria et al. (1986) had had depressive symptoms 1.5 to 36 years before onset of parkinsonism.

Relation between premorbid personality features and neuropsychological abnormalities found in Parkinson's disease

The neurochemical and morphological substrates of mental changes found in Parkinson's disease are still a matter of debate and speculations. It has been suggested that depression and bradyphrenia might be related to a depletion of meso-cortico-limbic dopamine projections while memory impairment and dementia could result from a loss of subcortico-frontal cholinergic projections (Agid et al., 1984). In addition dopamine deficiency in the caudate nucleus could interfere with "complex loop" function between basal ganglia and frontal association cortex and this might be related to the frontal lobe type of cognitive impairment found in the early stages of the disease.

If, as seems likely, the pathological brain process leading to clinical parkinsonism starts long before the onset of motor symptoms, premorbid personality traits peculiar to parkinsonians could be viewed as an early manifestation of their disease. They might be the earliest expression of some of the neuropsychological abnormalities found

Table 3. Possible premorbid correlates of neurobehavioural abnormalities in Parkinson's disease

Parkinson's disease	*? Pre-morbid traits*
— Set shifting difficulties	"Inflexibility" "Moral rigidity" "Obsessionality"
— Cognitive slowing	"Obsessional slowness" "Lack of spontaneity"
— Depression	"Introversion" "Shyness" "Depression"

after the onset of clinical symptoms; Table 3 summarizes this possible relationship. Interesting results along these lines have recently been reported by Lees (1989), who studied 12 young adult in-patients with obsessive compulsive behaviour and found a variety of neurological abnormalities, including difficulties in initiating some voluntary movements, carrying out two motor acts simultaneously and motor perseveration. Two patients had muscular rigidity and resting tremor was present in one. This was interpreted as suggesting the possibility that obsessional slowness could be a feature of Parkinson's disease.

Conclusions

Several lines of evidence support the old concept that patients with Parkinson's disease exhibit certain peculiar traits in their premorbid personality and behaviour. Current understanding has led us to diverge from the psychoanalysts' view in force in the first half of this century, when these features were seen as possible pathogenetic factors. Together with neuropsychological abnormalities found in the early stages of the disease these premorbid character traits could reflect slight changes in the brain neurotransmitter systems, which might possibly be present decades before onset of clinically manifest parkinsonism. Future studies should examine the risk of Parkinson's disease in psychiatric patient populations with depression and obsessive compulsive disorder and try to further define behavioural constellations which might occur as features of the disease.

References

Agid Y, Ruberg M, Dubois B, Javoy-Agid F (1984) Biochemical substrates of mental disturbances in Parkinson's disease. Adv Neurol 40: 211–218

Ajuriaguerra J de (1970) Etude psychopathologique des parkinsoniens. In: Ajuriaguerra J de, Gauthier G (eds) Monoamines noyaux gris centraux et syndrome de Parkinson. Masson, Paris, pp 327–351

Bihari K, Lees AJ (1987) Cigarette smoking, Parkinson's disease and ulcerative colitis. J Neurol Neurosurg Psychiatry 50: 635

Calne DB, Langston JW, Martin WRW, Stoessl AJ, Ruth TJ, Adam MJ, Pate BD, Schulzer M (1985) Positron emission tomography after MPTP: observations relating to the cause of Parkinson's disease. Nature 317: 246–248

Camp CD (1913) Paralysis agitans, multiple sclerosis and their treatment. In: White WA, Jeliffe SE (eds) Modern treatment of nervous and mental disease. Henry Kimpton, Philadelphia, pp 651–667

Canavan AGM, Passingham RE, Marsden CD, Quinn N, Wyke M, Polkey CE (1989) The performance on learning tasks of patients in the early stages of Parkinson's disease. Neuropsychologia 27: 141–156

Cohen-Booth G (1935) Paralysis agitans. Entstehungsbedingungen und Beeinflussungsmöglichkeiten. Nervenarzt 8: 69–83

Gibb WRG (1989) Dementia and Parkinson's disease. Br J Psychiatry 154: 596–614

Godwin-Austen RB, Lee PN, Marmot MG, Stern GM (1982) Smoking and Parkinson's disease. J Neurol Neurosurg Psychiatry 45: 577–581

Golbe LI, Cody RA, Duvoisin RC (1986) Smoking and Parkinson's disease. Search for a dose-response relationship. Arch Neurol 43: 774–778

Hornykiewicz O (1982) Brain neurotransmitter changes in Parkinson's disease. In: Marsden CD, Fahn S (eds) Movement disorders. Butterworth Scientific, London, pp 41–58

Hansch EC, Syndulko K, Cohen SN, Goldberg ZI, Potvin AR, Tourtellotte WW (1982) Cognition in Parkinson's disease: an event-related potential perspective. Ann Neurol 11: 599–607

Kessler II, Diamond KL (1967) Epidemiological studies of Parkinson's disease. 1. Smoking and Parkinson's disease. Am J Epidemiol 94: 16–25

Korten JJ, Ketterings K (1972) Anthropologische Aspekte der Parkinson'schen Krankheit. Nervenarzt 43: 201–205

Lees AJ (1989) The neurobehavioural abnormalities in Parkinson's disease and their relationship to psychomotor retardation and obsessional compulsive disorders. Behav Neurol 2: 1–11

Lees AJ, Smith E (1983) Cognitive deficits in the early stages of Parkinson's disease. Brain 106: 257–270

Mitscherlich M (1960) The psychic state of patients suffering from Parkinsonism. Psychosom Med 1: 317–324

Naville F (1922) Les complications et les séquelles mentales de l'encéphalite épidémique. Encéphale 17: 369–375

Poewe W, Gerstenbrand F, Ransmayr G, Plörer S (1983) Premorbid personality of Parkinson patients. J Neural Transm [Suppl 19]: 215–224

Poewe W, Gerstenbrand F, Karamat E, Schmidhuber-Eiler B (1989) The premorbid personality of patients with Parkinson's disease. In: Przuntek H, Riederer P (eds) Early diagnosis and preventive therapy in Parkinson's disease. Springer, Wien New York, pp 1–7

Riklan M, Weiner H, Diller L (1959) Somato-psychologic studies in Parkinson's disease. 1. An investigation into the relationship of certain disease factors to psychological function. J Nerv Ment Dis 129: 263–272

Royal College of Physicians (1977) Smoking and health. Pitman Medical, London

Rogers D, Less AJ, Smith E, Trimble M, Stern GM (1987) Bradyphrenia in Parkinson's disease and psychomotor retardation in depressive illness. Brain 110: 761–776

Sands IR (1942) The type of personality susceptible to Parkinson's disease. J Mt Sinai Hosp 9: 792–794

Snyder SH, D'Amato RJ (1986) MPTP: a neurotoxin relevant to the pathophysiology of Parkinson's disease. Neurology 36: 250–258

Santamaria J, Tolosa E, Valles A (1986) Parkinson's disease with depression: a possible subgroup of idiopathic parkinsonism. Neurology 36: 1130–1133

Todes CJ, Lees AJ (1985) The premorbid personality of patients with Parkinson's disease. J Neurol Neurosurg Psychiatry 48: 97–100

Taylor AE, Saint-Cyr JA, Lang AE, Kenny FT (1986) Frontal lobe dysfunction in Parkinson's disease. Brain 109: 845–883

Ward CD, Duvoisin RC, Ince SE (1984) Parkinson's disease in twins. Adv Neurol 40: 341–344

Wilson RS, Kasniak AW, Klawans HL, Garron DC (1980) High speed memory scanning in Parkinsonism. Cortex 16: 67–72

An antibody in the CSF of Parkinson's disease patients: summary of data and potential role as a diagnostic marker

P. M. Carvey[1], H. L. Klawans[1], L. C. Kao[1], A. Dahlström[2], and A. McRae[2, 3]

[1] Department of Neurological Sciences, Rush-Presbyterian St Lukes Medical Center, Chicago , IL, USA
[2] Institute of Neurobiology, Neuroscience Research Center of Göteborg, Sweden
[3] INSERM 259, Bordeaux, France

Summary

The CSF of Parkinson's disease patients was shown to possess an antibody (IgG) which immunocytochemically reacts with dopamine cells in the substantia nigra of the rat. This dopamine neuron antibody (DNAb) was also identified in the CSF of patients with possible nigral degeneration. In contrast, control patients or patients with neurologic disease which is not associated with nigral pathology, did not possess the DNAb in their CSF.

The data is most consistent with a hypothesis which suggests that the DNAb represents a secondary autoimmune response to nigral degeneration. As such, the DNAb may be useful as a diagnostic marker for Parkinson's disease and other disorders with nigral degeneration. Since nigral degeneration is thought to precede symptom expression by many years, the DNAb should theoretically be present in CSF prior to symptom expression and would thus represent an early, presymptomatic marker of Parkinson's disease.

Introduction

Immune processes have now been described in a variety of neurological disease states. The best characterized disorder is multiple sclerosis where CSF IgG can be observed to wax and wane with clinical symptoms (Johnson, 1980; Tourtellotte et al., 1980). Husby et al. (1976, 1977) have also described immunoglobulins directed at striatal tissue in the serum and CSF of patients with Huntington's chorea as well as Sydenham's chorea. Numerous investigators have now reported the existence of immunoglobulins in the serum and CSF of Alzheimer's disease patients which immunocytochemically react with various components of cholinergic neurons (Ishii and Haga, 1976; Filit et al., 1985; Chapman et al., 1986; McRae-Degueurce et al., 1987; Gaskin et al., 1987; Franceschi et al., 1988). There are immunoglobulin abnormalities in allergic encephalomyelitis (Gonatas, 1984), ataxia telangiectasia (McFarlin et al., 1972), myotonic dystrophy (McFarlin, 1984), as well as in patients exhibiting neuropsychiatric manifestations associated with systemic lupus erythematosus (Bluestein, 1984). Autoantibodies have been isolated which are directed at gabaergic synapses in stiff-man syndrome (Solimena et al., 1988, 1989) and Itagaki et al. (1987) have demonstrated that HLA class II positive microglia exist in the substantia nigra (SN) of patients with idiopathic Parkinson's disease (PD). Thus, the historical concept of the CNS as an immunologically privileged area must be reconsidered.

We have recently reported that the CSF of PD patients also contains an immunoglobulin (IgG) which reacts immunocytochemically with the SN as well as the ventral tegmental area (VTA) of the rat CNS (McRae-Degueurce et al., 1986 a, 1988 b; Carvey et al., 1988). This immunocytochemical reactivity is directed predominantly at cell bodies, although reaction with cell processes in the midbrain, as well as in the striatum, are occasionally observed. This immunocytochemical reactivity is significantly reduced by prior lesion of the SN using the DA neurotoxin, 6-hydroxydopamine. Incubation of dopamine (DA) neuron cultures with PD patient CSF similarly results in an immunocytochemical reactivity which appears directed at DA neurons (McRae et al., this volume). The antibody has thus been termed the DA-neuron antibody (DNAb), although immunologic reactivity directed at other cell types cannot be completely ruled out at this time.

We have also reported that the DNAb-positive CSF taken from patients undergoing adrenal medulla-to-brain transplantation, gradually loses its immunoreactivity in the months following surgery (McRae-Degueurce et al., 1988 b; Carvey et al., 1988). The time course of this disappearance is similar to the time course of clinical improvement observed in these patients (Penn et al., 1988). This may suggest that there is a relationship between the disappearance of the DNAb and clinical improvement in these patients. We will report here the DNAb results from a large series of patients which include patients with PD as well as appropriate controls. The results from these patients suggest that the DNAb may be useful as a biologic marker for PD.

Materials and methods

Patients

The CSF from 128 patients has been evaluated for the presence of the DNAb. Patients with PD (n = 65) were all moderately to severely affected and had a long standing history of disease which, unless otherwise stated, was responsive to levodopa therapy. With the exception of 11 patients which underwent adrenal medulla-to-brain transplantation, CSF samples were collected by lumbar puncture without stipulation of prior bed-rest or specific food consumption. CSF taken from transplant patients was collected by aspiration from the lateral ventricle at the time of surgery.

Lumbar CSF from 24 patients undergoing routine myelography served as non-neurologic disease controls. Aspirated ventricular CSF from 8 patients undergoing intraventricular shunting for the relief of elevated intracranial pressure, served as controls for the transplant patients. A total of 31 patients with various neurologic disorders were also evaluated for the presence of the DNAb in lumbar CSF. These included 8 patients with peripheral neuropathies, 4 with multiple sclerosis, 4 patients with movement disorders of unknown etiology, 3 patients with tumour or pseudo-tumour, 2 patients with Tourette's syndrome, and 1 patient each with olivopontocerebellar atrophy, spina bifida, Huntington's chorea, tardive dyskinesia, multifocal degeneration, polymyoclonus, post-anoxic chorea, spasmodic torticollus, cerebellar hemorrhage, and 1 patient with non-vascular locked-in syndrome secondary to aqueductal stenosis.

CSF samples were centrifuged for 15 minutes at 3,000 g. The CSF was drawn off and immediately frozen at $-60\,°C$. All samples were then transported to Sweden on dry ice where immunocytochemical reactivity was assessed (A.M.). Samples were always evaluated blind and numerous samples were evaluated twice under a different code name to insure reliability.

Immunocytochemical procedure

The procedure has been described in detail elsewhere (McRae-Degueurce et al., 1986 a, 1988 b). Briefly, rats were perfusion-fixed with a solution containing 5% glutaraldehyde and allyl alcohol in a cacodylate buffer. The brains were then removed, frozen, and coronally sectioned (10 μm) at the level of the substantia nigra. Consecutive sections were mounted on gelatinized slides and incubated with 2% normal goat serum for 30 minutes at room temperature. The sections were then incubated overnight at room temperature with patient CSF (1 : 1 with Tris/sodium metabisulfite buffer, pH 7.6) containing 1% goat serum and 0.5% Triton X-100. Control samples were incubated with biotinylated anti-human IgG serum excluding CSF. The sections were then processed for immunocytochemistry using the avidin-biotin-peroxidase complex.

Samples were judged immunopositive if reactivity was confined to the SN/VTA region. Samples exhibiting reactivity throughout the section were judged non-specific and termed immunonegative. It was observed that a greater percentage of the samples were judged non-specific when the CSF was not spun down prior to freezing. Freeze-fracture may thus contribute to non-specific markage. Sections not exhibiting any markage, were also termed immunonegative.

Results

Table 1 depicts the DNAb results from the 128 patients evaluated to date. CSF from 51 of the 65 PD patients (or 78.4%) have been designated DNAb positive using this procedure. This group includes 1 patient diagnosed with unresponsive PD whose CSF was judged immunonegative. In contrast, all 24 myelogram controls and all 8 ventricular CSF controls were judged immunonegative although 5 myelogram controls exhibited a non-specific markage.

Table 1 also depicts the DNAb results from the nonparkinsonian patients with other diagnosed neurologic diseases. Of the 31 patients

Table 1. Prevalence of DNAb in patients with probable, possible and not-probable substantia nigra degeneration

Diagnosis	# Studied	# Positive	% Positive	Nigral pathology
PD	65	51	74.4%	Probable
Extra pyramidal disorders	4	4	100.0%	Possible
Multi-focal degeneration	1	1	100.0%	Possible
Non-vascular locked in syndrome	1	1	100.0%	Possible
Peripheral neuropathy	8	0	0.0%	Not-probable
Multiple sclerosis	4	0	0,0%	Not-probable
Tumor/pseudo tumor	3	0	0.0%	Not-probable
Tourette's syndrome	2	0	0.0%	Not-probable
OPCA, spina bifida, HC, TD, spas. tort., polymyoclonus, post-anoxic chorea, cerebellar haemorrhage	1 each	0	0.0%	Not-probable
Myelogram	24	0	0.0%	Not-probable
Shunt patients	8	0	0.0%	Not-probable

OPCA Olivopontocerebellar atrophy; *HC* Huntington's chorea; *TD* tardive dyskinesia; *Spas. tort.* Spasmodic torticollis

evaluated, 6 were found to possess immunocytochemical reactivity in the midbrain region. Four of these patients had unusual movement or gait disorders of unknown etiology. Two of these undiagnosed movement disorders had very weak, but immunopositive reactions to SN/VTA region whereas 1 of the four patients had reactivity confined to the medial region of the SN only. The fifth DNAb positive, non-PD patient was an 8-year-old boy diagnosed as having multifocal degeneration with rigidity. The sixth DNAb positive, non-PD patient was a 22-year-old patient with acquired aqueductal stenosis secondary to viral meningitis who eventually developed levodopa responsive non-vascular locked-in syndrome (Rao and Costa, 1989). This patient exhibited cogwheel rigidity and had an abnormally low CSF homovanillic acid level of $11.63\,ng/ml$ (normal $HVA = 44.65 + 19.42$; $n = 218$). MRI of this patient revealed edema and encephalomalacia. Patients with other nonparkinsonian neurologic disorders ($n = 25$) were all immunonegative.

Discussion

Three clinical subgroups were examined for the presence of the DNAb in their CSF. These included, $1-$ patients with a definite history of PD, all of whom had probable nigral pathology; $2-$ patients with extrapyramidal movement disorders and possible nigral pathology; and $3-$ patients with various disorders without probable nigral pathology. A large percentage of the PD patients evaluated (78%) possessed an antibody (IgG) in their CSF which was immunoreactive to the cells of the SN of the rat. Six of the six patients (100%) with possible nigral pathology (based on the presence of extrapyramidal symptoms) were DNAb-positive. In contrast however, control CSF samples as well as a variety of neurologic disease states without probable nigral pathology, were all DNAb negative. These results extend our previous observations concerning the antibody and suggest that the DNAb is a common feature of PD and other disorders with SN degeneration (McRae-Degueurce et al., 1986 b, 1988 a, b, 1989). These results also support and extend the observations of Husby et al. (1976, 1977) as well as Itagaki et al. (1987), who also observed immunologic activity directed at CNS tissue in PD patients. There may thus be an autoimmune component in PD and the presence of the DNAb may represent a biologic marker of PD.

The hypothesized autoimmune process could represent a primary etiologic factor of PD or a secondary autoimmune component. Itagaki et al. (1987) observed HLA class II microglia phagocytosing tyrosine hydroxylase positive neurons in the brains of patients with idiopathic

PD as well as in the brains of patients with the amyotrophic lateral sclerosis-parkinsonian-dementia syndrome of Guam. The parkinsonian features of this syndrome, like MPTP-induced parkinsonism, are thought to result from exposure to a neurotoxin (Spencer et al., 1987).

Production of the DNAb as a secondary autoimmune process is also supported by the observation that patients with possible nigral involvement in their disease were also DNAb positive. These include a patient with multifocal degeneration exhibiting rigidity as well as a patient exhibiting non-vascular locked-in syndrome with cogwheel rigidity and subnormal CSF HVA, both of whom present clinical pictures consistent with nigral pathology. The 4 DNAb-positive patients with extrapyramidal movement disorders of unknown etiology may similarly have nigral involvement in their disease, although the pathologic picture in these patients is more difficult to predict and could only be established at autopsy. Immunologic involvement in these patients is, therefore, most consistent with the hypothesis that the DNAb is not an etiologic factor in PD, but rather, a secondary response to nigral degeneration, regardless of its cause. This would then suggest that the DNAb is not necessarily a biological marker for PD per se, but rather a biologic marker for nigral degeneration.

All PD patients however, would be expected to have nigral degeneration, with the possible exception of the levodopa non-responsive PD patient (this patient may have parkinsonian features secondary to striatal pathology which would be consistent with his DNAb-negative CSF). However, 20% of the PD patients thus far examined, were immunonegative. This false negative rate may simply represent a lack of sensitivity in the assay. Attempts to develop a traditional ELISA with its associated enhanced sensitivity, is currently under way. Alternatively, all PD patients may not exhibit an immunologic component to their disease. Immunonegative PD patients may therefore represent a subpopulation of PD. A third alternative would suggest that PD patients do not continuously produce the DNAb but that the production of the antibody may wax and wane as it does in multiple sclerosis (Tourtellotte et al., 1980).

A biologic marker of PD should also be specific for the SN. The markage pattern observed in PD patients however, also includes the VTA which is generally not thought to be involved in PD. However, recent studies by Riederer et al. (this volume) demonstrate decreased DA in the frontal cortical regions of PD patients. This would suggest projection cell loss and therefore VTA involvement as part of the pathologic picture of PD. If the DNAb is a secondary response to DA neuron degeneration, immunocytochemical reactivity to the VTA would therefore be anticipated. Prior to being utilized as a biological marker for PD, it would also need to be established whether or not cells (neurons and/or glia) in addition to DA neurons react with the DNAb. This has not been clearly established at this time. Examination of immunocytochemical reactivity in the locus ceruleus, which is involved in PD, are planned. However, the data collected thus far does suggest that the immunocytochemical reactivity observed in the SN is directed at DA neurons. This is supported by 1 — the observation that a lesion of the rat SN using 6 hydroxydopamine, significantly reduces the reactivity; 2 — the observation that DA neuron cultures react with the DNAb; and 3 — that the distribution of the reactivity of the DNAb is similar to the known distribution of dopaminergic neurons in the CNS.

If the DNAb represents a secondary autoimmune response to nigral degeneration, it would, theoretically, be present in the CSF prior to symptom expression. This is based on the commonly held assumption that nigral degeneration precedes symptom expression by many years, but that compensatory mechanisms within the DA system are able to prevent symptom expression until striatal DA content is less than 20% of normal (Hornykiewicz, 1979). The DNAb would then be truly considered an early marker for nigral degeneration implicating PD. It is also possible that, once the degenerative process of PD has advanced to the point where DNAb production occurs, the DNAb may contribute to the further progression of disease as our preliminary data suggests (McRae et al., this volume). This possible effect on disease progression would offer another potential therapy for PD involving immunosuppressive drugs.

The growing list of CNS degenerative disease states which have now been shown to possess antibodies directed at neuronal populations, ostensibly thought to be associated with symptom expression, may have far-reaching implications in our understanding of neurodegenerative disorders in general. Future work may establish that other neurodegenerative disease states induce neuron specific antibodies as appears to be the case in Alzheimer's disease, Huntington's chorea, stiff-man syndrome and PD. Procedures similar to those described in the present study would then be expected to be useful in elucidating the potential central locus of neurodegeneration in CNS disorders where pathology is thought to occur but has not yet been identified. And finally, consideration of the potential role of such autoantibodies in the progression of neurodegenerative disease states would be expected to result in a broader understanding of central disease process offering new potential therapeutic strategies.

38   P. M. Carvey et al.

Acknowledgements

The authors wish to thank Drs. C. G. Goetz, C. M. Tanner, W. Olanow, C. Shults, R. Wright, F. Morrell, D. Stefoski, R. Watts, J. W. Langston, and R. D. Penn for assisting us in gathering CSF samples. This work was supported by grants from the United Parkinson Foundation, The Boothroyd Foundation, and the Rush University Committee on Research.

References

Bluestein HG (1984) Antineuronal antibodies in the pathogenesis of neuropsychiatric manifestations of systemic lupus erythematosus. In: Behan P, Spreafico F (eds) Neuroimmunology. Raven Press, New York, pp 157–165

Carvey PM, Kroin JS, Zhang TJ, O'Dorisio TM, Yaksch TL, McRae A, Dahlstrom A, Kao LC, Goetz CG, Tanner CM, Shannon KM, Klawans HL (1988) Biochemical and immunochemical characterization of ventricular CSF from Parkinson's disease patients with adrenal medulla transplants. Neurology 38 [Suppl 1]: 144

Chapman J, Korczyn AD, Hareuveni M, Michaelson DM (1986) Antibodies to cholinergic cell bodies in Alzheimer's disease. In: Fisher A, Hanin I, Lachman C (eds) Alzheimer's and Parkinson's disease. Strategies for research and development. Plenum Press, New York, pp 329–336

Filit HM, Luine V, Reisberg B, Amador R, McEwen BS, Zabriske JB (1985) Studies of the specificity of autobrain antibodies in Alzheimer's disease. In: Hutton JT, Kenny AD (eds) Senile dementia. Alan R Liss, New York, pp 307–336

Franceschi M, Comola M, Nemni R, Pinto P, Iannaccone S, Smirne S, Canal N (1988) Neuron-binding antibodies in Alzheimer's disease and Down's syndrome. Neurology 38 [Suppl 1]: 285

Gaskin F, Kingsly BS, Fu SM (1987) Autoantibodies to neurofibrillary tangles and brain tissue in Alzheimer's disease. Establishment of Epstein-Barr virus-transformed antibody-producing cell lines. J Exp Med 165: 245–250

Gonatas NK (1984) Immunohistopathology of experimental allergic encephalomyelitis. In: Behan P, Spreafico F (eds) Neuroimmunology. Raven Press, New York, pp 113–126

Hornykiewicz O (1979) Compensatory biochemical changes at the striatal dopamine synapse in Parkinson's disease and limitations of L-dopa therapy. Adv Neurol 24: 275–281

Husby GL, van de Rijn I, Zabriskie JB, Abdin ZH, Williams RC (1976) Antibodies reacting with cytoplasm of subthalamic nuclei neurons in chorea and acute rheumatic fever. J Exp Med 144: 1094–1110

Husby GL, Davis LE, Wedege E, Kokmen E, Williams RC (1977) Antibodies to human caudate nucleus neurons in Huntington's chorea. J Clin Invest 59: 922–932

Ishii T, Haga S (1976) Immunoelectron microscopic localization of immunoglobulins in amyloid fibrils of senile plaques. Acta Neuropathol 36: 243–249

Itagaki S, McGeer PL, McGeer EG (1987) HLA-DR reactive microglia in Parkinson's disease. J Neuroimmunol 16 (1): 81

Johnson KP (1980) Cerebrospinal fluid and blood assays of diagnostic usefulness in multiple sclerosis. Neurology 30: 106–109

McFarlin DE (1984) Immunologic abnormalities associated with neurologic diseases. In: Behan P, Spreafico F (eds) Neuroimmunology. Raven Press, New York, pp 237–245

McFarlin DE, Strober W, Waldmann T (1972) Ataxia-telangiectasia. Medicine 51: 281–314

McRae-Degueurce A, Geffard M (1986 a) One perfusion mixture for immunocytochemical detection of noradrenaline, dopamine, serotonin and acetylcholine in the same rat brain. Brain Res 37: 217–219

McRae-Degueurce A, Gottfries CG, Karlsson I, Svennerholm L, Dahlstrom A (1986 b) Antibodies in the CSF of a Parkinson patient recognize neurons in rat mesencephalic regions. Acta Physiol Scand 126: 313–315

McRae-Degueurce A, Booj S, Haglid K, Rosengran L, Karlsson JE, Karlsson I, Wallin A, Svennerholm L, Gottfries C-G, Dahlstrom A (1987) Antibodies in cerebrospinal fluid of some Alzheimer's disease patients recognize cholinergic neurons in the rat central nervous system. Proc Natl Acad Sci 84: 9214–9218

McRae-Degueurce A, Klawans HL, Penn RD, Dahlstrom A, Tanner CM, Goetz CG, Carvey PM (1988 a) An antibody in the CSF of Parkinson's disease patients disappears following adrenal medulla transplantation. Neurosci Lett 94: 192–197

McRae-Degueurce A, Rosengren L, Haglid K, Booj S, Gottfries CG, Granerus AC, Dahlstrom A (1988 b) Immunocytochemical investigations on the presence of neuron-specific antibodies in the CSF of Parkinson's disease cases. Neurochem Res 13: 679–684

McRae A, Dahlstrom A, Klawans HL, Goetz CG, Tanner CM, Penn RD, Carvey PM (1989) Adrenal medulla transplantation in Parkinson's disease reduces the presence of a CSF antibody to the rat substantia nigra. Neurology 39 [Suppl 1]: 364

Penn RD, Goetz CG, Tanner CM, Klawans HL, Shannon KH, Comella CL, Witt TR (1988) The adrenal medullary transplant operation for Parkinson's disease: clinical observations in five patients. Neurosurgery 22: 999–1004

Rao N, Costa JL (1989) Recovery in non-vascular locked-in syndrome during treatment with sinemet. Brain Injury 3: 207–211

Solimena M, Folli F, Denis-Donini S, Comi GC, Pozza G, De Camilli P, Vicari AM (1988) Autoantibodies to glutamic acid decarboxylase in a patient with stiff-man syndrome, epilepsy, and type I diabetes mellitus. N Engl J Med 318: 1012–1020

Solimena M, Folli F, Pozza G, De Camilli P (1989) Autoantibodies directed against GABAergic synapses in stiff-man syndrome. Neurology 39 [Suppl 1]: 384

Spencer PS, Nunn PB, Hugon J, Ludolph AC, Ross SM, Roy DN, Robertson RC (1987) Guam amyotrophic lateral sclerosis-parkinsonian-dementia linked to a plant excitant neurotoxin. Science 237: 517–522

Tourtellotte WW, Potvin AR, Potvin HH, Ma BI, Baumhefner RW, Syndulko
    K (1980) Multiple sclerosis de novo CNS IgG synthesis: measurement,
    antibody profile, significance eradication, and problems. In: Bauer HJ,
    Poser S, Ritter G (eds) Progress in multiple sclerosis. Springer, Berlin
    Heidelberg New York, pp 106–110

Contribution of motor performance tests to the early diagnosis of Parkinson's disease

P. H. Kraus[1], P. Klotz[2], R. Steinberg[1], and H. Przuntek[1]

[1] Department of Neurology, St. Josef Hospital, University of Bochum, and
[2] Department of Neurology, University of Würzburg, Federal Republic of Germany

Summary

With the help of multivariate statistical methods, we analysed the results of a motor performance test series consisting of simple subtests. We were able to differentiate almost completely between parkinsonian patients (most of whom showed only slight or very slight clinical symptoms) and age-matched controls: we found only 2% false negative diagnoses and no false positives. Thus, impairment of fine motor skills is very frequent in the early stages of the disease and standardized motor performance tests can contribute to its detection. Since we established that a deterioration of the test results is highly age-correlated for these patients, we expect that follow-up testing will further contribute to improve selectivity.

Introduction

The diagnosis "Parkinson's disease" is only established when a constellation of typical clinical symptoms becomes visible. The prospect of new therapies with potential preventive effects makes it necessary to confirm the diagnosis as early as possible. Our approach to this problem is the examination of psychomotor performance using standardized apparative methods, because such tests have a high level of objectivity, reliability and validity. The examination of motor disorders by single apparative tests has been described in a number of studies in which significant differences between patient groups were demonstrated; see for example: Duvoisin (1971), Evarts et al. (1981), Flowers (1975). These tests reveal different aspects of the motoric disorders and some of them are applicable for the simple control of therapy. With the individual tests, however, there is too much overlap between the results from patients and from controls. Early diagnosis requires a reliable case-by-case answer. Since early diagnosis must also be a differential diagnosis, the methods to be developed must meet high requirements with respect to their selectivity (i.e. specificity and sensitivity).

We use a battery of simple subtests which are appropriate for differential diagnosis of motor disorders in diseases with multidimensional symptoms.

The results from the selected battery of tests provide much more information than those obtained by evaluating individual subtests. A simple example is the correlation between the results from the two hands, because the development of Parkinson symptoms is usually asymmetrical. It is also important to take into account the age dependence of fine motor ability (Kraus et al., 1987 a; Kraus and Przuntek, 1989). In line with the various constellations of clinical symptoms, various test profiles may be expected and these may also have differential diagnostic value.

Materials and methods

The performance test used in the present study consists of several subtests which were carried out separately for the right and left hand as described by Schoppe (1974).

Plugging: The subject is required to transfer 25 pins (diameter 2.5 mm, length 5 cm), individually and as quickly as possible, from a rack into a series of appropriate holes (diameter 2.8 mm) in a contact board. This test measures the time interval between the plugging in of the first and the last pin.

Tapping: The subject is required to tap on a contact board with a contact pencil as rapidly as possible. This test measures the number of contacts during two time intervals of 16 seconds each. As a measure of speed alteration (e.g., caused by fatigue) we use the difference between the two numbers as an additional parameter.

The *Steadiness 4.8 mm* test was performed in the smallest hole of the series in the board without support for the arm in action. A contact pencil had to be held for 32 seconds vertically in a hole, 4.8 mm in diameter, without touching either the rim or the bottom of the hole. This test measures the number and duration of contacts. The *Steadiness 8.5 mm* test was performed in the same manner using the hole 8.5 mm in diameter.

In the *Aiming* test the subject has to hit 20 contacts with the contact pencil. This test measures the total time required to perform the test, the number of hits, the number and duration of misses.

In the *Line tracing* test the patient has to follow a grooved path with a stylus as closely as possible. This test measures the total time required to perform the test, the number and duration of errors.

Controls and patients

The present study includes only controls and patients who are right-handed and aged between 40 and 85 years: 52 healthy controls (mean age 62.0 ± 12.0 years) and 107 patients (54 women and 53 men, average age 62.5 ± 8.6 years, average duration of disease 3.6 years, Hoehn and Yahr scale between stage 1 and 4). Patients were assessed as follows: 11% severe cases, 40% moderately severe cases, 49% slightly diseased (about 10% had only very discrete symptoms). The major symptoms were tremor in 13% of the patients and motor deficiency in 72%; in 15% of the cases tremor and motor deficiency were equivalent. Apart from recently diseased patients, all patients were under specific medication.

Statistical analysis was performed using the U-test according to Mann and Whitney for comparisons at the subtest level; the multivariate analysis included factor, cluster and discriminant analyses. The evaluation of the entire data (68 dimensions) was carried out with cluster analysis with squared Euclidean distances according to Ward using a 3 cluster solution. At the subtest level, evaluations did not take age dependence into account; discriminant analysis was carried out by using all the data after correction by age, allowing and including, additionally, average error times for subtests Steadiness, Aiming and Line tracing.

Results

The results of the subtests were compared using the U-Test according to Mann and Whitney because the parameters measured were not normally distributed. We found significant differences in all subtests except for Aiming, number of hits (right hand), and for Tapping-difference (left hand) (Table 1). There is, however, a marked overlap between the two groups (Fig. 1). The discriminating power at the subtest level is not sufficient for the purposes of early diagnosis, in particular because precisely those cases which are of most interest, i.e. those with few symptoms, are found in the overlap area. As expected in this group of mostly mildly diseased patients, the differences measured with the Steadiness test with the small hole (4.8 mm) were larger than those found with the 8.5 mm hole, because the floor effect with the larger hole leads to reduced resolution. The most marked differences were found for the subtest Plugging, whereas Tapping, which is used in different clinical scores, yielded relatively small differences.

First of all, we examined the battery of tests for internal consistency, making use of explorative statistical methods. The structuring of the variables by factorial analysis and the correlation between the individual subtests demonstrated that a useful combination of subtests had been selected, i.e., that indeed several different fine motor abilities were being measured.

Table 1. Evaluation at subtest level (Mann-Whitney-U-test)

Subtest	Right hand			Left hand		
	Mean rank		p =	Mean rank		p =
	Co (51)	PD (102)		Co (51)	PD (102)	
Steadiness 8.5, Errors	50.3	90.4	0.000*	51.8	89.6	0.000*
Steadiness 8.5, Error-time	52.8	89.1	0.000*	52.4	89.3	0.000*
Steadiness 4.8, Errors	50.4	89.7	0.000*	58.8	86.1	0.000*
Steadiness 4.8, Error-time	48.3	90.7	0.000*	51.6	89.7	0.000*
Tracing, Errors/Number	58.1	86.5	0.000*	63.4	83.8	0.007*
Tracing, Error-time	56.2	87.4	0.000*	62.5	84.3	0.004*
Tracing, Time	53.1	89.0	0.000*	55.5	87.8	0.000*
Aiming, Errors	53.1	89.0	0.000*	54.8	87.5	0.000*
Aiming, Hits	72.2	79.4	0.284 ns	63.2	83.2	0.003*
Aiming, Error-time	56.3	87.3	0.000*	53.5	88.1	0.000*
Aiming, Time	59.2	85.9	0.000*	64.7	82.5	0.019*
Tapping, Hits, 1st	92.8	69.1	0.002*	91.7	69.1	0.002*
Tapping, Hits, 2nd	89.8	70.6	0.012*	92.5	68.7	0.002*
Tapping, Difference	89.7	70.7	0.012*	77.6	76.0	0.837 ns
Plugging, Time	42.5	91.9	0.000*	45.8	89.1	0.000*

* Probability less significance level (0.05); *ns* not significant; *CO* controls; *PD* Parkinson's disease. Number of persons in parenthesis. With almost all the subtests, highly significant differences between the collectives were obtained

Explorative analysis at the inter-person level using cluster analysis revealed that the patients were grouped into interpretable clusters: for clarity purposes Fig. 2 shows a simple example for just three parameters for the right hand. In addition to the ability profile, the cluster analysis used here took into account the ability level.

Of particular interest are the last two groups shown in Fig. 2, both having moderate unsteadiness in the Steadiness (4.8 mm) test: the Tapping result of the one group is actually better than that of the healthy group whereas that of the other group is markedly worse. This is in good agreement with our observation that some patients with tremor can readily carry out arbitrary oscillatory movements whereas others are markedly impaired in their ability to produce such antagonistic alternating motions. The difference is not as marked in this simple three parameter model as when one actually measures the regularity of tapping (Kraus et al., 1987 a): although some patients

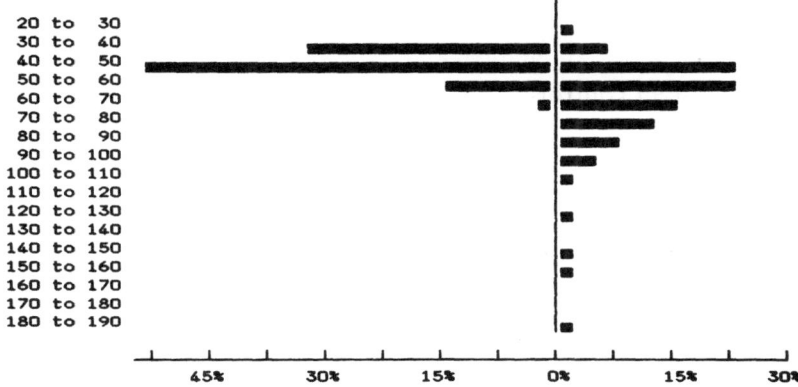

PLUGGING
TIME (/SEC), RIGHT HAND

Fig. 1. Plugging (right PD patients, left healthy controls): although there are highly significant differences between the collectives evaluated at the subtest level, precisely those cases of interest (few symptoms) lie in the overlap area. Therefore, the resolving power of an individual subtest is insufficient for the requirements of early diagnosis

MLS-Test-Profiles
of Parkinsonian Patients
51 Controls / 97 Patients

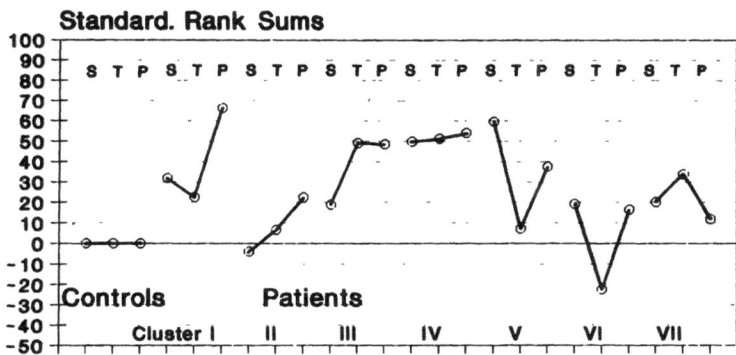

S: Steadiness, Number of Contacts, Right
T: Tapping, Number of Hits, Right
P: Plugging, Right

Fig. 2. With the 3-parameter model, 7 clusters of Parkinson patients with differing ability profiles are obtained

with tremor can still tap relatively rapidly, the regularity of this movement is enormously impaired.

Discriminant analysis allowed a satisfactory separation of the total patient group from the controls. The evaluation of the patient clusters (3 cluster solution) and the control group produced an optimal differentiation with only 2% false negative allocations and correct classification of all controls.

Discussion

The motivation for early diagnosis of Parkinson's disease is based principally on indications, mostly from animal experiments, that the course of the disease could possibly be influenced by medications which seem to have a protective effect. Confirmation of this hypothesis would then justify early therapeutic measures, for which an early diagnostic procedure would be required as quickly as possible.

To the question of therapy control, we studied the correlation between the results of the motor ability test series and the clinical classification of the patients according to the Webster Score (Webster, 1968) using data from a large multicentre study. This led to the conclusion that the fine motor ability test includes some parameters which cannot be obtained by clinical examination. At the early stages of Parkinson's disease clinical scores cannot be used to evaluate the motor tests because these scores were developed for therapy control and have their highest resolving power in quite a different range. In addition, their ability to differentiate is relatively coarse (mostly nominal and ranked scales) and rating is subjective. On the other hand, the advantages of apparative examinations are especially apparent at these stages: if the degree of difficulty of the test is appropriately chosen, the battery of tests emphasizes the differences most sensitively in the most interesting region, that is, at the border between healthy and diseased.

The fine motor ability test presented here is to be classified among the tests for pre-clinical diagnosis as a test for early symptoms, on the assumption that the impairment of fine motor ability is to be seen as a manifestation of the disease. On further development of the method with multidimensional total data registration and high resolution procedures, a contribution to diagnosis in the predictive sense can be expected even before recognizable clinical symptoms. There is at present no reliable method of predicting Parkinson's disease. A method such as that presented here for the measurement of fine motor ability could anticipate the time of diagnosis to such an extent that appropriate measures could be taken.

Even simple parameters from the battery of tests provide a good differentiation between the two groups. Correction of the data by patient's age, calculation of further parameters such as error times, and a clusterwise discriminant analysis appropriate for the problem yield optimal resolution.

The final evaluation of the established discriminant functions, however, cannot take place without the independent analysis of another sample population. The reliable classification of new individuals in the existing clusters is too difficult for routine use.

The fine motor ability test has been validated at the subtest level (Kraus et al., 1987 a); the corresponding validation after multivariate analysis has still to be carried out.

The results presented here are suitable, firstly as a model for assessing the possibilities of multivariate analysis of the battery of tests; secondly, for predicting the type of further developments.

This battery of tests can be made more powerful by the sensible addition of extra subtests which could take into account specific problems of differential diagnosis. We have already taken the first steps in this direction: for routine examinations, we use a modified battery of tests which also measures the regularity of rapid oscillatory movements (Kraus et al., 1987 a), and carry out apparative investigation of diadochokinesia movements as well (Kraus et al., 1987 b).

The results of our cross-sectional analysis of the age dependence of fine motor ability in the two groups indicate that follow-up examinations should bring a considerable improvement in resolution. In any case, the method presented here is not time-consuming and puts hardly any strain on the patient. Using the normal values that we have obtained, the examination can give its contribution as a screening test for the very early detection of motor disturbances. For the purpose of differential diagnosis, we expect a contribution from the differentiation of performance profiles for different diseases.

References

Duvoisin RC (1971) The evaluation of extrapyramidal disease. In: Ajuriaguerra J (ed) Monoamines, noyaux gris centraux et syndrome de Parkinson, Symposium Genève 1970. Masson, Paris, pp 313–325

Evarts EV, Teräväinen H, Calne DB (1981) Reaction time in Parkinson's disease. Brain 104: 167–186

Flowers KA (1975) Ballistic and corrective movements on an aiming task: intention tremor and Parkinsonian movement disorders compared. Neurology 25: 413–421

Kraus PH, Przuntek H (1989) Motor performance test. In: Przuntek H, Riederer P (eds) Early diagnosis and preventive therapy in Parkinson's disease. Springer, Wien New York, pp 75–82 (Key Topics in Brain Research)

Kraus PH, Klotz P, Fischer A, Przuntek H (1987 a) Assessment of symptoms of Parkinson's disease by apparative methods. J Neural Transm [Suppl 25]: 89–96

Kraus PH, Keck B, Klotz P, Przuntek H (1987 b) Computer aided analysis of diadochokinesia. Electroencephalogr Clin Neurophysiol 66 (5): 57

Schoppe KJ (1974) Das MLS-Gerät: Ein neuer Testapparat zur Messung feinmotorischer Leistungen. Diagnostica 20: 43–46

Webster DD (1968) Clinical analysis of the disability in Parkinson's disease. Modern Treatment, Hagerstown (Md) 5: 257–282

Electrophysiological methods in the early diagnosis of Parkinson's disease

C. H. Lücking, A. Hufschmidt, and G. Deuschl

Neurologische Klinik und Poliklinik der Universität, Freiburg i. Br.,
Federal Republic of Germany

Summary

Electrophysiological methods and motor performance tests are shown to be helpful to the early diagnosis of Parkinson's disease. The most sensitive out of five motor tests studied here is visuo-manual tracking. Tremor recordings allow to differentiate between physiological and pathological tremor. An enhancement of long-latency reflexes may precede the development of parkinsonian symptoms on the hitherto unaffected side in unilateral parkinsonism. The combination of motor function tests with tremor and long-latency reflex recordings is likely to be more sensitive than any single test.

Introduction

The clinical investigation and evaluation of the symptoms and complaints of parkinsonian patients are still the most important way to a correct diagnosis. At an early stage of the disease, however, the examiner is often misled by unspecific signs such as shoulder-arm pain, clumsiness and trembling or anxiety and depression. Even with the differential diagnosis of Parkinson's disease in mind, it is not always possible to arrive at a definite clinical decision. Most of the additional diagnostic procedures do not really contribute to clarify the diagnosis. In the near future, the PET-scan may turn out to be able to demonstrate the reduced metabolism in the nigrostriatal complex.

Electrophysiological methods play a major role in analyzing abnormal motor behaviour. Therefore, various attempts have been made in the past to study akinesia, tremor and rigidity more in-depth. There are only few investigations of the early stages of Parkinson's disease (Deutschl et al., 1989; Lücking et al., 1986).

The most disabling symptom for the patients, as a rule, is *hypo-*

or akinesia. In the past, various aspects of impaired motor function related to akinesia have been demonstrated. Reaction time to visual and auditory stimuli is prolonged in some patients, but the slowing is not dramatic and its correlation with the patient's clinical state is poor (Evarts, 1981). The additional delay observed in choice reaction time is even the same as in a normal population. Movement time seems to be a more reliable indicator of motor impairment (Hallett, 1985). Other tests which have proved to be sensitive include visuo-manual tracking (Flowers, 1978; Day et al., 1984) and the performance of sequential (Goldenberg et al., 1986) as well as simultaneous movements (Benecke, 1986). For the present study, a battery of very simple computerized tests of motor function was employed.

Combinations of different test have been used before as a diagnostic tool (Kraus and Przuntek, 1988).

Tremor in early parkinsonism may differ from that in the advanced stages in several aspects, namely in the frequency of resting tremor and in the amount and type of additional postural tremor. Classically, patients present a resting tremor which disappears when the involved muscles are voluntarily activated (type I). The main tremor frequency is between 4 and 5 Hz. In more than half of the patients the resting tremor is combined with a postural tremor, which can have the same frequency and reciprocal innervation pattern (type II). One quarter of the patients, however, exhibit postural tremor at a much higher frequency (6 to 10 Hz) with the characteristics of essential tremor (type III). A few patients show only an enhanced physiological tremor which may be detectable only by means of an accelerometer (type IV). Its main frequency is 8 to 12 Hz. Essential and enhanced physiological tremor can be differentiated by measuring the power spectrum of the accelerogram while weighting the arm or hand with an additional load (Hömberg et al., 1987; Deuschl et al., 1989). Under this condition, the frequency of physiological tremor decreases whereas in the essential tremor it does not.

Additional electrophysiological data may be provided by investigation of *long-latency reflexes* (LLR) evoked by electrical stimulation of the median nerve. In previous studies, it had been shown that in more than half of the patients with Parkinson's disease the early component LLR I is enhanced. As to the different tremor types, mainly type II tremor is characterized by an enhancement of LLR I. This is true for the early as well as the advanced stages of the disease.

Subjects and methods

The study is based on two indepedent groups of patients with early Parkinson's disease. Motor performance tests were carried out on 30 parkinsonian patients, 30 age-matched normal subjects and 6 patients with uni-lateral parkinsonian symptoms. Tremor and LLR recordings were performed on 24 patients (15 men, 9 women), 20 of whom had purely unilateral par-kinsonian symptoms. Their mean age was 58.7 years (range 49 to 78 years). The diagnosis of Parkinson's disease was accepted when at least two of the following main symptoms: resting tremor, bradykinesia, rigidity, or postural abnormality were detectable. The duration of illness was retrospectively as-sumed to be between 3 and 24 months (mean 15 months). At the time of examination the Webster score did not exceed 12 points in all patients.

Motor performance tests

The test battery consisted of:
1. Acoustic reaction time.
2. Visual reaction time.
3. Movement time: the patients had to lift their arm from the table and press a button placed at a distance of 50 cm to the left from the starting position of the hand. The task was started by an acoustic signal. Movement time as assessed here, therefore, includes an acoustic reaction time.
4. Diadochokinesis: the patients held a knob between their thumb and index finger and had to turn it to and fro (by supination/pronation) as rapidly as possible for ten seconds. The mean angular velocity of this movement was taken as an indicator of diadochokinesis.
5. Tracking: the patients were faced with an illuminated bar which crossed a TV screen at a steady rate from left to right and back. They had to keep up with this bar by a second bar which they could control by a potentiometer (pursuit tracking). As a parameter of tracking precision, the absolute distance between the target bar and the response bar was integrated over one trial.

For reaction time measurements, a button press was required as a re-sponse. The best value out of five trials was selected.

Tremor recording

Tremor was recorded by means of surface electrodes on two or three pairs of antagonistic muscles and by an accelerometer attached to the hand. Tremor frequency was determined from the accelerometer signal with an on-line Fast-Fourier-analyser (Nicolet Med. 999).

LLR recordings

Long-latency reflexes (and H-reflexes) were recorded from the thenar muscles by using surface electrodes after electrical stimulation of the median nerve. Details of the technique are published elsewhere (Deuschl et al., 1985).

Results

Motor performance

The patient group did significantly worse on all tests of motor performance. Figure 1 shows test scores of parkinsonian patients and normal controls. The data are normalized, so that the mean value of the control group is 100%. The difference in acoustic reaction time is significant only at a 2.5% level; all other differences are highly significant at levels of 0.5% and lower. The Fig. demonstrates that among the motor skills tested in this study tracking is the most sensitive in differentiating between normal subjects and parkinsonian patients.

Figure 2 displays a selection of results from 6 patients with unilateral or at least asymmetric Parkinson's disease. All patients were right-handed, and all showed their symptoms on the right side. This group was preferred for analysis, because their handedness can be expected to act to the advantage of the affected side. Otherwise, any asymmetries found could be explained partly by the subjects' hand dominance. The grey bars represent the individual scores on the symptomatic side. In this diagram, diadochokinesis is plotted as angular velocity. So, a slowing is symbolized by a grey bar which is shorter than the black one. The patients in this series exhibited remarkable differences in their test profile. Patient no. 5, for instance, is extremely bad at visual pursuit tracking. In diadochokinesis, however, he per-

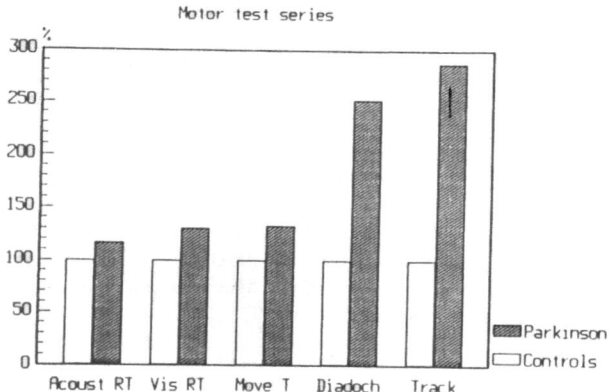

Fig. 1. Mean test scores of motor performance in 30 patients with early Parkinson's disease and 30 age-balanced controls. Normalized data (mean score of control group = 100%). For diadochokinesis, the reciprocal value of angular velocity is plotted. Significance levels, see text

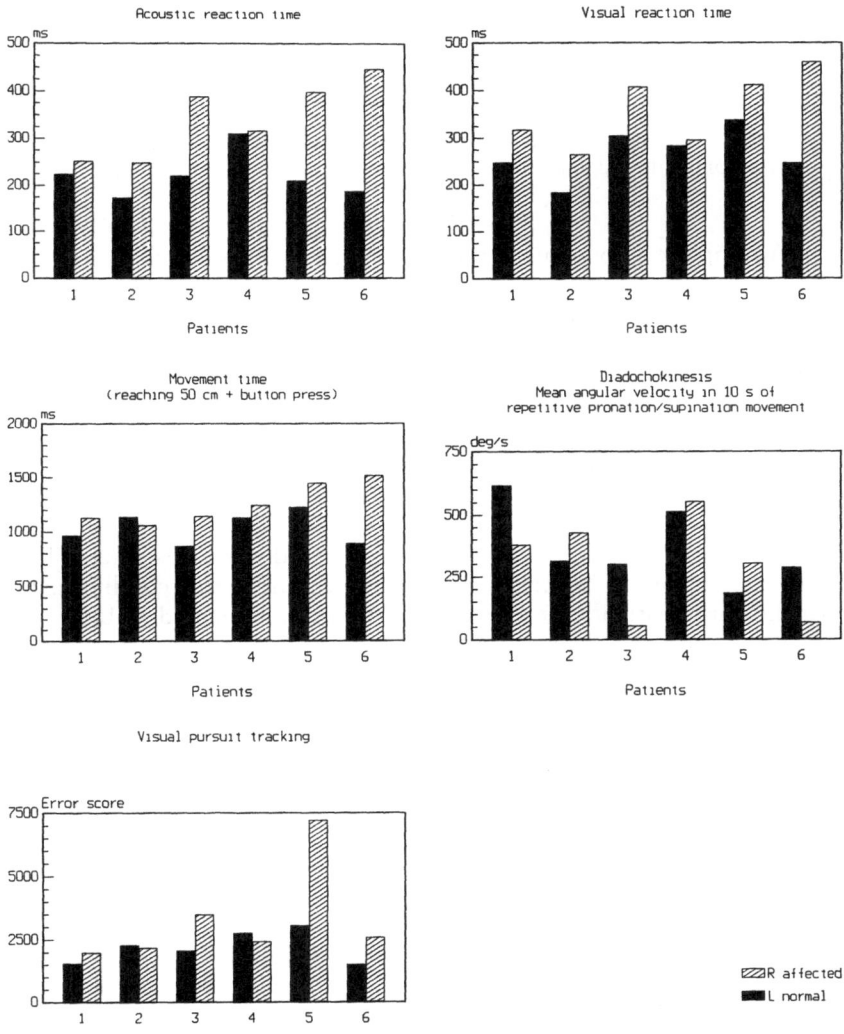

Fig. 2. Motor test scores in 7 patients with right-sided parkinsonian symptoms. *Black bars:* normal side; *grey bars:* affected side

forms faster with his affected hand. These two tests require a virtually identical motor output. The subjects have to turn the same knob, once to and fro as fast as they can, once to control their pursuit movements. The velocity of the hand movement required for tracking is much lower than and indeed far below the slowest movement recorded in any patient in spontaneous diadochokinesis. As another example, patient no. 6, who is the slowest of all in reaction time and movement time, performs quite well at the tracking task. These data on unilateral Parkinson's disease suggest that different patients exhibit different patterns of motor disability.

Tremor

Among the 24 patients in the second group, 21 presented a tremor. Seventeen had a resting tremor, five of them with a rather high frequency, between 6 and 9 Hz, which is not seen in advanced stages (Fig. 3). Sixteen patients showed a postural tremor with frequencies ranging between 5.5 and 8.5 Hz. This corresponds well to the postural tremor seen in patients with advanced parkinsonism. According to our classification in type I to IV, 6 patients belong to type I, 2 to type II, 9 to type III, and 4 had an enhanced physiological tremor (type IV). In some patients suffering from early Parkinson's disease without clinically visible tremor of the hands, the modification of the frequency pattern following successive loading of the hand was investigated. The frequency of physiological tremor decreases under this condition. In parkinsonian patients, however, the frequency remained stable even when the hand was loaded with 1 kg. We conclude that accelerometry may help to differentiate between physiological and pathological tremor even if they are clinically indistinguishable.

Long-latency reflexes

The normal reflex pattern consists of an early monosynaptic reflex (H-reflex) at a mean latency of 29 ms and two long-latency reflexes at

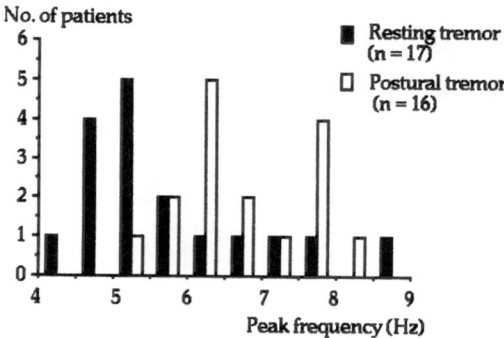

Fig. 3. Frequency range of resting tremor and postural tremor in parkinsonian patients. Nine patients had a combination of both tremor types

affected side   **unaffected side**

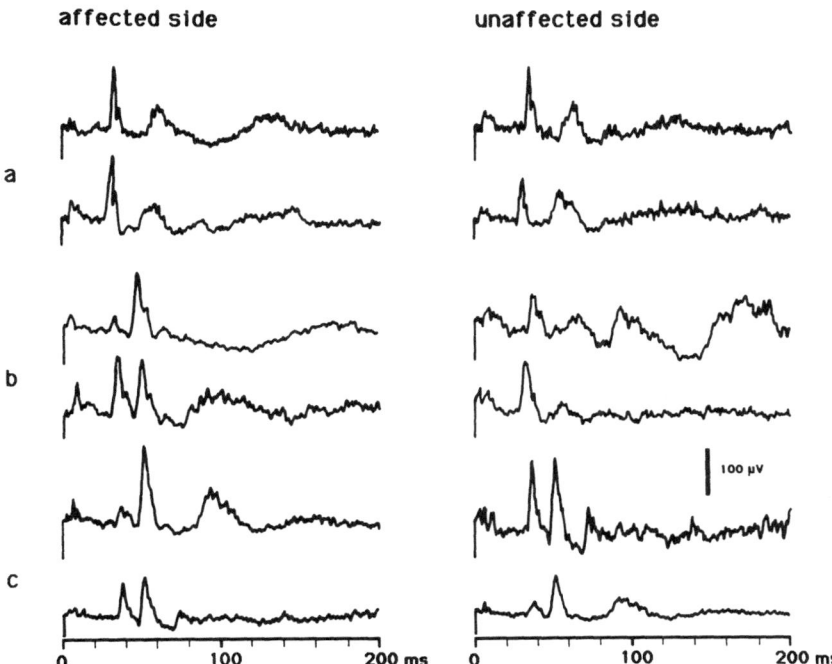

a

b

c

0   100   200 ms   0   100   200 ms

Fig. 4. Long-latency reflexes of 6 patients with unilateral Parkinson's disease recorded from both hands. Two patients with normal patterns with HR and LLR II (**a**). Enhanced LLR I on the clinically affected side only in two patients (**b**) and bilaterally enhanced LLR I in two patients (**c**)

around 50 ms and 60 ms, respectively (LLR I and II). The evaluation of long-latency reflexes centred on nine patients with unilateral parkinsonian symptoms. All of them had enhanced LLR on the affected side with a mean ratio between absolute LLR amplitude and baseline activity of 2.0 (normal: < 0.8). Interestingly, four patients exhibited a pathological LLR also in their clinically unaffected side. Examples are displayed in Fig. 4.

Conclusions

Our data on *motor performance* in early parkinsonism suggest that tests of akinesia are more sensitive if they require complex tasks rather than simple ones. This is illustrated by Fig. 1, in which the tests are arranged from right to left in an order of increasing complexity. The analysis of test scores in the group of unilateral parkinsonian subjects reveals that different patients exhibit different patterns of disturbances. For the early diagnosis of Parkinson's disease, therefore, a combination of motor functions tests will be more sensitive than any single one. Another conclusion is that our understanding of akinesia may develop from the concept of a unique symptom towards a set of basic disturbances which may combine in a variable fashion. This could be verified by demonstrating specific patterns of correlation and non-correlation between different motor abilities in the parkinsonian group. Such links are currently being investigated.

The findings on parkinsonian *tremor* presented in this study do not basically differ from previous data on patients with advanced Parkinson's disease. It became evident, however, that a pure resting tremor of higher frequency or a pure postural tremor of the essential type do not contradict the diagnosis of early parkinsonism. A slowing of tremor frequency is likely to occur along with the progression of the disease.

The value of *LLR recordings* as a diagnostic procedure is illustrated by the manifestation of an enhanced LLR I on the affected as well as on the clinically unaffected side of four patients with unilateral parkinsonism. In five patients, it was restricted to the affected side. Until now, no reliable data on LLR in the evolution of Parkinson's disease are available. However, the present finding of enhanced LLR I on the unaffected side of hemiparkinsonian patients shows that the altered reflex activity can precede the onset of clinical symptoms. So, this method may be important for the differential diagnosis in patients presenting an unusual tremor.

In contrast to biochemical, morphological and imaging procedures, electrophysiological methods record disorders of function which are at the basis of the symptoms and complaints of patients with Parkinson's disease. Therefore, further attempts should be made to develop quantitative methods for the documentation of motor performance to facilitate the diagnosis of movement disorders.

References

Benecke R, Rothwell JC, Dick JPR, Day BL, Marsden CD (1986) Performance of simultaneous movements in patients with Parkinson's disease. Brain 109: 739–757

Day BL, Dick JPR, Marsden CD (1984) Patients with Parkinson's disease can employ a predictive motor strategy. J Neurol Neurosurg Psychiatry 47: 1299–1306

Deuschl G, Lücking CH (1989) Tremor and electrically elicited long-latency reflexes in early stages of Parkinson's disease. In: Przuntek H, Riederer P (eds) Early diagnosis and preventive therapy in Parkinson's disease. Springer, Wien New York, p 103–110

Deuschl G, Schenck E, Lücking CH (1985) Long-latency responses in human thenar muscles mediated by fast conducting muscle and cutaneous afferents. Neurosci Lett 55: 361–366

Evarts EV, Teräiväinen H, Calne DB (1981) Reaction time in Parkinson's disease. Brain 104: 167–186

Flowers K (1978) Lack of prediction in the motor behaviour of parkinsonism. Brain 101: 35–52

Goldenberg G, Wimmer A, Auff E, Schnaberth G (1986) Impairment of motor planning in patients with Parkinson's disease: evidence from ideomotor apraxia testing. J Neurol Neurosurg Psychiatry 49: 1266–1272

Hallett M (1985) Quantitative assessment of motor deficiency in Parkinson's disease: ballistic movements. In: Delwaide PJ, Agnoli A (eds) Clinical neurophysiology in Parkinsonism. Elsevier, Amsterdam, p 351

Hömberg V, Hefter H, Reiners K, Freund HJ (1987) Differential effects of changes in mechanical limb properties on pyhsiological and pathological tremor. J Neurol Neurosurg Psychiatry 50: 568–579

Kraus PH, Przuntek H (1988) Motor performance test. In: Przuntek H, Riederer P (eds) Early diagnosis and preventive therapy in Parkinson's disease. Springer, Wien New York, pp 75–82

Lücking CH, Deuschl G, Strahl K, Schenck E (1986) Tremor im Früh- und Spätstadium der Parkinson-Krankheit. In: Fischer PA (Hrsg) Spätsyndrome der Parkinson-Krankheit. Editions Roche, Basel, S 99–113

Characterisation of Parkinson's disease using positron emission tomography

K. L. Leenders

PET Group, Paul Scherrer Institute, Villigen, Switzerland

Summary

Positron emission tomography (PET) can be applied in the study of the pathophysiology of Parkinson's disease (PD) and other conditions. An early diagnosis of PD should in principle be possible, since in this condition dopamine turnover is markedly decreased while dopamine D_2 receptor-density is generally unimpaired. In other neurodegenerative conditions accompanied by parkinsonism both "pre" and "post-synaptic" binding of tracers seems to be impaired.

In PD the loss of cells within the nigrostriatal pathway seems less outspoken when compared to the severe decrease of endogenous dopamine concentration.

Introduction

In recent years it has become possible to measure in vivo certain aspects of human striatal dopaminergic function using radio-labelled tracers and positron emission tomography (PET). Further validation and expansion of this method may lead to elucidation of the pathophysiology of brain disorders in which a disturbance of one or more neurotransmitter systems has been demonstrated.

Through PET studies it will now be possible to relate changes in neurotransmitter function to clinical features. Particularly, longitudinal studies starting in an early phase of the disease and using various types of tracers seem to be promising. Cross-sectional cohort studies are less suitable due to the rather small number of patients which can be currently scanned. This is not just caused by the relatively long duration of the scanning procedures; also the radiochemistry is often complicated and has to be performed immediately before a scan because of the short radioactive half-life of the radionuclides incorporated in the tracer-molecules. However, data handling and analysis are the most time-consuming aspect of measuring tissue function with PET. Reduction of count measurements into manageable units and conversion of time-activity curves into meaningful pharmacological or biochemical entities is a formidable task. It seems that the developments in this field are still in an early stage. The inevitably low patient throughput per scan laboratory, in combination with the small number of PET centres worldwide, makes it understand why accumulation of biological or clinical results with PET is a slow process.

Positron emission tomography (PET)

The compounds (ligands) used with PET are administered in trace amounts. A PET scanner is able to detect the uptake of these tracers into tissues, e.g. brain or heart, since a special type of radionuclide is incorporated as "label" in the tracer molecules. These radionuclides decay by emitting a positron, e.g. oxygen-15, carbon-11 or fluorine-18. A positron is a particle with the same mass as an electron but positively charged. The three main reasons for choosing this type of radionuclides are the following. First, they are nuclides of physiological atoms, which means that their incorporation into the required tracer molecules does not change, or only slightly, so the chemical properties of the tracer. Secondly, the short radioactive half-life (minutes to a few hours) allows administration of the tracer in a dose sufficient to obtain measurable signals while keeping the radiation dose low enough for human use. Thirdly, the characteristic physical features accompanying positron emission are at the basis of tomographical measurement of regional radioactivity.

Shortly after emission from a decaying nucleus a positron annihilates with an electron. This results in conversion of the masses of the positron and electron into two simultaneous high energy gamma rays (511 keV) travelling into opposite directions. The construction of most PET tomographs is such, that a ring of detectors surrounds the body. Simultaneous stimulation of two opposite detectors (coincidence event) by the two gamma rays resulting from emission of a positron allows exact determination of the direction from where the event took place. After collection of sufficient coincidence events (counts) within a certain time frame (seconds to minutes), the distribution of local radioactivity in the scanned cross-section (plane) can be calculated by standard tomographical reconstruction techniques. Thus for each region in the brain a time-activity curve can be determined in absolute units of radioactivity (microcurie per ml tissue). The buildup and washout of a tracer in a certain brain tissue region become more meaningful when they can be compared with the dose delivered to the brain via the arterial system. To obtain this information a series of blood samples is usually taken from a small indwelling radial artery cannula after administration of the tracer. From this a so-called arterial input curve is then derived. Whether the next step, namely calculation of a pharmacological or biochemical entity related to the tracer activity, can be achieved, depends on the specific properties of the tracer molecule. The mathematical models which are used for this purpose vary widely in complexity and various assumptions are waiting for validation.

Results and discussion

Pre-synaptic tracers

L-^{18}F-$fluoro$-$3,4$-$dihydroxyphenylalanine$ (^{18}F-dopa) is an analogue of L-dopa that can be used as a tracer for L-dopa transport from blood to brain, dopamine formation and subsequent conversion into metabolites (in striatum mainly HVA and DOPAC) (Garnett et al., 1983; Firnau et al., 1987; Leenders et al., 1986 a, b, c, 1988 b). Figure 1 illustrates the radioactivity distribution throughout the brain after ^{18}F-dopa administration to a healthy volunteeer and a patient with Par-

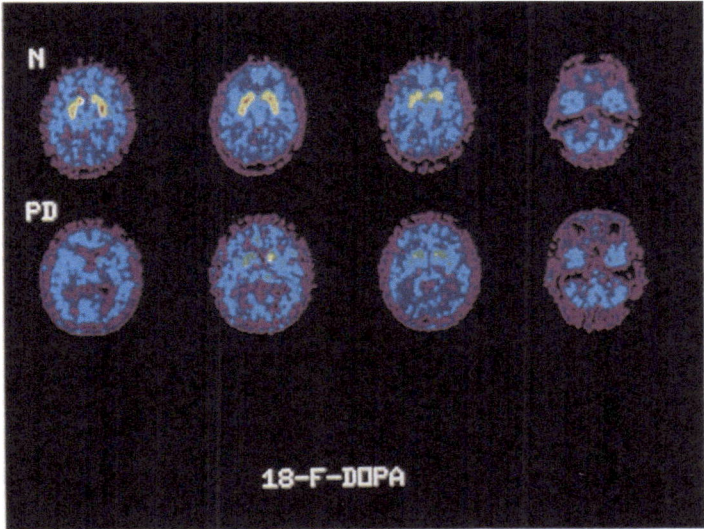

Fig. 1. L-^{18}F-6-fluorodopa uptake into brain measured in a healthy volunteer (upper row) and in a patient with Parkinson's disease (lower row). The images represent summated activity accumulated from one to two hours after tracer administration. Here 4 images are shown cutting the brain at four levels parallel to the orbitomeatal plane. The 3 left-hand side images cut through the striatum, whereas the right-hand side image is at a lower (cerebellar) level. Specific tracer uptake in the patient's striatal regions is markedly impaired compared to normal. The measurements were performed by the author using the PET tomograph at the MRC Cyclotron Unit, Hammersmith Hospital, London, UK

kinson's disease respectively. A typical time-activity graph is shown in Fig. 2.

L-dopa transport across the blood-brain barrier is an active, energy-dependent and strictly stereoselective process in competition with other large neutral amino acids (Leenders et al., 1986 c). ^{18}F-dopa is decarboxylated to ^{18}F-dopamine in the endothelial cells of brain capillaries and in the brain tissue itself, particularly in decarboxylase-rich regions like striatum. ^{18}F-dopamine is further metabolized into ^{18}F-HVA and ^{18}F-DOPAC, but Firnau and colleagues (Firnau et al., 1987) showed that in monkey brain during the first 1 to 1.5 hour after ^{18}F-dopa-administration, the radioactivity in striatum consisted predominantly of ^{18}F-dopamine. Since this metabolic pathway runs only in one direction, the sum of ^{18}F-dopamine, ^{18}F-HVA and ^{18}F-DOPAC formation must thus be determined by the regional dopadecarboxylation rate. In cerebral tissues other than the striatum, the O-methylated derivative was the major labelled compound.

In arterial plasma of subjects pretreated with carbidopa, the main metabolite after ^{18}F-dopa administration was found to be the O-methylated derivative (Boyes et al., 1986). Further advances in ^{18}F-dopa uptake measurements with PET may arise from using catechol-O-methyltransferase (COMT) inhibitors suitable for human use. Blocking of methylation in the periphery would improve the arterial input curve: the total radioactivity in plasma might possibly be equivalent to ^{18}F-dopa itself. Blocking of tissue methylation would result in all activity being derived from ^{18}F-dopamine or a metabolite beyond the decarboxylation step. The kinetic modelling of cerebral ^{18}F-dopa uptake to estimate regional dopamine formation would certainly become easier.

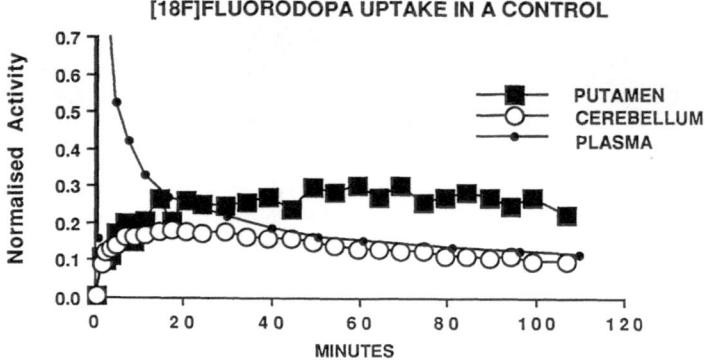

Fig. 2. Time-activity curve of L-^{18}F-6-fluorodopa uptake in putamen, cerebellum and arterial plasma in a healthy volunteer

The positive effect of a COMT inhibitor on ^{18}F-dopa uptake in rats has been demonstrated by Cumming et al. (1987).

[18]F-dopa uptake into brain can be expressed as a regional unidirectional influx constant K_i (Patlak et al., 1983, 1985). As expected, this index for presynaptic striatal dopa-decarboxylation activity is markedly decreased in patients with Parkinson's disease (Fig. 3). No overlap with controls is seen for the putamenal values, whereas some patients' caudate nucleus values are in the normal range. Mean values for putamen are about 40% of controls and for caudate nucleus about 85% of controls. The same percent decreases of dopadecarboxylase activity in the postmortem brain of parkinsonian patients were found by Nagatsu et al. (1979).

Nomifensine (NMF) binds specifically to catecholamine uptake sites on nerve terminals (Slater et al., 1984; Scatton et al., 1984). In the striatum specific nomifensine binding is virtually only related to the binding to dopaminergic nerve terminals. Unilateral lesions of the nigrostriatal dopaminergic pathway in rats produced a marked (about 80%) decrease of specific striatal binding of ^{3}H-nomifensine (Scatton et al., 1984).

Nomifensine labelled with the positron emitting radionuclide carbon-11 (^{11}C-NMF) has been applied in PET studies (Aquilonius et al., 1987; Leenders et al., 1988). A similar experiment, as reported by Scatton et al. (1984) in rats using autoradiography and ^{3}H-NMF, was performed on a Rhesus monkey using PET and ^{11}C-NMF (Leenders et al., 1988). PET scans were performed before and after administration

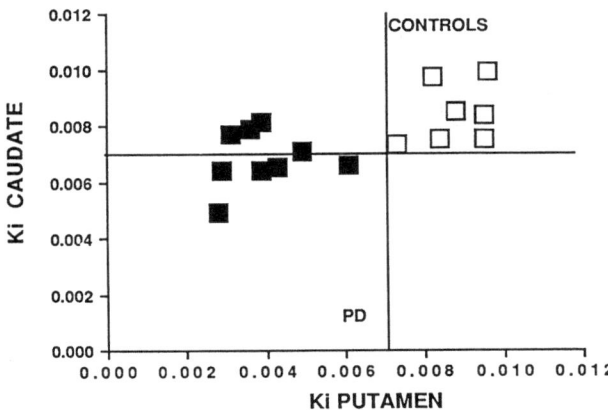

Fig. 3. L-^{18}F-6-fluorodopa uptake (K_i values) in putamen and caudate nucleus in a group of controls (n = 7) and parkinsonian patients (n = 9)

of MPTP, a neurotoxin specifically damaging or destroying dopaminergic neurons. MPTP was slowly infused as a solution (1.2 mg as a whole) through a catheter which was positioned into the right internal carotid artery via the femoral artery. Within two days left-sided akinesia developed, occasionally accompanied by marked dystonic postures of the left upper or lower limb. After apomorphine or levodopa akinesia disappeared and normal use of the limbs was observed. In addition, during about 45 minutes, rotation to the left occurred.

Striatal ^{11}C-NMF uptake 2 days after the lesion was normal on both sides. However, 9 days after the lesion the difference between striatal and non-dopaminergic brain tissue activity was reduced by about 80 to 90% in the right striatum, but remained unchanged on the left side. Apparently MPTP had been taken up by the nerve terminals in the right hemisphere and resulted in rapid functional but slower structural damage of the dopaminergic nigrostriatal pathway. Six weeks after the lesion the same reduction of ^{11}C-NMF uptake was seen in the right striatum, but after 5 months and 1 year about 50% of uptake had been restored. Clinically only mild left-sided hypokinesia was noticeable from several months after the lesion onwards.

Human studies using ^{11}C-NMF and PET have also been performed (Tedroff et al., 1988; Leenders et al., 1989). Figure 4 illustrates the time-activity curve of ^{11}C-NMF derived activity in a healthy subject's brain. The tracer concentrates in the striatum, but is also seen to accumulate in thalamic regions (not shown here). It is proposed here that striatal ^{11}C-NMF activity is determined by dopaminergic and

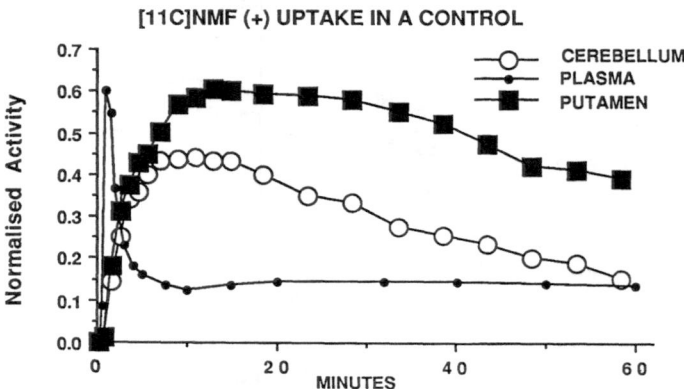

Fig. 4. Time-activity curve of ^{11}C-nomifensine uptake in putamen, cerebellum and arterial plasma in a healthy volunteer

thalamic ^{11}C-NMF activity by non-dopaminergic monaminergic uptake sites. An abundance of adrenergic innervation of the rat thalamus has been shown by Lindvall et al. (1974). As with ^{18}F-dopa (see above) striatal uptake of ^{11}C-NMF is likewise diminished in patients with Parkinson's disease (Fig. 5). Both putamenal and caudate nucleus values are now outside the normal range. Mean values of putamen and caudate nucleus were about 40% and 60% of controls respectively. This is not far from the values (30 and 50% respectively) of [3H]GBR 12935 (another substance binding to monaminergic uptake sites) binding in putamen and caudate nucleus in the post-mortem parkinsonian brain presented by Maloteaux et al. (1988).

Both ^{18}F-dopa and ^{11}C-nomifensine are thus found to be decreased by only about 40% in the putamen of parkinsonian patients. This is markedly less than the decreases of endogenous dopamine concentrations usually reported (5 − 15% of control values). These findings suggest that, although in Parkinson's disease most of the dopaminergic nerve cells of the nigrostriatal system may be defective in producing endogenous dopamine, still a considerable number of cells (and thus striatal nerve terminals) may structurally survive. This would then explain why at the initial stages of the disease exogenously administered L-dopa is clinically so effective: dopa can be decarboxylated and handled (retained) as dopamine in the still sizable number of nerve terminals. When the disease progresses, possibly accelerated by the influence of the regular flooding of the system by exogenous L-dopa, more nerve terminals are lost and therapeutic responses become more

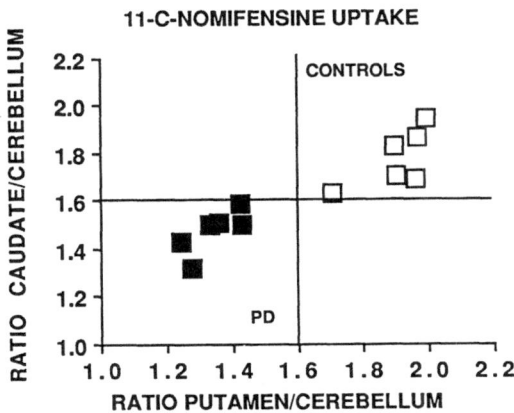

Fig. 5. Ratio of putamen or caudate nucleus to cerebellum values of ^{11}C-nomifensine uptake in controls (n = 6) and patients with Parkinson's disease (n = 6)

erratic, following more directly plasma L-dopa levels. Longitudinal studies comparing clinical progression of the disease with PET binding measures may provide an answer to this.

Post-synaptic tracers

Spiperone is a neuroleptic drug (butyrophenone) and is essentially a dopamine D_2 receptor antagonist. The radiolabelled analogue ^{11}C-methyl-spiperone (^{11}C-MSP) therefore binds predominantly to D_2 receptors in striatum where these receptors are highest in concentration (Fowler et al., 1986). However, also binding to serotonin receptors occurs, particularly in the cortical regions (Frost et al., 1987). ^{11}C-MSP has been used in man (Wagner et al., 1983; Wong et al., 1984) and Eckernäs et al. (1987) discussed the mathematical modelling associated with quantification of ^{11}C-MSP uptake.

Reports about its application in Parkinson's disease have been scarce so far (Leenders et al., 1985; Hägglund et al., 1987). Untreated parkinsonian patients showed similar striatal ^{11}C-MSP uptake compared to healthy controls. Levodopa drug treatment seemed to reduce ^{11}C-MSP uptake to some extent (Leenders et al., 1985), but the number of patients studied was small. These findings are in agreement with post-mortem results showing virtually no change in dopamine D_2 receptor densities in parkinsonian patients (Bokobza et al., 1984). The fact that, in this chronic disease, the post-synaptic dopaminergic system is apparently intact in the presence of a severe pre-synaptic lesion, "explains" why dopaminergic drug treatment is effective at all in Parkinson's disease. Patients with other chronic neurodegenerative diseases associated with parkinsonian features, like the Steele-Richardson-Olszewski syndrome, do not respond, or only slightly so, to levodopa therapy. In that condition impaired pre-synaptic dopaminergic function (Leenders et al., 1988) is accompanied by striatal dopamine D_2 receptor decreases (Baron et al., 1985), probably due to striatal neuronal cell loss.

Raclopride is a neuroleptic drug (substituted benzamide) and can also be radiolabelled to visualize dopamine receptor binding in human brain with PET (Farde et al., 1985). This compound is specific for dopamine D_2 receptors and quantitative analysis of these receptors can be achieved (Farde et al., 1986). ^{11}C-raclopride has not been extensively used in patients with Parkinson's disease yet, but one report showed normal striatal values before and after engraftment of homologous adrenal medulla tissue in 2 patients (Lindvall et al., 1987).

One Rhesus monkey has been studied using ^{11}C-raclopride before and after a unilateral nigrostriatal dopaminergic MPTP lesion which was produced by infusion of the neurotoxin through a right internal carotid artery infusion (Leenders et al., 1988). Two days after the lesion an increase of about 50% in specific tracer uptake was found in the lesioned striatum in the presence of a clinically impaired presynaptic dopaminergic function. Six weeks after the lesion, increased ^{11}C-raclopride and markedly decreased ^{11}C-nomifensine (see above) were found in the lesioned striatum, but normal values were observed on the unlesioned side. After 5 months and 1 year ^{11}C-raclopride uptake was normal again, while the pre-synaptic function was still impaired. This suggests that *acute* lesions of the nigrostriatal system can provoke a temporary increase of striatal dopamine D_2 receptor density. However, a *chronic* lesion of this system seems compatible with normal post-synaptic receptor density, at least in the absence of post-synaptic neuronal cell loss (see discussion of ^{11}C-MSP).

References

Aquilonius SM, Bergström K, Eckernäs SA, Hartvig P, Leenders KL, Lundqvist H, Antoni G, Gee A, Rimland A, Uhlin J, Långström (1987) In vivo evaluation of striatal dopamine reuptake sites using ^{11}C-nomifensine and positron emission tomography. Acta Neurol Scand 76: 283–287

Baron JC, Maziere B, Loc'h C, Sgouropoulos P, Bonnet AM, Agid Y (1985) Progressive supranuclear palsy: loss of striatal dopamine receptors demonstrated in vivo by positron tomography. Lancet ii: 1163–1164

Bokobza B, Ruberg M, Scatton B, Javoy-Agid F, Agid Y (1984) (3-H)-spiperone binding, dopamine and HVA concentrations in Parkinson's disease and supranuclear palsy. Eur J Pharmacol 99: 167–175

Boyes RE, Cumming P, Martin WRW, McGeer EG (1986) Determination of plasma [^{18}F]-6-fluorodopa during positron emission tomography: elimination and metabolism in carbidopa-treated subjects. Life Sci 39: 2243–2252

Cumming P, Boyes BE, Martin WRW, Adam M, Ruth T, McGeer EG (1987) Altered metabolism of [^{18}F]-6-fluorodopa in the hooded rat following inhibition of catechol-0-methyltransferase with U-0521. Biochem Pharmacol 36: 2527–2531

Eckernäs SA, Aquilonius SM, Hartvig P, et al (1987) Positron emission tomography (PET) in the study of dopamine receptors in the primate brain: evaluation of a kinetic model using ^{11}C-N-methyl-spiperone. Acta Neurol Scand 75: 168–178

Farde L, Ehrin E, Eriksson L, Greitz T, Hall H, Hedstrom CG, Litton JE, Sedvall G (1985) Substituted benzamides as ligands for visualisation of dopamine receptor binding in the human brain by positron emission tomography. Proc Natl Acad Sci USA 82: 3863–3867

Farde L, Hall H, Ehrin E, Sedvall G (1986) Quantitative analysis of D_2 dopamine receptor binding in the living human brain by PET. Science 231: 258–261

Firnau G, Sood S, Chirakal R, Nahmias C, Garnett ES (1987) Cerebral metabolism of 6-[F-18]Fluoro-L-dopa in the primate. J Neurochem 48: 1077–1082

Fowler JS, Arnett CD, Wolf AP, Shiue C-Y, MacGregor RR, Halldin C, Långström B, Wagner Jr HN (1986) A direct comparison of the brain uptake and plasma clearance of N-(^{11}C)methylspiroperidol and (^{18}F)N-methylspiroperidol in baboon using PET. Nucl Med Biol 13(3): 281–284

Frost JJ, Smith AC, Kuhar MJ, Dannals RF, Wagner Jr HN (1987) In vivo binding of ^{3}H-N-methylspiperone to dopamine and serotonin receptors. Life Sci 40: 987–995

Garnett ES, Firnau G, Nahmias C (1983) Dopamine visualized in the basal ganglia of living man. Nature 305: 137–138

Hägglund J, Aquilonius SM, Eckernäs SA, Hartvig P, Lundquist H, Gullberg P, Långström B (1987) Dopamine receptor properties in Parkinson's disease and Huntington's chorea evaluated by positron emission tomography using ^{11}C-N-methyl-spiperone. Acta Neurol Scand 75: 87–94

Leenders KL, Herold S, Palmer AJ, Turton D, Quinn N, Jones T, Frackowiak RSJ, Marsden CD (1985) Human cerebral dopamine system measured in vivo using PET. J Cereb Blood Flow Metab 5 [Suppl]: 517–518

Leenders KL, Frackowiak RJS, Quinn N, Marsden CD (1986a) Brain energy metabolism and dopaminergic function in Huntington's disease measured in vivo using positron emission tomography. Movement Disorders 1: 69–77

Leenders KL, Palmer AJ, Quinn N, Clark JC, Firnau G, Garnett ES, Nahmias C, Jones T, Marsden CD (1986b) Brain dopamine metabolism in patients with Parkinson's disease measured with positron emission tomography. J Neurol Neurosurg Psychiatry 49: 853–856

Leenders KL, Poewe WH, Palmer AJ, Brenton DP, Frackowiak RSJ (1986c) Inhibition of L-[^{18}F]fluorodopa uptake into human brain by amino acids demonstrated by positron emission tomography. Ann Neurol 20: 258–262

Leenders KL, Aquilonius SM, Bergström K, Bjurling P, Crossman AR, Eckernäs SA, Gee AG, Hartvig P, Lundqvist H, Långström B, Rimland A, Tedroff J (1988a) Unilateral MPTP lesion in a Rhesus monkey: effects on the striatal dopaminergic system measured in vivo with PET using various novel tracers. Brain Res 445: 61–67

Leenders KL, Frackowiak RJS, Lees AJ (1988b) Steele-Richardson-Olszewski syndrome. Brain energy metabolism, blood flow and fluorodopa uptake measured by positron emission tomography. Brain 111: 615–630

Lindvall O, Björklund A, Nobin A, Stenevi U (1974) The adrenergic innervation of the rat thalamus as revealed by the glyoxylic acid fluorescence method. J Comp Neurol 154: 317–348

Lindvall O, Backlund EO, Farde L, Sedvall G, Freedman R, Hoffer B, Nobin A, Seiger Å, Olson L (1987) Transplantation in Parkinson's disease: two cases of adrenal medullary grafts to the putamen. Ann Neurol 22: 457–468

Maloteaux JM, Vanisberg MA, Laterre C, Agid FJ, Agid Y, Laduron PM (1988) [3H]GBR 12935 binding to dopamine uptake sites: subcellular localization and reduction in Parkinson's disease and progressive supranuclear palsy. Eur J Pharmacol 156: 331–340

Nagatsu T, Kato T, Nagatsu I, Kondo Y, Inagaki S, Iizuka R, Narabayashi H (1979) Catecholamine-related enzymes in the brain of patients with parkinsonism and Wilson's disease. Adv Neurol 24: 283–292

Patlak CS, Blasberg RG (1985) Graphical evaluation of blood-to-brain transfer constants from multiple-time uptake data. Generalizations. J Cereb Blood Flow Metab 5: 584–590

Patlak CS, Blasberg RG, Fenstermacher JD (1983) Graphical evaluation of blood-to-brain transfer constants from multiple-time uptake data. J Cereb Blood Flow Metab 3: 1–7

Scatton B, Dubois A, Dubocovitch ML, Zahniser NR, Fage D (1984) Quantitative autoradiography of 3H-nomifensine binding sites in rat brain. Life Sci 36: 815–822

Slater P, Crossman AR (1984) Autoradiographic distribution of [3H]-nomifensine in brain. In: Linford-Rees W, Priest RG (eds) Nomifensine. A pharmacological and clinical profile. The Royal Society of Medicine, London, pp 15–19

Tedroff J, Aquilonius SM, Hartvig P, Lundquist H, Gee AG, Uhlin J, Långström B (1988) Monoamine re-uptake sites in the human brain evaluated in vivo by means of ^{11}C-nomifensine and positron emission tomography: the effects of age and Parkinson's disease. Acta Neurol Scand 77: 192–201

Wagner HN, Burns HD, Dannals RF, Wong DF, Langstrom B, Duelfer T, Frost JJ, Ravert HT, Links JM, Rosenbloom SB, Lukas SE, Kramer AV, Kuhar MJ (1983) Imaging dopamine receptors in the human brain by positron tomography. Science 221: 1264–1266

Wong DF, Wagner Jr HN, Dannals RF, Links JM, Frost JJ, Ravert HT, Wilson AA, Rosenbaum AE, Gjedde A, Douglass KH, Petronis JD, Folstein MF, Toung JKT, Burns HD, Kuhar MJ (1984) Effects of age on dopamine and serotonin receptors measured by positron tomography in the living human brain. Science 226: 1393–1396

Dopamine beta-hydroxylase and beta 2-microglobulin in cerebrospinal fluid: early markers in Parkinson's disease?

T. Nagastu[1], M. Mogi[2], M. Harada[2], and K. Kojima[3]

[1] Department of Biochemistry, Nagoya University School of Medicine, Nagoya,
[2] Department of Oral Biochemistry, Matsumoto Dental College, Shiojiri, and
[3] Hatano Research Institute, Food and Drug Safety Center, Hatano, Japan

Summary

Dopamine beta-hydroxylase (DBH) activity, DBH protein, and beta 2-microglobulin (B 2-MG), the light chain of the class 1 major histocompatibility complex (MHC), were found to be significantly reduced in cerebrospinal fluid (CSF) from parkinsonian patients. A significantly positive correlation was observed between DBH activity and DBH protein or B 2-MG. The results suggest a probable link between the changes in the immune system and those in the noradrenergic neurons in Parkinson's disease. It remains to be examined whether DBH or B 2-MG in CSF could be an early marker in Parkinson's disease.

Introduction

In parkinsonian brains, especially in the nigrostriatal regions, locus coeruleus and hypothalamus, tyrosine hydroxylase (TH), the biopterin (BP) cofactor, dopamine beta-hydroxylase (DBH), and phenylethanolamine N-methyltransferase (PNMT) are significantly decreased (Nagatsu et al., 1977, 1981, 1984). Of them, only DBH (Nagatsu, 1977) and BP are present in cerebrospinal fluid (CSF). The BP cofactor (Lovenberg et al., 1979), DBH activity (Matsui et al., 1981; Nagatsu et al., 1982) and DBH protein (Mogi et al., 1988) were found to be decreased in CSF from parkinsonian patients. Serum BP (Yamaguchi et al., 1983) and DBH activity (Nagatsu et al., 1982) were also significantly decreased in parkinsonian patients, but the decrease was not as marked as in CSF. Thus, the measure of BP and DBH activity or

DBH protein in CSF could be regarded as a more sensitive early marker of Parkinson's disease than their measure in serum.

We have recently found that beta 2-microglobulin (B 2-MG), a low molecular weight (11.8 kDa), non-glycosylated protein, which forms the light chain of the class I major histocompatibility complex (MHC-I), was decreased in the CSF of parkinsonian patients. Moreover, a significant positive correlation was observed between B 2-MG content and DBH activity in CSF from 45 patients (Mogi et al., 1989).

Materials and methods

Human B 2-MG, purified from human urine, was purchased from Calbiochem (La Jolla, CA, USA), anti B 2-MG IgG was from Dakopatts (Grostrup, Denmark).

The CSF was obtained from control patients without neurological diseases through lumbar puncture after local anaesthesia, and from parkinsonian patients. Each subject gave his fully informed consent to participate in the study. The CSF samples were all clear and no red blood cells were detected.

DBH activity in CSF was determined by HPLC with electrochemical detection (Suzuki et al., 1985). Tyramine was used as substrate and the DBH activities in CSF were expressed as μU (pmol of octopamine formed/min) per ml of CSF.

The DBH protein in CSF was measured by a newly established sandwich-type enzyme immunoassay (EIA) method using a solid phase (polystyrene beads) with β-D-galactosidase from E. Coli, as described by Mogi et al. (1984). Polystyrene beads with immobilized anti-DBH Fab' fragments were incubated in duplicate at 37 °C and shaken with 250 μl of standard DBH protein (purified human serum DBH) or CSF samples in a final volume of 500 μl with buffer G (0.01 M sodium phosphate buffer, pH 7.0, containing 0.3 M NaCl, 1 mM $MgCl_2$, 0.1% bovine serum albumin, 0.5% gelatin and 0.1% NaN_3). The beads were then incubated at 4 °C overnight with beta-D-galactosidase-labeled anti-DBH Fab'-fragments in buffer A. The beads were washed, and the beta-D-galactosidase activity bound to each bead was assayed with a fluorogenic substrate. 4-methylumbelliferyl-beta-D-galactoside.

B 2-MG in CSF was measured by the sandwich-type EIA which consisted of a solid phase (polystyrene beads) with immobilized anti-B 2-MG antibodies, and antibodies labeled with beta-D-galactosidase from E. Coli., as described by Mogi et al. (1984). Protein content was measured by the Bradford method (1976).

Results

In the EIA of DBH, DBH protein content versus bound beta-D-galactosidase activity was linear between 50 — 5000 pg of purified human serum DBH protein per tube. The limit of sensitivity, defined as the antigen level at which bound activity was 0 ± 2 SD, was 30 pg per tube. The anti-human pheochromocytoma DBH antibody showed no cross-reactivity with purified tyrosine hydroxylase from human adrenals, phenylalanine hydroxylase and dihydropteridine reductase from rat liver, and DOPA decarboxylase and phenylethanolamine N-methyltransferase from bovine adrenals, indicating that this EIA is specific for DBH.

We examined DBH activity and DBH content, as determined by the sandwich EIA in CSF from parkinsonian patients (Table 1). Parkinsonian patients had lower DBH content and activity than the control subjects. The mean CSF DBH content of parkinsonian patients was 16% of the value in control patients. The decrease in DBH content in parkinsonian patients paralleled the loss of DBH activity (19% of the controls). The severity of parkinsonian symptoms (Hoehn and Yahr, 1967) did not show a clear correlation with the degree of the decrease in DBH activity or DBH content (Table 1). The specific activity in CSF of controls was similar to that of parkinsonian patients; 1.14 U per mg of DBH in controls and 1.39 U per mg of DBH in parkinsonian patients. The specific activity in CSF DBH was similar to that in serum DBH from controls (1.40 U per mg of DBH). A significant positive correlation ($r = 0.79$) was observed between DBH activity and DBH protein measured by EIA in the CSF from 59 patients.

In the EIA of B 2-MG with anti-human urine B 2-MG antibodies, the standard urine B 2-MG content vs bound beta-D-galactosidase activity was linear from 3 to 300 pg of purified human B 2-MG per tube. The anti-human urine B 2-MG antibody showed no cross-reactivity with albumin from human sera indicating that this EIA is specific for B 2-MG. The limit of sensitivity was 5 pg per tube.

When we submitted the CSF from controls and parkinsonian patients to 2-D electrophoresis, and then to Western blot analysis using anti-B 2-MG IgG, those from patients with Parkinson's disease showed two major spots in the pH range $5.3 - 5.7$, with the same molecular weight (11.8 kDa) as previously reported (Gorevic et al., 1986). The pattern of B 2-MG in 2-D electrophoresis performed on CSF samples from controls was identical to that observed with parkinsonian CSF, suggesting that the B 2-MG protein in parkinsonian and control CSF may be identical and not a degraded form of the protein.

As shown in Table 2, parkinsonian patients showed a significantly lower B 2-MG concentration than control subjects. The mean B 2-MG content in CSF of parkinsonian patients was 35% of the value in controls.

There was no relationship between B 2-MG content and DBH activity in the CSF of control patients ($r = 0.01$), but a low correlation

Table 1. DBH activity and DBH content in CSF from control and parkinsonian patients

	Age (years)	N	DBH activity (μU/ml CSF)	DBH content (μg/ml CSF)	Specific activity (U/mg DBH)
Control patients	44 (18–65)	25	24.0 ± 3.7 (100%)	21.1 ± 3.1 (100%)	1.14
Parkinsonian patients	58 (32–78)	34	4.6 ± 0.7* (19%)	3.3 ± 0.7* (16%)	1.39
Stage 2		6	3.5 ± 1.2*	3.4 ± 1.6*	
Stage 3		14	5.2 ± 1.2*	3.9 ± 1.2*	
Stage 4		14	4.4 ± 0.9*	2.8 ± 0.8*	

Mean ± SEM; * significantly different from controls, $p < 0.005$
U μmol of octopamine formed per min

Table 2. Beta 2-microglobulin (B 2-MG) content in CSF from control and parkinsonian patients

	Age (years)	N	B 2-MG content (µg/ml of CSF)
Control patients	44 (18–87)	18	1.18 ± 0.11 (100%)
Parkinsonian patients	59 (40–78)	27	0.63 ± 0.09* (35%)
Stage 2		5	0.60 ± 0.20*
Stage 3		13	0.65 ± 0.15*
Stage 4		9	0.62 ± 0.13*

Mean ± SEM; * significantly different from controls, $p < 0.005$

between B 2-MG content and DBH activity seems to exist in CSF of parkinsonian patients ($r = 0.25$). Furthermore, a significantly positive correlation ($r = 0.87$) was observed between B 2-MG content and DBH activity in CSF from 45 control and parkinsonian patients.

The values of CSF B 2-MG of the two patients who were not given levodopa (0.50 and 0.56 µg/ml CSF) were similar to those of the other parkinsonian patients treated with levodopa.

Discussion

Both B 2-MG content and DBH content or activity in CSF were significantly lower in parkinsonian patients than in controls (Tables 1 and 2). Both B 2-MG and DBH contents did not show any clear correlation with the severity of the disease, but the values were significantly low at a relatively early stage (stage 2).

The decrease in DBH content and activity is in line with the diminished DBH activity in parkinsonian brains (Nagatsu et al., 1977; Nagatsu, 1984), and may reflect a lowered noradrenergic activity in the parkinsonian brain.

The significance of the decrease in CSF B 2-MG in parkinsonians is not clear, but may suggest a probable link between an immunological change and the change in noradrenergic neurons in Parkinson's disease. A poor relationship between serum and CSF B 2-MG levels was reported by Starmans et al. (1977). B 2-MG levels in serum do not seem to be determined by renal clearance (Wibell et al., 1973). It is likely that B 2-MG levels in CSF can be determined by its production in the brain, independently of any peripheral changes. Thus, the decrease in B 2-MG content in CSF from parkinsonian patients may be due to decreased release of B 2-MG into CSF from the brain.

Table 3. Changes in proteins and enzyme activities in the CSF from parkinsonian patients

CSF proteins		Changes	Reference
B 2-MG	protein	decreased	Mogi et al. (1989)
DBH	protein	decreased	Mogi et al. (1988)
	activity	decreased	Nagatsu et al. (1982)
Proline endopeptidase	activity	decreased	Hagihara and Nagatsu (1987)
Dipeptidyl peptidase IV	activity	no change	Hagihara et al. (1987)
Acetylcholinesterase	activity	no change	Ruberg et al. (1986)
Butyrylcholinesterase	activity	no change	Ruberg et al. (1986)
Dipeptidyl peptidase II	activity	increased	Hagihara et al. (1987)

Table 3 shows changes in various proteins and enzyme activities in CSF. We have recently found an increased activity of dipeptidyl peptidase II in CSF from parkinsonian patients without any significant changes in dipeptidyl peptidase IV activity (Hagihara et al., 1987 a). In contrast, proline endopeptidase activity was decreased in CSF from parkinsonian patients, although serum activity did not change (Hagihara and Nagatsu, 1987). No changes in acetylcholinesterase or butyrylcholinesterase activity was observed in CSF of parkinsonian patients (Ruberg et al., 1986). Therefore, the decrease in CSF DBH and B 2-MG may be specific to Parkinson's disease, and not due to a general decrease in CSF components.

The possibility of using CSF DBH and B 2-MG as early markers in Parkinson's disease should be further examined.

Acknowledgements

This investigation was supported by grants from Grant-in-Aid for Scientific Research on Priority Areas, Ministry of Education, Science and Culture to T. Nagatsu, and from the Human Health Science Foundation to K. Kojima.

References

Bradford MM (1976) Rapid and sensitive method for quantitation of microgram quantities of proteins utilizing the principle of protein-dye binding. Anal Biochem 72: 248–254

Gorevic PD, Munoz PC, Casey TT, DiRaimondo DR, Stone WJ, Prelli FC, Rodrigues MM, Poulik MD, Frangione B (1986) Polymerization of intact beta 2-microglobulin in tissue causes amyloidosis in patients with chronic hemodialyses. Proc Natl Acad Sci USA 83: 7908–7912

Hagihara M, Nagatsu T (1987) Post-proline cleaving enzyme in human cerebrospinal fluid from control patients and parkinsonian patients. Biochem Med Metabol Biol 38: 387–391

Hagihara M, Mihara R, Togari A, Nagatsu T (1987) Dipeptidyl-aminopeptidase II in human cerebrospinal fluid: changes in patients with Parkinson's disease. Biochem Med Metabol Biol 37: 360–365

Hoehn MM, Yahr MD (1967) Parkinsonism: onset, progression and mortality. Neurology 17: 427–442

Lovenberg W, Levine RA, Robinson DS, Ebert M, Williams AC, Calne DB (1979) Hydroxylase cofactor activity in cerebrospinal fluid of normal subjects and patients with Parkinson's disease. Science 204: 624–626

Matsui H, Kato T, Yamamoto C, Fujita K, Nagatsu T (1981) Highly sensitive assay for dopamine-β-hydroxylase activity in human cerebrospinal fluid by high-performance liquid chromatography-electrochemical detection: properties of the enzyme. J Neurochem 37: 289–296

Mogi M, Kojima K, Nagatsu T (1984) Detection of inactive or less active forms of tyrosine hydroxylase in human adrenals by a sandwich enzyme immunoassay. Anal Biochem 138: 125–132

Mogi M, Harada M, Kojima K, Inagaki H, Kondo T, Narabayashi H, Arai T, Teradaira R, Fujita K, Kiuchi K, Nagatsu T (1988) Sandwich enzyme immunoassay of dopamine-β-hydroxylase in cerebrospinal fluid from controls and parkinsonian patients. Neurochem Int 12: 187–191

Mogi M, Harada M, Kojima K, Adachi T, Narabayashi H, Fujita K, Naoi M, Nagatsu T (1989) Beta 2-microglobulin decrease in cerebrospinal fluid from parkinsonian patients. Neurosci Lett 104: 241–246

Nagatsu T (1977) Dopamine-β-hydroxylase in blood and cerebrospinal fluid. Trends Biochem Sci 2: 217–219

Nagatsu T, Kato T, Numata (Sudo) Y, Ikuta K, Sano M, Nagatsu I, Kondo Y, Inagaki S, Ikuta R, Hori A, Narabayashi H (1977) Phenylethanolamine N-methyltransferase and other enzymes of catecholamine metabolism in human brain. Clin Chim Acta 75: 221–232

Nagatsu T, Yamaguchi T, Kato T, Sugimoto T, Matsuzaki S, Akino M, Nagatsu I, Iizuka R, Narabayashi H (1981) Biopterin in human brain and urine from control and parkinsonian patients: application of a new radioimmunoassay. Clin Chim Acta 109: 305–311

Nagatsu T, Wakui Y, Kato T, Fujita K, Kondo T, Yokochi F, Narabayashi H (1982) Dopamine beta-hydroxylase activity in cerebrospinal fluid of parkinsonian patients. Biomed Res 3: 95–98

Nagatsu T, Yamazaki T, Rahman MK, Trocewicz J, Oka K, Hirata Y, Nagatsu I, Narabayashi H, Kondo T, Iizuka R (1984) Catecholamine-related enzymes and the biopterin cofactor in Parkinson's disease and related extra-pyramidal diseases. Adv Neurol 40: 467–481

Ruberg M, Rieger F, Villageois A, Bonnet AM, Azid Y (1986) Acetylcholinesterase and butyrylcholinesterase in frontal cortex and cerebrospinal fluid of demented and non-demented patients with Parkinson's disease. Brain Res 362: 83–91

Starmans JJP, Vos J, Van Der Helm HJ (1977) The beta 2-microglobulin content of the cerebrospinal fluid in neurological disease. J Neurol Sci 33: 45–49

Suzuki H, Yata J, Kojima K, Nagatsu T (1985) Simple and sensitive assay of dopamine-β-hydroxylase in human cerebrospinal fluid by high-performance liquid chromatography with electrochemical detection. J Chromatogr 341: 176–181

Wibell L, Evrin PE, Berggard I (1973) Serum beta 2-microglobulin in renal disease. Nephron 10: 320–331

Yamaguchi T, Nagatsu T, Sugimoto T, Matsuura S, Kondo T, Iizuka R, Narabayashi H (1983) Effects of tyrosine administration on serum biopterin in normal controls and patients with Parkinson's disease. Science 219: 75–77

Urinary dopamine sulfate conjugates in Parkinson's disease

E. Kienzl[1], K. Eichinger[2], K. Jellinger[1], W. Kuhn[3], G. Fuchs[4], W. Danielczyk[5], W. Wesemann[6], and P. Riederer[7]

[1] Ludwig-Boltzmann-Institute of Clinical Neurobiology, Lainz-Hospital, Vienna,
[2] Division of Chromatography and Spectroscopy, Institute of Organic Chemistry, Technical University, Vienna, Austria
[3] Department of Neurology, University of Würzburg,
[4] Parkinson Clinic, Wohlfahrt, Federal Republic of Germany
[5] Department of Neurology, Lainz Geriatric Hospital, Vienna, Austria
[6] Department of Neurochemistry, Institute of Physiology II, University of Marburg,
[7] Clinical Neurochemistry, Department of Psychiatry, University of Würzburg School of Medicine, Federal Republic of Germany

Summary

Dopamine-3-O-sulfate (DA-3-O-S) and dopamine-4-O-sulfate (DA-4-O-S) have been shown to be important end-products of L-dopa metabolism. Therefore, when measured in urine samples of patients with Parkinson's disease (PD), they may give indications of disorders in the peripheral metabolism of catecholamines. In addition, information about the reliability of DA sulfation after L-dopa therapy may be of significance in assessing its role in the elimination of DA from the peripheral nervous system.

Although DA-3-O-S seems to be the predominant sulfo-conjugate in urine, it is neither changed in PD compared to controls with or without other neurological disorders nor in depression syndrome. By contrast, DA-4-O-S is significantly decreased in de novo PD patients. However, a similar reduction is notable in cases of other neurological disorders. In depressed patients, the loss of this compound is less pronounced as compared to de novo PD patients. Treatment with combined L-dopa therapy increases primarily DA-3-O-S, while changes in DA-4-O-S are only marginal.

It can be concluded that urinary DA-3-O-S cannot be used as a marker for PD, while DA-4-O-S is significantly reduced in a variety of neurological disorders and, in particular, in PD. Further studies are necessary to elucidate its role as possible peripheral marker to distinguish preclinical PD and depression syndrome.

Introduction

In plasma more than 95% of dopamine (DA) is present as its sulfated forms DA-3-O-sulfate (DA-3-O-S) and DA-4-O-sulfate (DA-4-O-S) (range $17 - 48$ pmol/ml); (Van Loon, 1983; Yamamoto et al., 1985; Scott and Elchisak, 1987), and considerable amounts of DA-sulfates are excreted in urine. The ratio of free to total DA in urine was found to be approximately the same as in plasma (Rutledge and Hoehn, 1973: free DA $94 \pm 21 \mu g/24$ h, total DA $659 \pm 93 \mu g/24$ h; Westerink and ten Kate, 1986: free DA 1267 ± 192 nmol/24 h, conjugated DA 6448 ± 3478 nmol/24 h). Since urinary excretion under physiological conditions directly reflects the plasma levels of DA-3-O-S and DA-4-O-S (Yamamoto et al., 1985), urinary assays may provide in vivo information on central nervous system (CNS) functions and dysfunctions, e.g. in neurological disorders.

Both extraneuronal and neuronal sources of the DA-sulfates have been postulated. However, the origin of DA-sulfates is not clearly defined. Sources of sulfation are the chromaffin cells of the adrenal medulla, paraganglia, peripheral autonomic dopaminergic nerves and enterochromaffin cells or mast cells storing DA. An increase in these DA metabolites is especially observed after food intake (Van Loon, 1983). Conjugation of DA in CNS in the proximity of synapses may also play a role in the physiological regulation of dopaminergic neurotransmission and in the inactivation of DA formed from L-dopa. To elucidate the role of DA-sulfates as metabolic products and physiologically active substances, many studies dealing with the determination of both DA-sulfates in blood, urine and cerebrospinal fluid (CSF), and of phenolsulfotransferase (PST) activity have been published. However, despite the great interest in the analysis of DA-sulfates, until recently, no generally accepted analytical method existed. Therefore, the values of concentrations of these substances reported by various groups differed widely (Rutledge and Hoehn, 1973; Jenner and Rose, 1973; Foldes and Meek, 1974; Bronaugh et al., 1975; Arakawa et al., 1979; Renskers et al., 1980; Buu et al., 1981; Elchisak and Carlson, 1982; Elchisak, 1983 a, b, 1987; Yamamoto et al., 1985; Scott and Elchisak, 1987). Most authors were not able to distinguish between DA-3-O-S and DA-4-O-S (Fig. 1). By contrast, our improved analytical method combines sensitivity of detection with unequivocal substance identification and simplicity (Kienzl and Eichinger, 1988). Using this method we investigated the 24 h excretion pattern of the two DA-sulfates in urine of controls, de novo parkinsonian patients, PD patients with or without L-dopa therapy, a group of persons with different neurologic diseases and patients with major depression, in order to

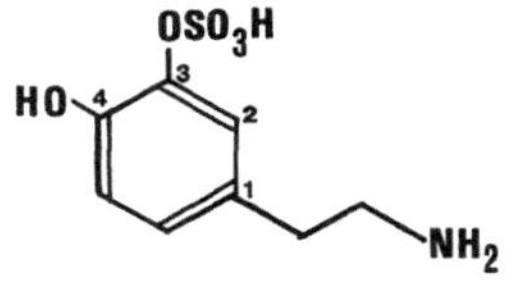

Fig. 1. Chemical structures of DA-3-O-S (dopamine-3-O-sulfate) and DA-4-O-S (dopamine-4-O-sulfate)

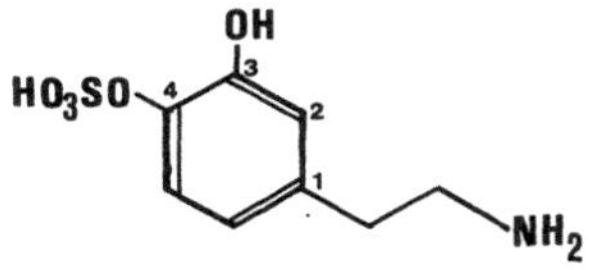

study the contribution of DA-sulfoconjugation and its significance as a marker of pathophysiological processes.

Patients and methods

Twenty four-hour urine samples were obtained from various groups of patients and a control group:

1) Controls, free of neurological, psychiatric or other organic diseases, included 9 females and 6 males aged 57 ± 2.9 years (mean ± sem).

2) A group of 16 drug naive patients with Parkinson's disease (PD), aged 63.6 ± 2.9 years (11 females, 5 males) hospitalized for the first time. These patients were diagnosed as having both early and late onset PD (range: 36 to 77 years) of stage I and II according to Hoehn and Yahr (1967).

3) Another group of patients with PD (n = 19; 12 females, 7 males; aged 69.9 ± 2.5 years) received antiparkinson therapy without L-dopa or L-dopa plus a peripheral decarboxylase inhibitor. Treatment consisted of amantadine sulfate and anticholinergics or the partial dopamine agonist "terguride". The duration of the disease was 5.9 ± 4.4 years. Most of these patients were at stage II – III (Hoehn and Yahr).

4) The majority of PD patients were treated with L-dopa plus the peripherally acting decarboxylase inhibitor benserazide (Madopar® 62.5 mg/day up to 6 times 125 mg/day) and amantadine sulfate (n = 21; 16 females and 5 males; age 73.1 ± 1.5 years, duration of disease: 7.7 ± 4.4 years; Hoehn-Yahr stage II – IV).

5) Controls and PD patients were compared with a group of patients suffering from a variety of other neurological disorders, including multiple sclerosis (n = 3), polyneuropathy (n = 2), cerebral infarction (n = 2) and single cases of myasthenia gravis, arthritis, cerebral atrophy and Wilson's

disease. All (4 males, 7 females, aged 60.7 ± 3.3 years) were inpatients at a hospital for chronic diseases. They were treated for their respective disease according to usual drug strategies.

6) In addition, a group of outpatients (6 females, 2 males, aged 45.1 ± 4.2 years) with major depression diagnosed according to DSM III was studied.

Chemicals

1-Heptanesulfonic acid (sodium salt) was purchased from SIGMA (St. Louis, MO, USA); Titriplex III (Ethylenediaminotetraacetic acid, EDTA), glacial acetic acid, hexane (analytical grade) and water for chromatography (LiChrosolv) were supplied by MERCK (Darmstadt, FRG). Chemicals for synthesis were reagent grade and were purchased from MERCK (Darmstadt, FRG) with the exception of dopamine hydrochloride which was obtained from SIGMA (St. Louis, MO, USA).

Arylsulfatase was a partially purified sulfatase type VI (4,2 units/mg protein) from Aerobacter aerogenes in 50% glycerol-0.01 M TRIS solution, pH 7.5, obtained from SIGMA (St. Louis, MO, USA).

Sample preparation

The specimens were the 24 h urines, kept frozen at $-30\,°C$ after being acidified to about pH 2 with 25% hydrochloric acid. For HPLC-analysis 2 ml urine were extracted twice with 1 ml hexane to remove lipids and 100 µl glacial acetic acid was added. Prior to dilution (1 : 10) with the mobile phase the sample was pressure filtered through a 0.45 µm ACRO LC 25 filter (Gelman Sciences, Ann Arbor, MI, USA). Prior to injection the samples were diluted once more 1 : 10 with the mobile phase. No further sample preparation was necessary. All procedures were carried out at room temperature. Because of the stability of the DA-sulfate-isomers, time of storage, extraction and analysis could be neglected.

For enzymatic hydrolysis a sulfatase solution was diluted 1 : 1000 with a standard solution (1.3 g/ml DA-4-O-S and 0.85 g/ml DA-3-O-S in 0,01 M TRIS-buffer, pH 7.1). The degradation was performed in a water bath at 26 °C. The reaction was stopped by addition of glacial acetic acid and sudden freezing. For analysis the samples were thawed and diluted with the mobile phase.

Standard preparation

Reaction of anhydrous dopamine hydrochloride with sulfuric acid (conc.) gave DA-4-O-S. A mixture of chlorsulfonic acid and pyridine in chloroform was found to be an excellent reagent for the synthesis of a nearly equimolar mixture of DA-3-O-S and DA-4-O-S (Kienzl and Eichinger, 1988).

HPLC-System

The reverse phase HPLC determinations were carried out using a solvent delivery module Model 5000 and a Coulochem 5100 A detector with a 5010 analytical cell (ESA, Bedford, MA, USA) connected to a recording integrator CI-10 (LDC/Milton-Roy, Riviera Beach, FL, USA). Potentials were + 0.50 V (conditioning cell), − 0,35 V (detector 1) and + 0,45 V (detector 2). Stationary phase was a 250 × 4,6 mm S 5 ODS (5 µm) column (Spherisorb, Norwalk, CT, USA). The mobile phase (pH 3.6) consisted of a solution of 0.8 ml (6 mmol) anhydrous acetic acid, 250 mg Na-1-heptanesulfonate and 80 mg EDTA in 992 ml water (LiChrosolv, Merck). The flow rate was 0.7 ml/min. Prior to use the mobile phase was filtered through a 0.22 µm membrane filter (Sartorius, Göttingen, FRG). Peak identifications were performed with authentic standards. Retention times were 13.1 minutes for DA-4-O-S and 13.8 minutes for DA-3-O-S.

Results

DA-3-O-S

The average excretion of DA-3-O-S in controls (1.73 mg/24 h, n = 15) was not considerably higher than that in de novo PD cases (1.03 mg/24 h, n = 16) or in individuals with various neurological disorders or depression (Table 1). A nearly tenfold increase in urinary DA-3-O-S output only was found in PD patients treated with Madopar® (range: 3.99 − 16.2 mg/24 h). Likewise, in PD cases treated with amantadine-sulfate in combination with terguride or anticholinergic drugs without L-dopa therapy an appreciable increase in the amount of excreted DA-3-O-S isomer could be detected (range: 0.70 − 17.8 mg/24 h).

DA-4-O-S

Drug naive PD patients showed a markedly decreased excretion of DA-4-O-S (0,30 mg/24 h, n = 16) (Table 1). However, the extreme

Table 1. Mean urinary 24 h excretion values of DA-3-O-S and DA-4-O-S in controls and patients with neurological disorders

Group	DA-4-O-S mg/24 h mean ± SEM	DA-3-O-S mg/24 h mean ± SEM	Ratio DA-4-O-S DA-3-O-S
1 Controls	1.62 ± 0.42	1.73 ± 0.45	1.15 ± 0.20
2 PD de novo	0.30 ± 0.05	1.03 ± 0.15	0.29 ± 0.03
3 PD no L-DOPA	1.07 ± 0.18	4.23 ± 0.94	0.36 ± 0.05
4 PD + L-DOPA	2.50 ± 0.28	9.44 ± 0.77	0.26 ± 0.02
5 Controls: div. neurol. disord.	0.49 ± 0.11	1.40 ± 0.32	0.33 ± 0.05
6 Depression	0.63 ± 0.31	1.67 ± 0.27	0.38 ± 0.05

low levels in de novo PD were not significantly distinguishable from the decreased DA-4-O-S levels in other neurological disorders or depression (Tables 1, 2, 3). Moreover, when L-dopa or another parkinsonian therapy without L-dopa was administered, DA sulfoconjugation in the 4-O position of the aromatic ring was increased (Table 1).

Ratio

The predominant isomer of the excreted DA-sulfoconjugates in both PD and in all neurological disorders examined is DA-3-O-S. The ratio DA-4-O-S/DA-3-O-S is 1.15 in healthy subjects and is significantly different from that of patients with PD (Table 1, 2). The ratio is 0.29 for de novo PD (Table 1). No kind of anti-Parkinson therapy was able to change the different pattern of the distribution between DA-4-O-S and DA-3-O-S excreted in 24 h urine (Table 1). However, patients with other neurological disorders had a similar low ratio (Tables 1, 2, 3).

Table 2. Excretion of DA-4-O-S and DA-3-O-S (mg/24 h) in human urine

Group	DA-4-O-S	DA-3-O-S	Ratio
PD de novo Controls	***	n. s.	***
PD de novo Controls: Neurol. Disord.	n. s.	n. s.	n. s.
PD de novo Depression	**	*	n. s.
Controls Controls: Neurol. disorders	**	n. s.	**
Depression Controls	*	n. s.	*
Depression Controls: Neurol. disorders	n. s.	n. s.	n. s.

*** $p < 0.001$
** $p < 0.01$
* $p < 0.05$
n. s. non-significant
Statistical evaluation: significance according to U-test

Table 3. Excretion of DA-4-O-S and DA-3-O-S (mg/24 h) in human urine

Group	DA-4-O-S	DA-3-O-S	Ratio
PD + L-DOPA No L-DOPA	***	***	n. s.
PD + L-DOPA PD de novo	***	***	n. s.
PD + L-DOPA Depression	***	***	**
PD + L-DOPA Controls	n. s.	***	***
PD no L-DOPA PD de novo	***	**	n. s.
PD no L-DOPA Depression	n. s.	n. s.	n. s.
PD no L-DOPA Controls	n. s.	***	***

*** $p < 0.001$
 ** $p < 0.01$
n. s. non-significant
Statistical evaluation: significance according to U-test

It is obvious that there was much inter-individual variation in the amount of excreted DA-sulfates. Also daily fluctuations in concentrations were notable. Measurements of both DA-sulfates on three consecutive days gave mean deviations of about 46%. By contrast, the calculated ratio did not show any variability; its time course remained constant.

To assess any effects of age on conjugated DA, a series of linear correlations were calculated. We found no age dependency in the excretion of DA-sulfates (data not given).

Usually synthetic DA-sulfate conjugates are chemically stable. DA-3-O-S is also metabolically inert when incubated with a solution of the arylsulfohydrolase from Aerobacter aerogenes. By contrast, as shown in Fig. 2, DA-4-O-S is metabolically labile in that it can be hydrolyzed by a sulfatase under physiological conditions.

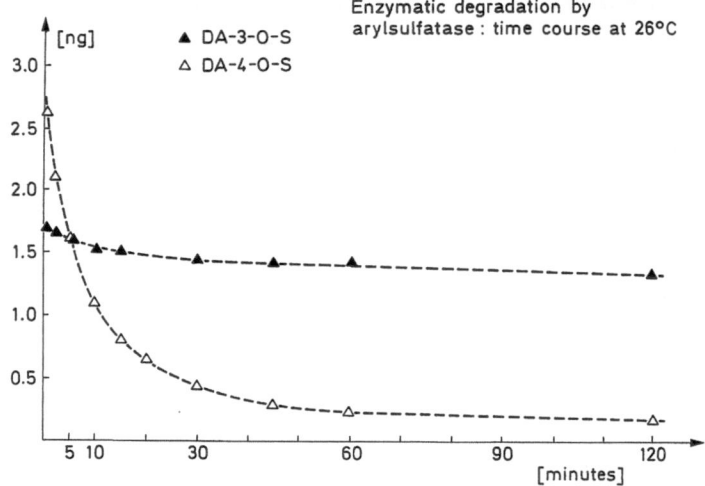

Fig. 2. Time course of the enzymatic hydrolysis of a synthesized mixture of DA-3-O-S and DA-4-O-S. (Method and substance preparation are described in the text)

Discussion

The complex clinical signs and symptoms of PD, including motor disability, mental disorders and endocrine disturbances, lead to believe that this is a generalized disease not limited to distinct areas of the brain (Barbeau, 1969). Reduction of the DA synthesizing enzyme tyrosine hydroxylase (TH) in striatum is paralleled by its reduction in adrenal medulla (Riederer et al., 1978), by loss of DA in the urine (Barbeau, 1961) and by a significant loss of DA, NA and adrenaline in biopsied adrenal medullary tissue obtained during surgical transplantation (Liebermann, pers. comm.), as well as in post-mortem adrenal medullary tissue of PD (Carmichael et al., 1988). Furthermore, Lewy bodies, the anatomical markers of PD, have been found in the adrenal medulla, autonomic ganglia and parasympathetic enteric nerve plexuses of patients with PD (f. rev. see Jellinger, 1989). There is also evidence of disturbed autonomic functions, e.g. orthostatic hypotension in L-dopa treated or untreated PD patients (Birkmayer and Neumayer, 1972). In addition, low plasma renin activity in PD (Barbeau, 1975) was suggested to be related to possible central or peripheral dopaminergic receptors (Sullivan et al., 1973). Indeed, some evidence of important peripheral functions of DA has been given (Van Loon, 1983). These include a possible role in neurotransmission in sympathetic ganglia, altering blood flow and renin release, modulation of aldosterone and sodium secretion, and mediation of certain stress responses. In human controls, peripheral sympathetic activity is associated with increase in plasma DA concentrations that parallels the increase in plasma NA and adrenaline (A). However, both the source and regulation of plasma and urinary DA concentrations remain poorly understood. Nevertheless, drug-induced alterations of DA concentrations in the periphery may reflect neurotransmitter mechanisms regulating sympathetic activity (Van Loon et al., 1979; Van Loon, 1980). Free DA disappears from the circulation in $1-3$ minutes, while the half-life of sulfated DA is about 2 hours (Corneille et al., 1983). Therefore, investigation of urinary excretion of DA-sulfoconjugates may serve as valuable markers for peripheral dopaminergic activity. Although the origins of DA-sulfate, which constitutes the largest portion of total circulating catecholamines, remain obscure (Kuchel et al., 1985), there are studies indicating some relationship between central and peripheral DA-metabolism (Tyce et al., 1986). It has been suggested that these metabolites in plasma or urine may mirror the metabolism in CNS only when a severe alteration of the dopaminergic CNS system is present (Elsworth et al., 1986; Bacopoulos et al., 1979). Moreover, a positive relationship between CSF and plasma DA-sulfates, NA and HVA was found (Kuchel et al., 1985; Tyce and Rorie, 1982), suggesting unidirectional movement of sulfated catecholamines and HVA from brain to blood (Kuchel et al., 1985).

Another unresolved problem is the putative role of sulfoconjugation in brain, important with regard to PD. PST is the synthesizing enzyme and can contribute by up to 15% to the total DA-metabolism. Sulfate donor is 3'-phosphoadenosine-5'-phosphosulfate (Renskers et al., 1980) (Fig. 3). PST is present at least in two forms in many human tissues and has also a wide heterogeneous distribution in rat and human brain (Foldes and Meek, 1974; Rivett et al., 1982; Weinshilboum, 1986). Brain PST is localized in kainic acid sensitive neurons intrinsic to the striatum, whereas sulfoconjugation was not affected by treatment with 6-hydroxydopamine, indicating a separate metabolic cytosolic compartment (Rivett et al., 1984; Buu, 1985). Studies on rat brains revealed that the synthesis of DA-sulfates in the hypothalamus, striatum and brain stem was induced only after inhibition of MAO activity by pargyline. The increase in DA-sulfate was significantly correlated with the degree of MAO inhibition (Buu, 1985). However, DA-sulfates themselves are believed to be resistant to MAO activity (Renskers et al., 1980). The role of catecholamines sulfoconjugation in relation to deamination by MAO and O-methylation by COMT in CNS and the periphery is not clear. This question is important with

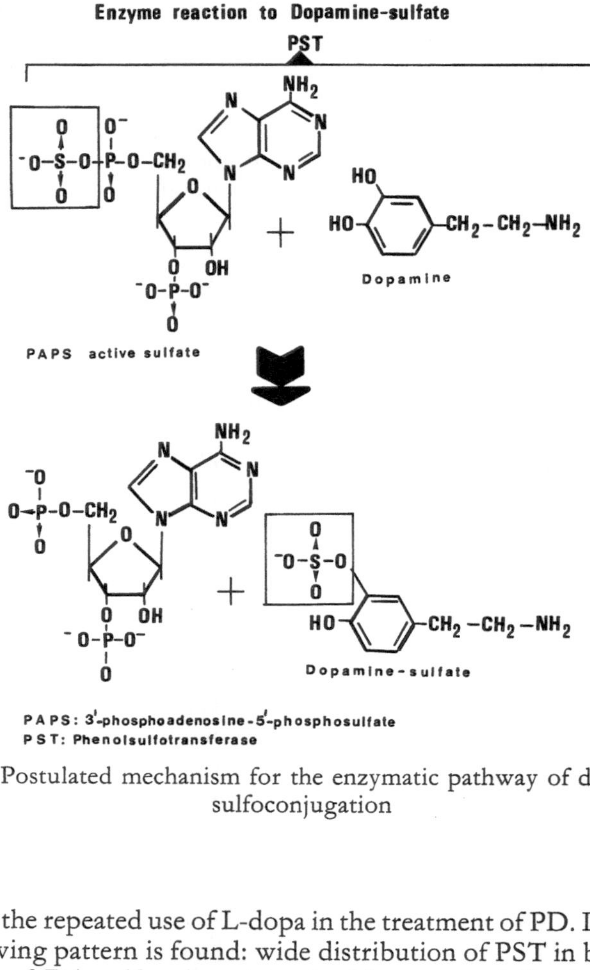

Fig. 3. Postulated mechanism for the enzymatic pathway of dopamine sulfoconjugation

regard to the repeated use of L-dopa in the treatment of PD. In humans, the following pattern is found: wide distribution of PST in brain, high excretion of DA-sulfate in urine of subjects, especially those under-going L-dopa therapy, and extremely low levels of DA-sulfate in brain (Sofic, pers. comm.). In blood, PST is found exclusively in platelets. Human platelets were suggested to be a good peripheral model for PST activity. Platelet PST activity was increased in PD patients treated with L-dopa; its activity was correlated to the dosage and duration of L-dopa treatment (Glover et al., 1983). These findings might explain the enhanced formation of sulfoconjugation under L-dopa therapy. However, the level of endogenous inorganic sulfate is normally rate-limiting in the formation of arylsulfates in vivo (Krijgsheld et al., 1982). To some extent a sulfate transfer from amantadine-sulfate to DA with regard to an increased sulfate pool may explain the higher DA-sulfate values in PD patients without any L-dopa therapy.

The amount of L-dopa utilized as neurotransmitter by dopaminergic neurons and related basal ganglia accounts for only a small portion of the therapeutical dosage, while circulating L-dopa is taken up by various tissues including kidney, liver, heart, dopaminergic and sympathetic systems where it is decarboxylated to DA (Bacopoulos et al., 1979). It is suggested that DA may be a neurotransmitter or cotransmitter in the autonomic nervous system (Lackovic et al., 1982). According to that theory, DA in plasma closely reflects autonomic nervous system activity (Van Loon, 1980). In addition to adrenal sympathetic nerves and, possibly, kidney, only a small amount of DA may enter plasma from brain (Van Loon et al., 1979). It is widely accepted that the DA-conjugates are irreversible metabolic end-products similar to the products of MAO and COMT. However, there is some evidence that DA-sulfates could possess storage functions for further retransformation into the free form by means of an arylsulfatase system (Demassieux et al., 1987). In the present study we tried to set up a model for the enzymic hydrolysis of DA-sulfate-esters by an arylsulfatase system. An increased rate of hydrolysis was found for DA-4-O-S, whereas DA-3-O-S seemed to be much less affected (Fig. 3). Also under pathophysiological conditions DA-4-O-S seems to be more accessible to degradation than DA-3-O-S. There is also evidence that DA-sulfates may be active themselves in reacting with vascular α-adrenergic receptors in decreasing blood pressure when infused into dogs (Scott and Elchisak, 1983).

In conclusion the results of our study provide evidence that particularly DA-4-O-S is significantly involved in PD, resulting in an altered ratio of DA-4-O-S and DA-3-O-S. Furthermore, in many earlier investigations the use of precise clinical criteria failed to distinguish depression from the onset of common symptoms in PD. The higher values of DA-4-O-S in major depression compared to the lower ones in de novo PD and in various neurological disorders might be an additional tool for differential diagnosis. Therefore, estimation of the DA-sulfate pattern in the periphery may be a valuable indicator of dopaminergic activity and may serve as future marker for neuropsychiatric disorders related to dysfunctions of the dopaminergic system.

References

Arakawa Y, Imai K, Tamura Z (1979) High-performance liquid chromatographic determination of dopamine sulfoconjugates in urine after L-dopa administration. J Chromatogr 162: 311–318

Bacopoulos NG, Hattox SE, Roth RH (1979) 3,4-Dihydroxyphenylacetic acid and homovanillic acid in rat plasma: possible indicators of central dopaminergic activity. Eur J Pharmacol 56: 225–236

Barbeau A (1961) Biochemistry of Parkinson's disease. Excerpta Medica, ICS 38: 152–153

Barbeau A (1969) Parkinson's disease as a systemic disorder. In: Gillingham FJ, Sonaldson ML (eds) 3rd Symposium on Parkinson's disease. Edinburgh, Livingstone, pp 66–73

Barbeau A (1975) Pathophysiology of the oscillations in performance after long-term therapy with L-dopa. In: Birkmayer W, Hornykiewicz O (eds) Advances in parkinsonism. Editions Roche, Basel, pp 424–434

Birkmayer W, Neumayer E (1972) Die moderne medikamentöse Behandlung des Parkinsonismus. Z Neurol 202: 257–280

Bronaugh RL, Mattox SE, Hoehn MM, Murphy RC, Rutledge CO (1975) The separation and identification of dopamine 3-O-sulfate and dopamine 4-O-sulfate in urine of parkinsonian patients. J Pharmacol Exp Ther 195: 441–452

Buu NT (1985) Dopamine sulfoconjugation in the rat brain: regulation by monoamine oxidase. J Neurochem 45: 470–476

Buu NT, Duhaime J, Savard C, Truong L, Kuchel O (1981) Presence of conjugated catecholamines in rat brain: a new method of analysis of catecholamine sulfates. J Neurochem 36(2): 769–772

Carmichael SW, Wilson RJ, Brimijoin WS, et al (1988) Decreased catecholamine levels in the adrenal medulla of patients with Parkinson's disease. N Engl J Med 319: 254–256

Corneille L, Lachance S, Demassieux S, Carrière S (1983) Turnover of free and conjugated serum catecholamines during haemodialysis. Clin Invest Med 6: 11–17

Demassieux S, Bordeleau L, Gravel D, Carrière S (1987) Catecholamine sulfates: end-products or metabolic intermediates? Life Sci 40: 183–191

Elchisak MA (1983 a) Determination of conjugated compounds by liquid chromatography with electrochemical detection using postcolumn hydrolysis. J Chromatogr 255: 475–482

Elchisak MA (1983 b) Determination of dopamine-O-sulfate and norepinephrine-O-sulfate isomers and serotonine-O-sulfate by high-performance liquid chromatography using dual-electrode electrochemical detection. J Chromatogr 264: 119–127

Elchisak MA (1987) Analytical techniques for the determination of phenolic amine neurotransmitter conjugates. Life Sci 41: 913–916

Elchisak MA, Carlson JN (1982) Method for analysis of dopamine sulfate isomers by high-performance liquid chromatography. Life Sci 30: 2325–2336

Elsworth JD, Leahy DJ, Roth RH, Redmond jr DE (1986) Homovanillic acid concentrations in brain, CSF and plasma as indicators of central dopamine function in primates. J Neural Transm 68: 51–62

Foldes A, Meek JL (1974) Occurrence and localisation of brain phenosulfotransferase. J Neurochem 23: 303–307

Glover V, Lees AJ, Ward C, Stern GM, Sandler M (1983) Platelet pheno-sulfotransferase activity in Parkinson's disease. J Neural Transm 57: 95–102

Hoehn MM, Yahr MD (1967) Parkinsonism: onset, progression and mortality. Neurology 17: 427–442

Jellinger K (1989) Pathology of Parkinson's syndrome. In: Calne DB (ed) Handbook of experimental pharmacology, vol 88. Springer, Berlin Heidelberg New York, pp 47–112

Jenner WN, Rose FA (1973) Studies on the sulphation of 3,4-dihydroxy-phenylethylamine (dopamine) and related compounds by rat tissues. Biochem J 135: 109–114

Kienzl E, Eichinger K (1988) Preparation of dopamine 3-O-sulfate and do-pamine 4-O-sulfate as reference substances and high-performance liquid chromatographic trace determination. J Chromatogr 430: 263–269

Krijgsheld KR, Scholtens E, Mulder GJ (1982) The dependence of the rate of sulfate conjugation on the plasma concentration of inorganic sulfate in the rat in vivo. Biochem Pharmacol 31: 3997–4000

Kuchel O, Hausser C, Buu NT, Tenneson S (1985) CSF sulfoconjugated catecholamines in man: their relationship with plasma catecholamines. J Neural Transm 62: 91–97

Lackovic Z, Relja M, Neff NH (1982) Catabolism of endogenous dopamine in peripheral tissues: is there an independent role for dopamine in peripheral neurotransmission? J Neurochem 38: 1453–1458

Renskers KJ, Feor KD, Roth JA (1980) Sulfation of dopamine and other biogenic amines by human brain sulfotransferase. J Neurochem 34: 1362–1368

Riederer P, Rausch WD, Birkmayer W, Jellinger K, Seemann D (1978) CNS modulation of adrenal tyrosine hydroxylase in Parkinson's disease and metabolic encephalopathies. J Neural Transm [Suppl 14]: 121–131

Rivett AJ, Eddy BJ, Roth JA (1982) Contribution of sulfate conjugation, deamination and O-methylation to metabolism of dopamine and nor-epinephrine in human brain. J Neurochem 39: 1009–1016

Rivett AJ, Francis A, Whittemore R, Roth JA (1984) Sulfate conjugation of dopamine in rat brain: regional distribution of activity and evidence for neuronal localisation. J Neurochem 42: 1444–1448

Rutledge CO, Hoehn MM (1973) Sulfate conjugation and L-dopa treatment of parkinsonian patients. Nature 244: 447–450

Scott MC, Elchisak MA (1983) Cardiovascular effects of dopamine sulfate in anesthetized dogs. Fed Proc 42: 1363

Scott MC, Elchisak MA (1987) Direct measurement of dopamine-O-sulfate in plasma and cerebrospinal fluid. J Chromatogr 413: 17–23

Sullivan JM, Nakano KK, Tyler HR (1973) Plasma renin activity during levodopa therapy. Significance of long- and short-term treatment. J Am Med Assoc 224: 1726–1727

Tyce GM, Rorie DK (1982) Conjugated dopamine in superfusates of slices of rat striatum. J Neurochem 39: 1333–1339

Tyce GM, Messick JM, Yaksh TL, Byer DE, Danielson DR, Rorie DK (1986) Amine sulfate formation in the central nervous system. Fed Proc 45(8): 2247–2253

Van Loon GR (1980) Abnormal catecholamine mechanisms in hypothalamic-pituitary disorders. Metabolism 29 [Suppl 1]: 1198–1202

Van Loon GR (1983) Plasma dopamine: regulation and significance. Fed Proc 42: 3012–3018

Van Loon GR, Sole MJ, Bain J (1979) Effects of bromocriptine on plasma catecholamines in normal men. Neuroendocrinology 28: 425–434

Weinshilboum RM (1986) Phenol sulfotransferase in humans: properties, regulation and function. Fed Proc 45: 2223–2228

Westerink BHC, ten Kate N (1986) 24 h excretion patterns of free, conjugated and methylated catecholamines in man. J Clin Chem Clin Biochem 24: 513–519

Yamamoto T, Yamatodani A, Nishimura AM (1985) Determination of dopamine-3- and -4-O-sulfate in human plasma and urine by anion-exchange high-performance liquid chromatography with fluorimetric detection. J Chromatogr 342: 261–267

Salsolinol and the early detection of Parkinson's disease

P. Dostert[1], M. Strolin Benedetti[1], and G. Dordain[2]

[1] Farmitalia Carlo Erba, Research and Development — Erbamont Group, Milan, Italy
[2] Hôpital Nord, Service de Neurologie, Clermont-Ferrand, France

Summary

In a previous study, the daily urinary excretion of total (R + S) salsolinol was shown to be significantly lower in de novo parkinsonian patients than in healthy volunteers.

Here, we report on the urinary concentrations and daily urinary excretion of (R)- and (S)-salsolinol in an additional sample of 12 healthy subjects and 9 de novo parkinsonians. The urinary concentrations of both enantiomers of salsolinol were lower than the limit of detection in the urine of the parkinsonians. (R)-Salsolinol was detectable in the urine of 8 healthy subjects, whereas (S)-salsolinol was detectable in the urine of 3 only. Since 1,2-dehydrosalsolinol was present in the urine of parkinsonian patients, it is suggested that salsolinol bioprecursors such as 1,2-dehydrosalsolinol and 1-carboxysalsolinol might be additional markers for the early detection of Parkinson's disease.

Introduction

The biosynthesis of salsolinol (1-methyl-1,2,3,4-tetrahydro-6,7-isoquinolinediol) has been suggested to occur in humans by condensation of dopamine either with acetaldehyde (Robbins, 1968) or with pyruvic acid followed by oxidative decarboxylation, to give 1,2-dehydrosalsolinol (3,4-dihydro-1-methyl-6,7-isoquinolinediol), and subsequent reduction (Brossi, 1982). Salsolinol possesses an asymmetric center at C-1 and exists as R and S enantiomers. There is convincing evidence that (R)-salsolinol is the predominant, if not the only enantiomer present in the urine of healthy subjects (Dostert et al., 1987; Strolin Benedetti et al., 1989 a), at least in the absence of salsolinol-containing food or chronic alcoholic beverage intake. Moreover, the presence of 1,2-dehydrosalsolinol has recently been established in the urine of healthy subjects (Dostert et al., 1990), supporting Brossi's hypothesis.

In man, the presence of salsolinol was first established in the urine of parkinsonian patients on L-dopa medication (Sandler et al., 1973). In another study, using a gas chromatography-mass spectrometry (GC-MS) method, the cerebrospinal fluid concentrations of total (R + S) salsolinol were found to be lower, although not significantly different, in de novo parkinsonians than in controls (Dordain et al., 1984). The daily urinary excretion of total salsolinol was also measured in the two groups after hydrolysis with β-glucuronidase-arylsulphatase, as well as in the parkinsonian patients after 7 days of L-dopa therapy. In the non-treated parkinsonians, the daily urinary excretion of salsolinol was found to be significantly lower than in controls, being 39 ± 7 and 197 ± 54 nmol/day (mean $\pm$ S.D.), respectively. Salsolinol excretion was dramatically increased in patients under L-dopa therapy $(1409 \pm 148$ nmol/day).

The significant decrease in salsolinol excretion in de novo parkinsonian patients compared to controls noted in the above-mentioned preliminary study encouraged us to further study the daily urinary output of salsolinol in an additional sample of de novo parkinsonians and controls. Since a method to measure both enantiomers separately had meanwhile been made available (Pianezzola et al., 1989), the urinary concentrations of (R)-and (S)-salsolinol were determined in this study.

Material and methods

Study design

Twelve healthy volunteers, 11 men and 1 woman aged 39.0 ± 7.8 years (mean $\pm$ S.D.) and 9 patients, 3 women and 6 men aged 69.8 ± 11.6 years (mean $\pm$ S.D.), diagnosed as parkinsonians entered the study (Table 1). The patients had never been treated with L-dopa with the exception of subjects 15 and 17, who had stopped L-dopa therapy 6 months and 45 days before the beginning of the study, respectively. None of the patients was on medicines known to change tissue dopamine concentrations. Intake of food and beverages known to contain appreciable amounts of salsolinol (Strolin Benedetti et al., 1989 b) was forbidden 24 hours before and during urine collection. Alcoholic beverage intake was also strictly limited. The 24-h urines were collected in the presence of semicarbazide and $NaHSO_3$ and stored at $-20\,°C$ until analysis. In the case of healthy volunteers 1 to 6, urines were collected in two fractions: from 9 a.m. to 5 p.m. and from 5 p.m. to 9 a.m.

Salsolinol extraction and analysis

Urine underwent hydrolysis with β-glucuronidase-arylsulphatase (Helix pomatia, 37 °C, 16 h). Salsolinol extraction from urine brought to pH 8.5 was

Table 1. Characteristics of the healthy volunteers and parkinsonian patients

Healthy volunteers			Parkinsonian patients			
Subjects	Sex	Age (years)	Subjects	Sex	Age (years)	Hoehn and Yahr (grade)
1	m	44	13	m	64	2
2	m	49	14	m	56	3
3	m	49	15	m	88	3
4	m	41	16	m	83	4
5	m	48	17	m	70	4
6	m	34	18	f	53	1
7	m	39	19	f	75	2
8	m	42	20	f	65	1
9	f	31	21	m	74	2
10	m	26				
11	m	33				
12	m	32				

carried out as described by Strolin Benedetti et al. (1989 a), using phenyl-boronic acid cartridges and a subsequent elution of the absorbed salsolinol with acetic acid and methanol. Derivatization and HPLC analysis of the R and S enantiomers of salsolinol with electrochemical detection were performed according to Pianezzola et al. (1989). The limit of detection of each enantiomer was 14 pmol/ml urine. The precision of the method was determined intra- and inter-day; it proved to be 12 and 13% for the R and S enantiomers respectively, expressed as coefficient of variation.

Results

In all the parkinsonian patients, the urinary concentrations of (R)- and (S)-salsolinol were found to be lower than the limit of detection (14 pmol/ml).

The urinary concentrations and daily excretion of (R)- and (S)-salsolinol in the healthy volunteers are given in Table 2. In subjects 1 to 6, when present, (S)-salsolinol was only detectable in the fraction 5 p.m. − 9 a.m. Therefore, the urinary concentrations of (S)-salsolinol were determined in this fraction, whereas the urinary concentrations of (R)-salsolinol were determined from the 24-h urines. (R)-Salsolinol was detected in the urine of 8 out of the 12 healthy volunteers and (S)-salsolinol was measurable in the urine of 3 subjects only. In the healthy subjects, the daily urinary excretion of total salsolinol was found to vary from 0 to 135 nmoles.

Table 2. Urinary concentrations (UC) and daily urinary excretion (DUE) of (R)- and (S)-salsolinol in 12 healthy volunteers

Subjects	UC (pmol/ml)		DUE (nmoles)		
	R	S	R	S	R + S
1	39.2	—	38.0	—	38.0
2	53.3	14.0	88.4	12.6	101.0
3	60.5	—	47.2	—	47.2
4	38.5	—	35.5	—	35.5
5	60.4	18.0	48.3	7.2	55.5
6	103.3	23.3	124.0	11.2	135.2
7	—	—	—	—	—
8	19.0	—	39.7	—	39.7
9	—	—	—	—	—
10	—	—	—	—	—
11	—	—	—	—	—
12	16.2	—	14.5	—	14.5

Discussion

In this additional study, the urinary concentrations of (R)- and (S)-salsolinol were lower than the limit of detection in the 6 patients. By contrast, in the preliminary study (Dordain et al., 1984) the urinary concentrations of total (R + S) salsolinol were found to be measurable (estimated limit of detection 10 pmol/ml) in the urine of all the 6 de novo parkinsonians. It is worth noting that the patients of the second study had apparently a more severe parkinsonism than the patients of the first study, as evaluated by Hoehn and Yahr scale.

In 4 out of the 12 healthy volunteers, the urinary concentrations of (R)-salsolinol were also lower than the limit of detection. It is to note that the HPLC method used in this study was different from the GC/MS method used in the Dordain study, where (R)- and (S)-salsolinol were not measured separately. In addition, in one of these four volunteers, subject 7, neither the R nor the S enantiomer of salsolinol was detectable in urine after administration of Madopar (3 × 62.5 mg/day) for 7 days (Dostert et al., 1990), suggesting that the biosynthesis of salsolinol is apparently impaired in some individuals. Conversely, urine of subjects 9, 10 and 11 contained substantial amounts of both (R)- and (S)-salsolinol after administration of Madopar. Urinary concentrations of 1,2-dehydrosalsolinol ranging from 121.5 to 312.4 pmol/

ml were found in the urine of subjects 7 to 12 (Dostert et al., 1990), showing the lack of correlation between the presence of salsolinol and that of 1,2-dehydrosalsolinol in human urine. Detectable amounts of 1,2-dehydrosalsolinol were also found in the urine of most parkinsonian patients (Dostert et al., unpublished results). Therefore, it may be wondered whether salsolinol bioprecursors such as 1,2-dehydrosalsolinol and/or 1-carboxysalsolinol, the condensation product of dopamine and pyruvate, might provide additional markers for the early detection of Parkinson's disease at least as reliable as salsolinol itself.

References

Brossi A (1982) Mammalian TIQ's: products of condensation with aldehydes or pyruvic acid? Prog Clin Biol Res 90: 123–133

Dordain G, Dostert P, Strolin Benedetti M, Rovei V (1984) Tetrahydroisoquinoline derivatives and parkinsonism. In: Tipton KF, Dostert P, Strolin Benedetti M (eds) Monoamine oxidase and disease. Prospects for therapy with reversible inhibitors. Academic Press, London, pp 417–426

Dostert P, Strolin Benedetti M, Dedieu M (1987) Ratio of enantiomers of salsolinol in human urine. Pharmacol Toxicol 60 [Suppl 1]: 13

Dostert P, Strolin Benedetti M, Bellotti V, Allievi C, Dordain G (1990) Biosynthesis of salsolinol, a tetrahydroisoquinoline alkaloid, in healthy subjects. J Neural Transm 81: 215–223

Pianezzola E, Bellotti V, Fontana E, Moro E, Gal J, Desai DM (1989) Determination of the enantiomeric composition of salsolinol in biological samples by high-performance liquid chromatography with electrochemical detection. J Chromatogr Biomed Appl 495: 205–214

Robbins JH (1968) Possible alkaloid formation in alcoholism and other diseases. Clin Res 16: 554

Sandler M, Bonham Carter S, Hunter KR, Stern GM (1973) Tetrahydroisoquinoline alkaloids: in vivo metabolites of L-dopa in man. Nature 241: 439–443

Strolin Benedetti M, Bellotti V, Pianezzola E, Moro E, Carminati P, Dostert P (1989 a) Ratio of the R and S enantiomers of salsolinol in food and human urine. J Neural Transm 77: 47–53

Strolin Benedetti M, Dostert P, Carminati P (1989 b) Influence of food intake on the enantiomeric composition of urinary salsolinol in man. J Neural Transm [Gen Sect] 78: 43–51

MR imaging of putamenal iron predicts response to dopaminergic therapy in parkinsonian patients

C. W. Olanow[1], M. Alberts[2], W. Djang[2], and J. Stajich[2]

[1] University of South Florida, Department of Neurology, Tampa, FL, and
[2] Duke University, Department of Medicine, Durham, NC, USA

Summary

We have previously demonstrated abnormal signal attenuation in the putamen consistent with increased iron on high field strength (1.5 Tesla) MR scans in patients with Parkinson Plus syndromes. We now describe 31 therapeutically naive parkinsonian patients who underwent MR imaging prior to entering a prospective randomized blinded study comparing bromocriptine to Sinemet as primary treatment. Patients were followed for a mean of 35 months at which time the MR scans were reviewed by a blinded radiologist and compared to the clinical scores recorded by blinded observers. Ten patients were considered to have abnormal scans due to increased signal attenuation in the putamen. Total Parkinson score in these patients deteriorated from baseline by 38.8% while those with normal MR scans improved by 33% (p < 0.0001). 60% of patients with abnormal MR scans had a poor response while 75% of those with normal MR scan had a good response to drug treatment. Eight patients were suspected by the blinded observers of having evolved into a Parkinson Plus syndrome and seven of these had abnormal signal attenuation in the putamen on the initial MR scan. We propose that signal attenuation in the putamen consistent with increased iron reflects striatal damage, predicts a poor response to dopaminergic therapy and can be identified early in the course of the illness.

Introduction

Using high field strength (1.5 Tesla) MR imaging, we previously demonstrated that patients with Parkinson Plus syndromes frequently have signal attenuation in the putamen which equals or exceeds that in the globus pallidus (Fig. 1) (Drayer et al., 1986 b). Normal adults demonstrate decreased signal on high field strength MR in specific brain regions (substantia nigra, red nucleus, globus pallidus, dentate

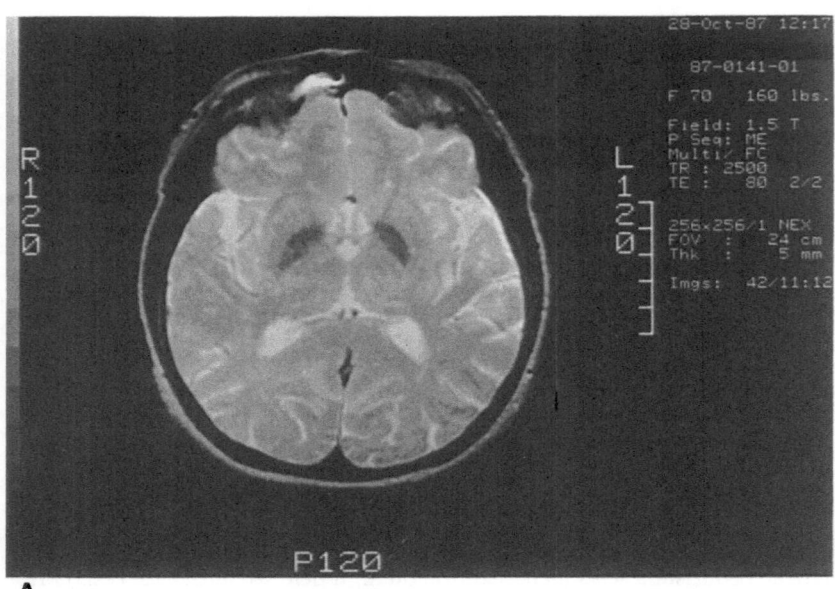

A

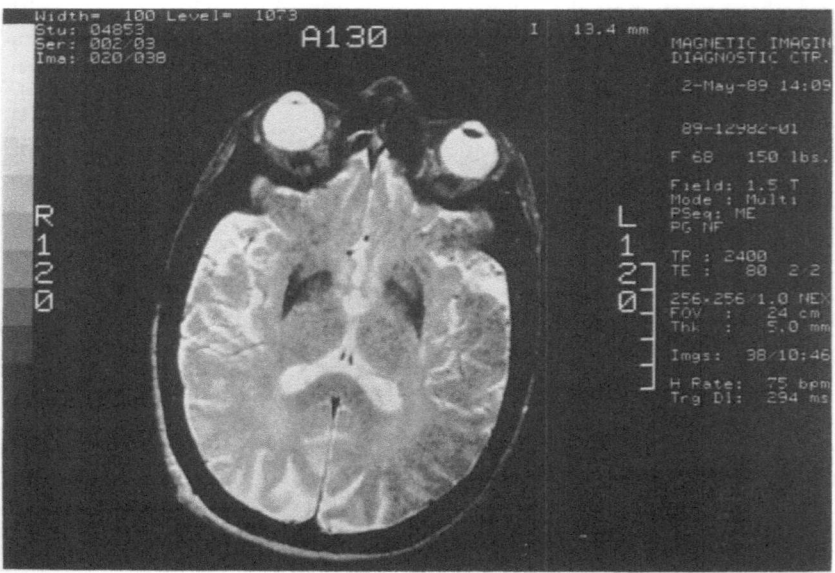

B

nucleus of the cerebellum) and anatomic studies using Perls stain for ferric iron demonstrate that these regions correlate precisely with areas of iron accumulation (Drayer et al., 1986 a). In normal individuals these areas of signal attenuation on MR scan are not observed until approximately one year of age corresponding with the appearance of iron on Perls stain (Olanow et al., 1989). In the globus pallidus, decreased signal is first observed in the medial portion and may be difficult to distinguish from decreased signal due to myelinated fibers in the adjacent internal capsule. Thereafter, signal attenuation in the globus pallidus becomes more pronounced through age 30, remains at a relative plateau until the fifth or sixth decade, and becomes more pronounced in the later years of life (Fig. 2). MR detection of signal attenuation in the putamen in normal subjects is not recognized until the third or fourth decade. It is first seen in the posterolateral portion of the putamen and increases markedly poorer the fifth decade (Fig. 2). In some elderly controls signal attenuation in the putamen approaches that observed in the globus pallidus, but in normal individuals signal attenuation in the putamen does not exceed that in the globus pallidus (Olanow et al., 1989). This pattern is in keeping with the accumulation of iron in brain specific regions (Hallgren and Sourander, 1958). We have proposed that signal attenuation in the putamen which exceeds that observed in the globus pallidus is a pathological state and when observed in a parkinsonian patient suggests underlying striatal degeneration and a Parkinson Plus syndrome (Olanow and Drayer, 1987).

Parkinsonian patients have an increased incidence of abnormal signal attenuation in the putamen compared to controls (Olanow and Drayer, 1987). Abnormal signal attenuation in the putamen in these patients correlates with an accelerated rate of disease progression, a poor response to dopamine replacement therapy and a clinical picture suggestive of a Parkinson Plus syndrome (Olanow and Drayer, 1987). In an analysis to be reported elsewhere, we found that 90% of patients clinically diagnosed as having Parkinson Plus syndromes can be dis-

---

Fig. 1. A High field strength MR image through basal ganglia of a normal adult. Note that signal attenuation is almost exclusively confined to the region of the globus pallidus and that little if any signal attenuation is noted in the putamen. **B** High field strength MR image through the basal ganglia of a patient with a Parkinson Plus syndrome. Note prominent signal attenuation in the putamen (particularly posterolateral region) which exceeds that in the globus pallidus

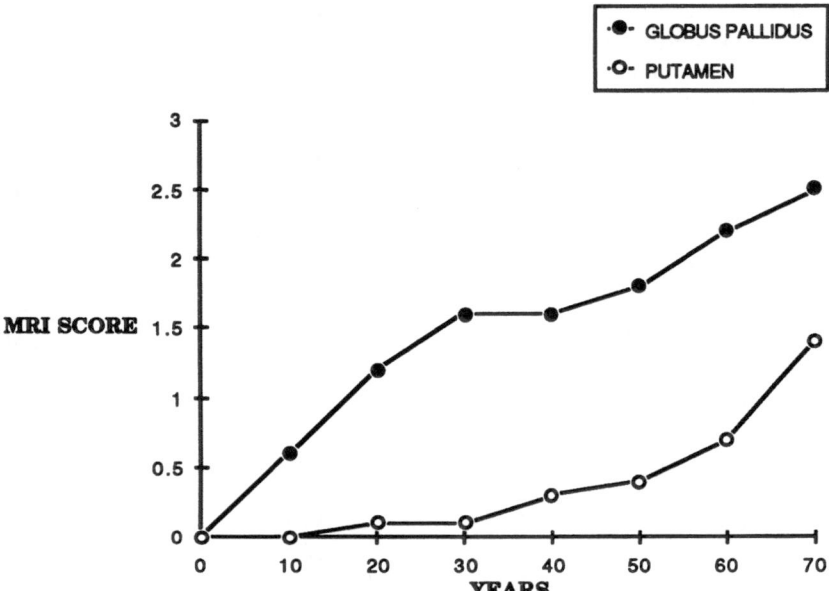

Fig. 2. MRI determination of iron score by decade in the globus pallidus and putamen in a normal population. Note that the iron score is 0 at birth and gradually increases with age. The globus pallidus shows a more rapid rate of accumulation in earlier years and putamen a more rapid rate in later years. In normal elderly patients, signal attenuation in the putamen may approach that seen in the globus pallidus, rarely equals and does not exceed the signal attenuation in the globus pallidus

tinguished from patients diagnosed as having idiopathic Parkinson's disease by MR scan.

We feel that abnormal signal attenuation in the putamen in patients with a Parkinson Plus syndrome reflects underlying striatal damage which accounts for their lack of response to dopamine replacement therapy. Patients with idiopathic Parkinson's disease have degeneration of neurons in the substantia nigra but relative preservation of the neurons and dopamine receptors in the striatum and thus retain the capacity to respond to dopamine replacement therapy. By contrast, patients with Parkinson Plus syndromes have degeneration of striatal neurons and may thus lack dopamine receptors necessary for a response to dopamine replacement therapy. As clinical distinction between these two groups of patients may be extremely difficult, particularly early in the course of their disease, signal attenuation in the putamen seen on MR scan may serve as a marker of striatal degeneration which predicts the response to L-dopa therapy.

To better assess the role of MR imaging as an early marker for Parkinson Plus syndromes in patients with early parkinsonism, we have designed a blinded prospective study to compare MR abnormalities in the putamen with the course and response to treatment of previously untreated Parkinson patients.

Patients and methods

Patients were selected from the Movement Disorders Clinic at Duke University. All had clinical features compatible with early Parkinson's disease and none had received prior dopamine replacement therapy. Inclusion criteria included ages 40 − 75, disease duration of less than 5 years, and Hoehn-Yahr stage of 1 to 3. Patients of either sex were included. Exclusion criteria included secondary parkinsonism due to drug, trauma, tumour, encephalitis, etc., clinically significant medical illness and prior dopamine replacement therapy. Prior to initiation of therapy, MR studies were performed on a high field strength (1.5 Tesla) MR system (General Electric). Both T 1 and T 2 weighted spin echo pulse sequences were employed. The T 1 partial saturation images were obtained using a repetition time (TR) of 500 ms and an echo delay (TE) of 20 ms. The T 2 spin echo series used a TR of 2,000 ms and TE of 100 ms (SE 2,000/100). An image matrix of 128 × 256 and a single signal average (2 excitations) were used routinely for axial images. Thin sections were obtained through the region of the basal ganglia. Intermediate weighted images, with a TR of 2,000 ms and TE of 30 ms (SE 2,000/30), were obtained routinely.

Patients included in this study were also participants in a prospective randomized single-blind double observer study comparing Sinemet to bromocriptine in previously untreated parkinsonian patients (Olanow et al., 1987). As part of that study, patients were assigned to either Sinemet or bromocriptine and treated with the lowest dose of study medication that would provide them with a clinically satisfactory response. Patients randomized to bromocriptine could receive Sinemet as an adjunct after 12 months. Clinical scores and response to drug therapy were evaluated by blinded investigators using a modified Columbia scale, modified Northwestern Disability Scale and Hoehn-Yahr stage. An estimate of response to therapy and accuracy of clinical diagnosis was made at the conclusion of the study by the blinded evaluators who were unaware of the MR results.

After a mean follow-up of 35 months, a visual analysis of MR images was performed by a blinded neuroradiologist (WD) who was unaware of the clinical history of any patient. "Iron score" was evaluated on the SE 2,000/100 image using a simplified quantitative scoring system (Table 1). Iron score was graded in the globus pallidus, caudate, putamen, thalamus, red nucleus, substantia nigra, dentate nucleus and superior colliculus and these results will

Table 1. Iron score as demonstrated by MR imaging

0	No detectable signal attenuation
1	Trace signal attenuation
2	Mild but definite signal attenuation
3	Moderately prominent signal attenuation
4	Markedly prominent signal attenuation

be reported separately. In this study emphasis was placed on the iron score in the putamen. The putamen was considered to be abnormal if signal attenuation equalled or exceeded that observed in the neighbouring globus pallidus and the neuroradiologist was asked to grade each scan as normal (MR−) or abnormal (MR+) based on this criteria. Statistical analysis to compare MR findings with clinical response was performed using an analysis of variance (ANOVA).

Results

Thirty-one previously untreated parkinsonian patients underwent MR imaging as part of the above protocol. 21 were judged to have normal signal attenuation in the putamen (MR−) and 10 were judged to have abnormal signal attenuation in the putamen (MR+). Baseline data in the two groups are illustrated in Table 2. Treatment data in the two groups after a mean follow-up of 35 months is illustrated in Table 3. Total Parkinson score was improved from baseline by 33% in patients with MR− scans while patients with MR+ scans deteriorated by 38.8% in comparison to their baseline level (Fig. 3). Patients

Table 2. Patient data

	MR +	MR −
Number	10	21
Sex		
male	7	8
female	3	13
Mean age (years)	71.7	58.9*
Mean duration PD (mos.)	13.4	20.
Baseline total PD score (max 450)	122	88.3

* $p < 0.01$

Table 3. Treatment data

	MR +	MR −
	35.4	34.8
Total PD score-baseline	122.0	88.3
Total PD score-final visit	169.3	59.2*
% Improvement in PD score from baseline	− 38.8	+ 33.0*
Mean sinemet dose (mg. L-dopa)	533.3	363.2**
Mean bromocriptine dose (mg.)	16.5	20.8

  * p < 0.0001
** p < 0.01

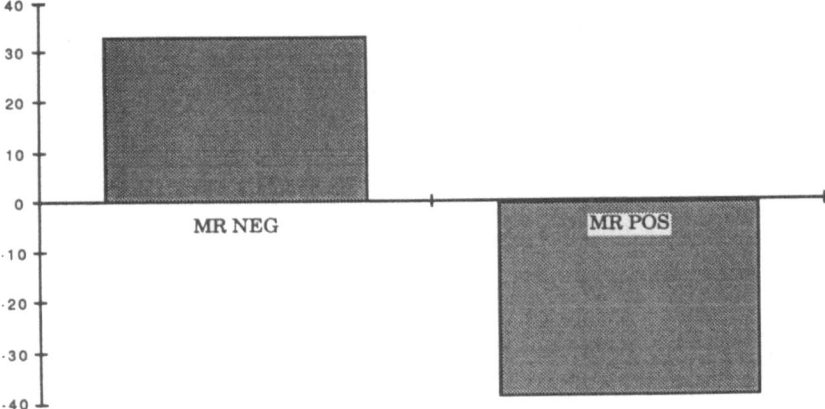

Fig. 3. Percent change in total Parkinson score from baseline in patients with MR − and MR + scans. Note that those with MR − scans improved by 33% while those with MR + scans deteriorated by 38.8%

with MR + scans received a daily dose of Sinemet which was 47% greater than that of patients with MR − scans. At the final visit, the response to dopamine replacement therapy was rated by the blinded investigators and the results are noted in Table 4. 90% of patients with MR + scans were judged to have had a poor or fair response while 76% of patients with MR − scans had a good response. Eight patients

Table 4. Clinical response

	MR +	MR −
Poor	6 (60%)	1 (5%)
Fair	3 (30%)	4 (19%)
Good	1 (10%)	16 (76%)

were considered to have developed a Parkinson Plus syndrome based on a clinical picture dominated by impairment of speech and postural reflexes and a lack of response to levodopa replacement therapy. Seven of these patients had MR + scans. Two had eye finding suggestive of Progressive Supranuclear Palsy and one had orthostatic hypotension suggesting Shy-Drager syndrome.

Clinical benefits were comparable in the groups randomized to Sinemet and to Parlodel (which could be supplemented with Sinemet) and will be reported elsewhere. There was no difference in the distribution of MR + and MR − scans in either treatment group. An analysis of tremor, rigidity, bradykinesia, gait, postural stability and speech at the time of presentation was not significantly different in patients with MR + and MR − scans. Mean iron scores for globus pallidus, substantia nigra, red nucleus and dentate nucleus did not serve to distinguish the two groups. Five patients with MR + scans and four with MR − scans had repeated studies and no differences from the initial scan were detected after a mean of 35 months.

Discussion

We demonstrate in a prospective fashion that high field strength MR imaging of the putamen in previously untreated parkinsonian patients predicts their response to dopamine replacement therapy. Ten of 31 previously untreated Parkinson patients had MR changes in the putamen in which signal attenuation was equal to, or more pronounced than, that present in the neighbouring globus pallidus. These patients as a group had a markedly poorer response to dopamine replacement therapy than patients who had a normal MR pattern in the putamen. 60% of MR + patients had a poor response to levodopa and 70% developed a Parkinson Plus syndrome. By contrast, only one patient (5%) with an MR − scan had a poor response to levodopa therapy. These data suggest that MR imaging performed prior to the introduction of treatment predicts the response to replacement therapy and the likelihood of developing a Parkinson Plus syndrome.

At the time of clinical presentation, patients with MR + scans were older, had a shorter duration of disease and had more severe parkinsonism than patients with MR − scans. Increasing age is associated with increased iron and more prominent signal attenuation in the putamen and the differences in the age of the two groups may have influenced the MR appearance. However, aging does not preclude a good response to levodopa and the pattern of an MR + scan is rarely seen in normal controls and we do not think that age alone accounts for these observations (Olanow et al., 1989). No presenting clinical feature predicted treatment response or outcome as accurately as high field strength MR imaging of the putamen.

Eight of 31 patients (26%) in our study were suspected of having a Parkinson Plus syndrome after three years of follow-up. Three patients with MR + scans continue to be clinically indistinguishable from patients with idiopathic Parkinson's disease but it is possible that they might develop Parkinson Plus syndromes and that the incidence will increase. While none of our patients has come to autopsy for formal histopathologic confirmation of diagnosis, this study is in keeping with the idea that it may be difficult to distinguish patients with early Parkinson's disease from patients with Parkinson Plus syndromes and that the incidence of Parkinson Plus syndromes may be higher than what has been previously appreciated.

These MR studies suggest that patients with parkinsonism can be divided into "nigral" and "striatal" types, which can be distinguished early in the course of the illness. These two types may have different etiologies and the ability to differentiate them may be of importance in designing specific therapy. MR differentiation may permit greater diagnostic accuracy in selecting patients for studies of Parkinson's disease or Parkinson Plus syndromes, thereby allowing studies to be performed with greater reliability using fewer patients and at a lesser cost.

Signal attenuation in the putamen of Parkinson Plus patients is likely to be due to the accumulation of iron. MR changes consistent with increased iron have also been observed in the substantia nigra of patients with Parkinson's disease and Parkinson Plus syndromes (Olanow and Drayer, 1987; Braffman et al., 1988). Earle reported increased brain iron in parkinsonian patients using x-ray fluorescent spectroscopy (Earle, 1968). More recently there have been reports which confirm that iron is increased in the substantia nigra of patients with Parkinson's disease and in the substantia nigra and striatum of patients with Progressive Supranuclear Palsy (Dexter et al., 1989 a, c). The significance of the iron accumulation is unknown. Iron may simply be a nonspecific marker of a degenerative process as similar changes have been observed in other degenerative disorders (Drayer et al., 1987). There is, however, considerable interest in the possibility that iron may contribute to the pathogenesis of Parkinson's disease and Parkinson Plus syndromes. Iron is a transition metal which catalyzes the formation of hydrogen peroxide and the generation of free hydroxyl radicals (Halliwell and Gutteridge, 1985). That it may play such a role in Parkinson's disease is supported by the observation that brain ferritin is reduced in patients with Parkinson's disease (Dexter et al., 1989 a). Iron bound to ferritin is inactive. The finding of decreased ferritin coupled with increased iron suggests an alteration in iron regulation resulting in the increased availability of iron in a molecular form which can facilitate oxidation reactions with the generation of free radicals. Further concern regarding a possible role for oxidation reactions in the etiology of Parkinson's disease is derived from observations of decreased glutathione and increased lipid peroxidation in the substantia nigra of Parkinson's disease patients and clinical studies which suggest that the antioxidant, deprenyl, may have protective effects (Perry et al., 1982; Dexter et al., 1989 b; Birkmayer et al., 1985).

MRI is a non-invasive tool which is now widely available and can differentiate parkinsonian patients with and without abnormal signal attenuation in the putamen. This appears to categorize Parkinson patients into a "striatal" and a "nigral" form and to predict the rate of disease progression, the response to therapy, and the likelihood of developing a Parkinson Plus syndrome. In vivo mapping of brain iron may provide important clues as to the pathogenesis of these disorders and further studies to delineate the significance of the pathological accumulation of brain iron are indicated.

References

Birkmayer W, Knoll J, Riederer P, Youdim MBH, Hars V, Marton J (1985) Increased life expectancy resulting from addition of l-deprenyl to madopar treatment in Parkinson's disease: a long term study. J Neural Transm 64: 113–127

Braffman BH, Grossman RI, Goldberg HI, Stern MB, Hurtig HI, Hackney DB, Bilaniuk LT, Zimmerman RA (1988) MR imaging of Parkinson's disease with spin echo and gradient echo sequences. Am J Nucl Rad 9: 1093–1099

Dexter DT, Carayon A, Jenner P, Agid F, Agid Y, Marsden CD (1989 a) Increased nigral iron and ferritin levels in Progressive Supranuclear Palsy. Br J Pharmacol [Suppl]: 382 P

Dexter DT, Carter CJ, Wells FR, Javoy-Agid F, Agid Y, Lees A, Jenner P, Marsden CD (1989 b) Basal lipid peroxidation in substantia nigra is increased in Parkinson's disease. J Neurochem 52: 381–389

Dexter DT, Wells FR, Lees AJ, Agid F, Agid Y, Jener P, Marsden CD (1989 c) Increased nigral iron content and alterations in other metal ions occurring in brain in Parkinson's disease. J Neurochem 52: 1830–1836

Drayer BP, Burger P, Darwin R (1986 a) Magnetic resonance imaging of brain iron. Am J Nucl Rad 7: 373–380

Drayer BP, Olanow W, Burger P, Johnson GA, Herfkens R, Riederer S (1986 b) Parkinson Plus syndrome: diagnosis using high field MR imaging of brain iron. Radiology 159: 493–498

Drayer B, Burger P, Herwitz B, Dawson D, Cain J (1987) Reduced signal intensity on MR images of thalamus and putamen in multiple sclerosis. Am J Nucl Rad 8: 357–363

Earle KM (1968) Studies on Parkinson's disease and including x-ray fluorescent spectroscopy of formalin-fixed brain tissue. J Neuropathol Exp Neurol 27: 1–14

Hallgren B, Sourander P (1958) The effect of age on the nonhemin iron in the human brain. J Neurochem 3: 41–51

Halliwell B, Gutteridge JMC (1985) Oxygen radicals in the nervous system. Trends Neurol Sci 8: 22–26

Olanow CW, Drayer BP (1987) Brain iron: MRI studies in Parkinson syndromes. In: Fahn S, Marsden CD, Calne D, Goldstein M (eds) Recent developments in Parkinson's disease. Macmillan Healthcare, Flurham Park, pp 135–143

Olanow CW, Alberts MJ, Stajich J, Burch G (1987) A randomized blinded study of low dose bromocriptine vs. low dose carbidopa/levodopa in untreated Parkinson patients. In: Fahn S, Marsden CD, Calne D, Goldstein M (eds) Recent developments in Parkinson's disease. Macmillan Healthcare, Flurham Park, pp 201–208

Olanow CW, Holgate RC, Murtaugh R, Martinez C (1989) MR imaging in Parkinson's disease and aging. In: Calne DB, Comi G, Crippa D, Horolwski R, Trabucci M (eds) Parkinsonism and aging. Raven Press, New York, pp 156–164

Perry TL, Godin DV, Hansen S (1982) Parkinson's disease: a disorder due to nigral glutathione deficiency? Neurosci Lett 33: 305–310

The neurotoxic component in Parkinson's disease may involve iron-melanin interaction and lipid peroxidation in the substantia nigra

M. B. H. Youdim and D. Ben-Shachar

Rappaport Family Research Institute, Department of Pharmacology, Technion, Haifa, Israel

Summary

There are indications of oxidative stress in Parkinson's disease (PD). This can be attributed to the selective increases of iron (III) due to its decompartmentalization and lipid peroxidation and a decrease of glutathione (GSH) oxidation in substantia nigra (SN). The net effect would be accumulation of H_2O_2 and the formation of free cytotoxic hydroxyl (OH·) radical from the interaction of iron either with H_2O_2 or melanin or both, to drive the Fenton reaction. Identification of two high-affinity binding sites for Fe^{3+} on dopamine-melanin indicates the iron binding potency of melanin. This interaction does not result in prevention but rather in potentiation of iron-induced lipid peroxidation, a process selectively inhibited by iron chelators. Thus, H_2O_2 and melanin are thought to participate in altering the redox state of iron between its two valencies with resultant formation of OH·. This hypothesis can explain the vulnerability and selectivity of melaninized nigrostriatal dopamine neurons to degeneration in PD. Determination of OH· or lipofusin in the CSF, as a by-product of lipid peroxidation, may be a valid approach to early diagnosis of PD.

Introduction

Attempts at identifying the etiology and an early marker of Parkinson's disease (PD) have failed, mainly due to the lack of explanation for the vulnerability of melaninized dopamine neurons in the substantia nigra (SN) (Hornykiewicz, 1988). Current opinion supports the view of the involvement of a neurotoxic event, resulting from endogenously or exogenously derived cytotoxins (Tanner, 1989). However, at best the evidence for such cytotoxins is rather tenuous (Baker, 1989). Although the MPTP (N-methyl-4-phenyl-1,2,3,6-tetrahydropyridine) animal model of PD has proved to be a valuable tool, it is of limited use since it does not totally mimic the pathological conditions of idiopathic PD (Kinemuchi et al., 1987; Riederer and Youdim, 1987). In the absence of identifying a neurotoxin in the environment of basal ganglia, other factors more immediate to SN must be considered. Recent advances in the chemical pathology of PD have implicated the vulnerability of melaninized nigrostriatal dopamine neurons to the process of oxidative stress as a result of altered oxygen free radical formation and disposition (Youdim, 1988; Youdim et al., 1989). Ultimately, the balance between the latter two could be the crucial factor in the degeneration of nigrostriatal dopamine neurons.

In the mammalian brain little attention has been given to the formation of oxygen-derived free radicals and functional biochemical changes related to the formation of free radicals or any agent that might attenuate the formation of free radicals in brain tissue (Halliwell and Gutteridge, 1985). A free radical is an atom or molecule that has an unpaired electron. They are produced as a consequence of a variety of metabolic processes and have been associated with lipid peroxidation of cellular membrane and cell degeneration (Fitzsimons, 1979; Wilson, 1979; Triggs and Willmore, 1984). There are a number of disorders in which in vivo oxygen-derived free radical formation is implicated; these include brain and heart ischaemia, rheumatoid arthritis, inflammation, trauma, intracellular toxicity of lung by chemicals (e.g. Paraquat) and radiation (Wilson, 1979; Halliwell and Gutteridge, 1986; Kohen and Cherin, 1985; Halliwell, 1989). In ageing and Parkinson's disease, a case has been made for oxygen-derived free radicals and this has been attributed to age-related changes in free radical scavenger enzyme, glutathione peroxidase or altered increased activity of oxidative enzymes (e.g. monoamine oxidase) capable of producing significant amounts of hydrogen peroxide (Donaldson and Barbeau, 1985; Cohen, 1985; Spina and Cohen, 1988, 1989; Youdim et al., 1989). The latter processes appear particularly appropriate for the striatum since this tissue has an extremely high metabolic activity and is sensitive to endogenous or exogenous insults.

Brain iron and Parkinson's disease

Oxygen free radicals are most often initiated by some agents. In the above disorders, the direct involvement of free iron or an abnormality of iron metabolism has more often been cited in oxygen free radical formation and initiation of tissue damage than any other agent. Although the direct participation of iron in oxygen free radical formation can no longer be questioned, its exact mechanism of action remains controversial (Wilson, 1979; Braughler et al., 1986; Minotti and Aust, 1987). Nevertheless it is accepted that lipid peroxidation resulting from the formation of oxygen free radical, the cytotoxic hydroxyl (OH·) radical, is dependent on the alteration of the redox state of iron between its two valencies and the presence of H_2O_2 (Wilson, 1979; Halliwell and Gutteridge, 1986; Halliwell, 1989). Thus, the possible abnormality of iron metabolism in PD has not escaped notice and more recently it has received considerable attention (Youdim, 1988; Riederer et al., 1989; Youdim et al., 1989). This is particularly important since regional iron distribution in mammalian brain (human, monkey and rats) is uneven and unique, as compared to other metals (Youdim, 1985; Riederer et al., 1989). Of all the brain regions the highest iron content is present in the substantia nigra and globus pallidus (Riederer et al., 1989), two brain regions known for their neurodegeneration in PD and Haller Varden Spatz disease (HVS), respectively (Youdim, 1985). The substantially increased accumulation of iron in globus pallidus of HVS brain has been known for some time and confirmed on several occasions (Yehuda and Youdim, 1988). One outstanding feature of globus pallidus in HVS is the demyelination and denervation of the neurons. The most compelling finding for PD is the recent confirmation by us (Sofic et al., 1988; Riederer et al., 1989) and others (Drayer et al., 1986; Dexter et al., 1989 a) of the study by Earle (1968) reporting the increased iron content in parkinsonian SN. These findings, together with reports of the diminished availability of glutathione (GSH) and ascorbate in parkinsonian brains (Perry and Young, 1986; Riederer et al., 1989) the necessary components for disposition of H_2O_2 by glutathione peroxidase, are supportive of the oxidative stress in PD. Free iron is known to be capable of reducing tissue GSH. Thus the increased availability of H_2O_2, resulting from oxidative deamination and auto-oxidation of dopamine, can, in essence, participate in altering the valencies of Fe^{2+} between its two states and drive the Fenton reaction (Wilson, 1979). The resultant feature of this reaction is the formation of OH· and increased lipid peroxidation (Braughler et al., 1986; Minotti and Aust, 1987). In particular, we have noted a significantly greater increase of Fe^{3+} as compared to Fe^{2+} in parkinsonian SN, inasmuch as the ratio of Fe^{2+}/Fe^{3+} changes from roughly $3:1$ in control SN to $1:1$ in SN of PD (Sofic et al., 1988; Riederer et al., 1989). This hypothesis is fully compatible with the property of iron to induce lipid peroxidation (Braughler et al., 1986; Minotti and Aust, 1987). Indeed one recent study has shown increased basal lipid peroxidation selectively in SN and not in other brain regions in PD (Dexter et al., 1989 b). Confirmation of this finding is essential.

In summary, it is now apparent that all the elements necessary to induce selective oxidative stress in parkinsonian SN exist (a) elevated availability of H_2O_2 due to increase of MAO-B activity with ageing and auto-oxidation of dopamine to melanin (Konradi et al., 1986); (b) reduction in glutathione (GSH) (see Riederer et al., 1989), the rate limiting cofactor of glutathione peroxidase and the main enzymatic pathway for disposition of H_2O_2 in the brain (Spina and Cohen, 1988, 1989); (c) increased availability of iron and its altered valencies by melanin (Riederer et al., 1989; Youdim et al., 1989; Ben Shachar et al., 1989) and (d) increase of basal lipid peroxidation (Dexter et al., 1989 a).

Selectivity of substantia nigra dopamine neuron degeneration iron-melanin interaction

If increased iron participates in the oxidative stress due to siderosis of SN (Youdim et al., 1989), it may be asked (a) how is iron transported into the brain and why should the SN, which normally accumulates high iron content, be endowed with a further increase of iron in PD, (b) where is the iron deposited, intra- or extraneuronally, in melaninized dopamine neurons (c) in what form and how is it bound (d) can iron-induced cytotoxicity be used as an index of early diagnosis in PD by measuring free hydroxyl (OH·) radical in vivo and (e) would iron chelators be useful for retarding the progression of neurodegeneration. The answers to all these questions are not easily available but progress towards their resolution is at hand. In the final analysis, the processes by which only melanin-containing dopamine neurons are vulnerable to degeneration (Hirsch et al., 1988) must be considered.

It has always been inferred and generally accepted that melanin in melaninized cells (tissue) is a scavenger of radicals and protective to these cells (Masson et al., 1960; Pilas et al., 1988). However, the role of melanin in pathological as compared to normal conditions requires a thorough reassessment. This is especially relevant with regard to its role in PD. It is the pigmented (Mann and Yates, 1983) and the melanin-containing dopamine neurons of SN which degenerate in PD (Hirsch et al., 1988). This specificity suggests that melanin itself may participate in the neurodegeneration of dopamine neurons. Under certain conditions melanin, instead of being a radical scavenger, can be highly effective as a promoter of oxygen free radicals, especially because of its ability to generate cytotoxic OH· in the presence of divalent metals, such as iron. This has been attributed to the avidity by which melanin binds iron more tenuously than any other metal (Bruenger et al., 1967; Larsson and Tjalve, 1979; Potts and Au, 1976). The metal binding property of melanin has been known for many years. Neuromelanin or melanin formed from auto-oxidation of dopamine or noradrenaline possesses many carboxyl, phenolic, hydroxyl and quinoid groups with strong affinity for metals and drugs (Larsson and Tjalve, 1979). The iron binding activity of melanin can result in promotion of the Fenton reaction, via the reduction of Fe^{3+} to Fe^{2+} in the presence of H_2O_2 (Masson et al., 1960; Pilas et al., 1988). The net effect is the liberation of OH· and possible lipid peroxidation. Indeed chelated iron is considered to be even more effective in radical formation (Pilas et al., 1988). By contrast Fe^{2+} bound by melanin does not participate in OH· formation (Pilas et al., 1988).

Increase of Fe^{2+} and Fe^{3+} in SN of parkinsonian brain (Sofic et al., 1988; Riederer et al., 1989) cannot be attributed to ferritin or haemosiderin-iron, since they are not increased as much as that of total iron. Therefore iron is either decompartmentalized and is available in a free state or it is accumulated in a bound form, so that it would be available for alteration of its redox state.

The degeneration of nigrostriatal dopamine neurons would not explain the substantially increased iron in this region, since MPTP treated monkeys and mice do not show alteration of iron content in the striatum (Youdim et al., 1988). Consideration should be given to either increased uptake or decreased efflux of iron in the brain resulting from specific changes in blood-brain barrier (BBB) in the SN. At the present there is no firm evidence that BBB is affected in different brain regions of PD. Although transferrin and transferrin receptors have been identified in the capillary endothelial cells, forming the BBB in different brain regions, it is apparent that the processes by which iron is taken up and deposited in the brain differ significantly from those in the small intestine and liver. Thus in both human and rat brain, the regions (SN, globus pallidus, dente gyrus, interpeduncular nuclei, caudate nucleus, putamen, red nucleus) with high iron content have extremely low transferrin receptors (Hill, 1988). In the absence of a significant amount of ferritin the most obvious candidate for complexing free iron in the SN is the selective presence of melanin in this region. Indeed we have recently described for the first time the presence of two high affinity binding sites for Fe^{3+} on dopamine melanin. These sites have relative K_D's values of 13 and 200 nM and B_{max} values of 1.37 and 17.4 mmol/mg melanin, respectively (Ben Shachar et al., 1989). Comparison of these data with those available in the literature shows that the dopamine-melanin has a higher affinity and greater capacity for binding iron than other metals and drugs including MPTP and MPP^+ (Larsson and Tjalve, 1979; D'Amato et al., 1988). The iron binding capacity of melanin is concentration- and pH-dependent. The optimal conditions are 3 µg melanin/ml and pH 6.5. The examination of a large number of drugs with the possible ability to displace $^{59}Fe^{3+}$ from dopamine melanin has shown that only iron-chelators and psychotropic drugs that chelate iron are the most potent inhibitors (IC_{50} $60-200$ nM). Other drugs, including MPTP, MPP^+ and chloroquine, which are able to bind to melanin do not affect $^{59}Fe^{3+}$ even at 2000 nM concentration (Ben Shachar et al., 1989). Therefore, the binding of iron is specific and not mutually shared by other agents capable of binding to dopamine melanin. By contrast, dopamine itself promotes the binding of $^{59}Fe^{3+}$ to dopamine-melanin. At first this finding would appear to be at odds with the ability of dopamine-melanin to bind iron. However, it is possible that in the presence of iron more dopamine is auto-oxidized to melanin, thus increasing the latter's polymeric resin nature. On the other hand, dopamine may allosterically modify the structure of melanin for greater iron binding. In any event this finding itself may be crucial and relevant to the chemical pathology of neurodegeneration in Parkinson's disease. The fact that both iron and melanin increase in the substantia nigra with ageing, suggests that their stoichiometry in relationship to lipid peroxidation induced dopaminergic neurodegeneration may be relevant (Fig. 1) (Ben-Shachar et al., 1989).

There is no doubt that free or chelated iron can promote the formation of OH· and induce lipid peroxidation of cell membranes in vitro. This has been used as evidence for its in vivo cytotoxicity. However, direct in vivo evidence for lipid peroxidation and free radical formation is hard to come by due to the lack of appropriate in vivo

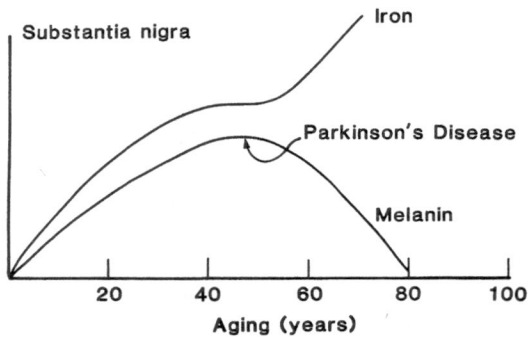

Fig. 1. Depiction of the time course of iron and melanin deposition in the human substantia nigra (SN) with ageing. The increase of iron in parkinsonain SN could alter the ratio of iron melanin, resulting in neurotoxicity induced lipid peroxidation and neurodegeneration of melanin containing neurons with the resultant disappearance of melanin

methodology. In any event we have confirmed (Ben Shachar et al., 1989) the lipid peroxidation promoting property on iron by measuring malondialdehyde (MDA) formation using rat and human cortical homogenate preparations (Triggs and Willmore, 1984). Melanin alone, at low concentration (3 μg/ml), is inhibitory but at higher concentrations (> 10 μg/ml) is an even more potent inhibitor of MDA formation. By contrast interaction of Fe^{3+1} with melanin does not result in inhibition of MDA formation, but rather, in potentiation of iron induced lipid peroxidation, which can be fully inhibited by specific iron chelators, desferrioxamine, o-phenanthroline and 8-hydroxyquinoline (Fig. 2). These findings may go some way to support our hypothesis regarding selective promotion of lipid peroxidation in the SN of parkinsonian brain by the increased availability of iron-melanin complex (Youdim et al., 1989). Thus, it is more than possible that at certain concentrations of endogenous iron, melanin can be protective against lipid peroxidation. However, when the stoichiometry of iron to melanin changes in favour of iron, lipid peroxidation takes place (Fig. 2). A similar process could also occur in PD (Fig. 1). As a consequence of lipid peroxidation induced neurodegeneration, the melanin in the dopamine neurons would eventually disappear. The ability of iron chelators, desferrioxamine, o-phenanthroline and chlorpromazine

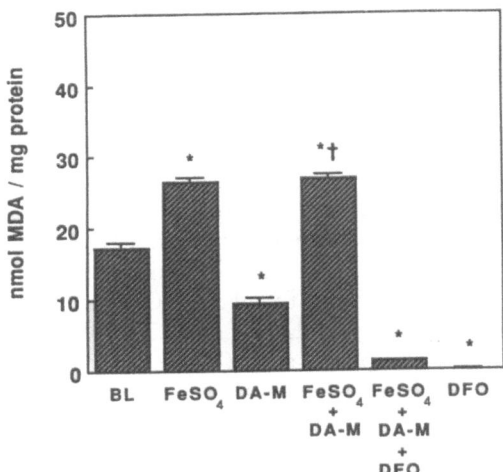

Fig. 2. Lipid peroxidation properties of iron and melanin alone or in combination as measured by the appearance of malondialdehyde (MDA). * p < 0.01, † p < 0.001. *BL* basal level; *DA-M* dopamine melanin (3 μg/ml); FeSO₄, 100 μM; *DFO* desferrioxamine (100 μM)

to inhibit the binding of iron to melanin, and iron-melanin complex induced lipid peroxidation, predict iron chelators as an important possible class of radical scavenger therapeutic drugs for the CNS. However, although desferrioxamine is a useful drug for the treatment of iron overload in the peripheral tissues, it does not cross the blood-brain barrier. One class of iron chelators such as the 21-aminosteroid U74006F with the ability to cross the BBB (Braughler et al., 1989; Braughler et al., 1988) has now been developed and is being used with promising results for treatment of trauma and ischaemia (Hall and Yonkers, 1988; Hall, 1988). Consideration should also be given to other lipophilic iron chelators including 3,4-dihydroxybenzoic acid and lipoic acid.

Consequences of iron "overload" in substantia nigra

It is well established that iron overload in tissues can result from a number of causes, the common feature of which in man is the limitation of iron elimination from the body. Thus, any increase in tissue iron accumulation, no matter by what process, cannot be compensated by an increase in iron excretion, especially in brain with its low iron turnover (Ben Shachar et al., 1986). The ultastructural and biochemical patterns in iron overload show increased displacement of ferritin (Chrichton, 1979) from one site to another resulting in availability of free iron. Clearly the risk that iron can become decompartmentalized in such a situation is increased and production of free radicals can therefore lead to tissue damage, resulting from lipid peroxidation of membrane lipids. Free iron can also increase the rate of lipid oxidation by catalyzing the decomposition of lipid hydroperoxides. Skin pigmentation is one common feature of iron overload. It is due to increased melanin production, which is thought to arise from a free-radical mechanism similar to that involved in the formation of melanin by sunlight. The iron in the skin catalyzes free radical generation and amplifies melanin formation (Crichton, 1979). Therapy of iron overload with iron chelators (e.g. 3,4-dihydroxybenzoic acid) results in depigmentation of skin (Peterson et al., 1974; Alexander, 1960). By analogy, the dense melaninization of SN dopamine neurons, resulting from auto-oxidation of dopamine, may initiate the formation of oxygen free radicals and H_2O_2 in response to decomparmentalized free iron. It is the Fe^{2+} form of iron which is the most potent free-radical generator and thus the lipid peroxidation potential of iron is greater when melanin is present since melanin-like ascorbate can regenerate Fe^{2+} from Fe^{3+} (Pilas et al., 1988). We believe this hypothesis can also explain parkinsonism in managanese miners (Donaldson and

Barbeau, 1986). Manganese, like iron, can change its valency between two states and be bound to melanin. Furthermore its accumulation in the striatum can result in the reduction of glutathione peroxidase and glutathione (GSH) (Liccione and Maines, 1988), the main pathway of H_2O_2 decomposition in the brain.

In vivo detection and measurement of oxygen radical induced lipid peroxidation in Parkinson's disease

The involvement of oxygen free-radical reactions has been suggested and demonstrated in a number of diseases (Halliwell, 1987). However, most of the evidence has been obtained from in vitro studies. Free-radical reactions may either play a role in the pathology of the disease or be a by-product of cellular injury and not be related to onset or initiation of the disease. The fact that MPTP does not induce oxygen free-radical reaction, a phenomenon recently reported to occur selectively in the substantia nigra from parkinsonian brains (Dexter et al., 1989 b), is indicative of altered pathology due to radical formation. In the last few years a number of novel methods for detection of oxygen radical reactions in vivo has been developed. These procedures depend on the conversion of substances (e.g. benzene, salicylate, aspirin and uric acid) by the free radical reaction to specific metabolites which can then be detected either in the urine or blood serum (Halliwell et al., 1988). The implications in PD are clear, where oxidative stress has been suggested to occur on several occasions, including the present discussion. Such techniques may be applicable for examining free radicals in the CSF.

References

Alexander P (1960) Protection of macromolecules in vitro against damage by ionizing radiation. In: Hollander A (ed) Radiation protection and recovery. Pergamon Press, New York

Barker R (1989) Environmental toxins and Parkinson's disease. Trends Neurosci 12: 182

Ben Shachar D, Ashkenazi R, Youdim MBH (1986) Long-term consequences of early iron deficiency on dopaminergic neurotransmission in rats. Int J Dev Neurosci 4: 81–88

Ben Shachar D, Riederer P, Youdim MBH (1989) Iron-melanin interaction in the substantia nigra as the neurotoxic component of Parkinson's disease (submitted)

Braughler JM, Duncan LA, Chase RL (1986) The involvement of iron in lipid peroxidation importance of ferric to ferrous ratio in initiation. J Biol Chem 261: 10282–10289

Braughler JM, Pregenzer JF, Chase RL, Ducan LA, Jacobson EJ, McCall JM (1987) Novel 21-aminosteroids as potent inhibitors of iron-dependent lipid peroxidation. J Biol Chem 262: 10438–10440

Bruenger FW, Stover BJ, Atherton DR (1967) The incorporation of various metal ions in vivo and in vitro produced melanin. Radiat Res 32: 1–12

Cohen G (1985) In: Lajtha A (ed) Handbook of neurochemistry, vol 4. Plenum Press, New York, pp 315–330

Crichton RR (1979) Interaction between iron metabolism and oxygen activation. In: Oxygen free radicals and tissue damage. Ciba Foundation Symposium No 65 New Ser. Excerpta Medica, Amsterdam, pp 57–76

D'Amato RJ, Lipman ZP, Snyder SH (1986) Selectivity of the Parkinson neurotoxin MPTP: toxic metabolite MPP+ binds to neuromelanin. Science 231: 987–991

Dexter DT, Wells FR, Agid F, Agid Y, Lees AJ, Jenner P, Marsden CD (1989 a) Increased nigral iron content in post-mortem parkinsonian brain. J Neurochem 52: 1830–1836

Dexter DT, Carter CJ, Wells FR, Javoy-Agid F, Agid Y, Lees A, Marsden CD (1989 b) Basal lipid peroxidation in substantia nigra is increased in Parkinson's disease. J Neurochem 52: 381–389

Donaldson J, Barbeau A (1985) Managanese neurotoxicity: possible clues to aetiology of human brain disorders. In: Gabay S, Harris J, Ho BT (eds) Metal ions in neurology and psychiatry. Alan R Liss, New York, pp 259–285

Drayer BP, Olanow W, Burger P, Johnson GA, Herfkens R, Riederer P (1986) Parkinson Plus syndrome: diagnosis using high field MR imaging of brain iron. Radiology 159: 493–498

Earle KM (1968) Trace metals in parkinsonian brains. J Neuropathol Exp Neurol 27: 1–14

Fitzsimons DW (1979) Oxygen free radicals and tissue damage. Ciba Foundation Symposium No 65 New Ser. Excerpta Medica, Amsterdam

Hall ED (1988) Effects of the 21-aminosteroid 474006F on post-traumatic spinal cord ischaemia in cats. J Neurosurg 68: 462–465

Hall ED, Yonkers PA (1988) Attenuation of postischaemic cerebral hypoperfusion by the 21-aminosteroid 474006F. Stroke 19: 340–344

Halliwell B (1987) Oxidants and human disease: some new concepts. FASEB J 1: 358–364

Halliwell B (1989) Oxidants and the central nervous system: some fundamental questions. Acta Neurol Scand (in press)

Halliwell B, Grootveld M, Harparkash K, Fagerheim I (1988) Aromatic hydroxylation and uric acid degradation as methods for detecting and measuring oxygen radicals in vitro and in vivo. In: Rico-Evans C, Halliwell B (eds) Free radicals methodology and concepts. Richelieu Press, London, pp 33–59

Halliwell B, Gutteridge JMC (1986) Oxygen free radicals and iron in relation to biology and medicine: some problems and concepts. Arch Biochem Biophys 246: 501–514

Halliwell B, Gutteridge JMC (1988) Oxygen radicals and the nervous system. Trends Neurosci 8: 22–26

Hill JM (1988) In: Youdim MBH (ed) Brain iron: neurochemical and behavioural aspects. Taylor and Francis, London, pp 1–24

Hirsch E, Graybeil AM, Agid YA (1988) Melaninized dopaminergic neurons are differentially susceptible to degeneration in Parkinson's disease. Nature 334: 345–348

Hornykiewicz O (1988) Neurochemical pathology and the etiology of Parkinson's disease: basic facts and hypothetical possibilities. Mt Sinai J Med 55: 11–20

Kinenuchi H, Fowler CJ, Tipton KF (1987) The neurotoxicity of 1-methyl-4-phenyl-1,2,3,6-tetrahydropyridine (MPTP) and its relevance to Parkinson's disease. Neurochem Int 11: 359–373

Kohen R, Chevion M (1986) Paraquat toxicity is enhanced by iron and reduced by desferrioxamine in laboratory mice. Biochem Pharmacol 34: 1841–1843

Konradi C, Riederer P, Youdim MBH (1986) Hydrogen peroxide enhances the activity of monoamine oxidase type B and not type A. A pilot study. J Neural Transm [Suppl] 22: 61–73

Larsson B, Tjalve H (1979) Studies on the mechanism of drug binding to melanin. Biochem Pharmacol 28: 1181–1187

Liccione JJ, Maines MD (1988) Selective vulnerability of glutathione metabolism and cellular defense mechanism in rat striatum to manganese. J Pharmacol Exp Ther 247: 156–161

Mann DM, Yates PO (1983) Possible rate of neuromelanin in the pathogenesis of Parkinson's disease. Mech Aging Dev 21: 193–203

Masson HS, Ingram DJ, Allen B (1960) The free radical property of melanins. Arch Biochem Biophys 86: 225–230

Minotti G, Aust SD (1987) The requirement for iron (III) in the initiation of lipid peroxidation by iron (II) and hydrogen peroxide. J Biol Chem 262: 1098–1104

Perry TL, Young VN (1986) Idiopathic Parkinson's disease, progressive supranuclear palsy and abnormal glutathione metabolism in the substantia nigra of patients. Neurosci Lett 67: 269–274

Peterson CM, Graziano JH, Grady RW, Jones RL, Cerami A (1974) Clinical evolution of 2,3-dihydroxybenzoic acid as an oral chelating drug. Am Soc Hematol 168: 101 (Abstract)

Pilas B, Sana T, Kalyanaraman B, Swartz HM (1988) The effect of melanin on iron associated decomposition of hydrogen peroxide. Free Rad Biol Med 4: 285–293

Potts AM, Au PC (1976) The affinity of melanin for inorganic ions. Exp Eye Res 22: 487–491

Riederer P, Sofic E, Rausch WD, Schmidt B, Reynolds GP, Kellinger K, Youdim MBH (1989) Transition metals, ferritin, glutathione and ascorbic acid in parkinsonian brain. J Neurochem 52: 515–520

Riederer P, Youdim MBH (1987) MPTP-induced dopaminergic neurotoxicity. A useful model in the study of Parkinson's disease? Neurochem Int 11: 379–381

Sofic E, Riederer P, Heinsen H, Beckmann H, Reynolds GP, Hebenstreit G, Youdim MBH (1988) Increased iron (III) and total iron content in postmortem substantia nigra of parkinsonian brain. J Neural Transm 74: 199–205

Spina MB, Cohen G (1988) Exposure of school synaptosomes to L-dopa increases levels of oxidized gluathione. J Pharmacol Exp Ther 247: 502–507

Spina MB, Cohen G (1989) Dopamine turnover and glutathione oxidation. Implications for Parkinson's disease. Proc Natl Acad Sci USA 86: 1398–1400

Tanner CM (1989) The role of environmental toxins in the etiology of Parkinson's disease. Trends Neurosci 12: 49–59

Triggs WJ, Willmore LJ (1984) In vivo lipid peroxidation in rat brain following intracortical Fe^{++} injection. J Neurochem 42: 976–980

Wilson RL (1979) Hydroxyl radicals and biological damage in vitro: what relevance to in vivo. In: Oxygen free radicals and tissue damage. Ciba Foundation Symposium No 65. New Ser. Excerpta Medica, Amsterdam, pp 19–42

Yehuda S, Youdim MBH (1988) Brain iron deficiency: biochemistry and behaviour. In: Youdim MBH (ed) Brain iron neurochemical and behavioural aspects. Taylor and Francis, London, pp 89–114

Youdim MBH (1985) Brain iron metabolism in relation to dopaminergic neurotransmission. In: Lajtha A (ed) Handbook of neurochemistry, vol 10. Plenum Press, New York, pp 731–735

Youdim MBH (1988) Iron in the brain: implications for Parkinson's and Alzheimer's diseases. Mt Sinai J Med 55: 97–101

Youdim MBH, Ben-Shachar D, Riederer P, Melamed E (1988) MPTP neurotoxicity and brain iron. In: MPTP-induced neurotoxicity Symposium, Cagliari (Abstract)

Youdim MBH, Ben Shachar D, Riederer P (1989) Is Parkinson's disease a progressive siderosis of substantia nigra resulting in iron and melanin induced neurodegeneration? Acta Neurol Scand (in press)

Round table on Parkinson's disease

Participants: P. Riederer (moderator), D. B. Calne, T. Caraceni, S. Garattini, M. Gonce, M. Przuntek, U. K. Rinne.

Riederer: May I suggest that we discuss questions of particular interest in Parkinson's disease, and to concentrate this discussion on three topics: first, basic research, for example is there any new evidence from experimental work that might lead to new drug development; second, early diagnostic test batteries; and third, therapeutic strategies, such as the use of dopaminergic agonists in the very early stages of Parkinson's disease, as suggested by Prof. Rinne.

Youdim: Well, iron does interest me very much. I think we can learn a lot from desferrioxamine, a major drug for treatment of iron overload and paraquat toxicity, and from its specific ability to chelate iron. I mentioned the compounds that are now being produced by Upjohn, the 21-aminosteroids (lazaroids) which get across the brain and, at least in the animal models of ischaemia and trauma, seem to protect the brain. We have some preliminary data, using these 21-aminosteroids in the 6-hydroxydopamine model, which show that these compounds protect against 6-hydroxydopamine-induced lesions. Also, many years ago Gerald Cohen showed that antioxidants did protect against 6-hydroxydopamine. So, I think that this may be of future interest. As you know, inhibition of MAO should prevent the formation of hydrogen peroxide. Because of the role of iron-II in producing hydroxyl radicals from hydrogen peroxide, iron chelators might also contribute to decrease lipid peroxidation.

Corsini: What about the MPTP story?

Youdim: The MPTP story is not obvious and I forgot to mention that we have examined the effect of MPTP on iron. MPTP-treated monkeys and mice did not show any change in iron metabolism; secondly, I think that it has now been established that MPTP doesn't cause lipid peroxidation, which is now being shown in idiopathic Parkinson's disease. We are going to look at iron chelators, to see whether they can actually prevent MPTP-induced parkinsonism; however, there may be some sort of radicals formed with MPTP.

Fariello: I just take an issue. As I said before, MPTP is parkinsonism. It may not be Parkinson's disease, but it is a form of parkinsonism. Taking a look at Dr. Langston's patients, if one didn't know that they had MPTP, one would say they have Parkinson's disease. So, MPTP is Parkinson. One thing Prof. Youdim: according to your data, shouldn't MPTP protect against Parkinson's disease with regard to the melanin effect on iron?

Youdim: We have shown that MPTP did not displace iron from melanin. Why should it protect? It may be that there are different sites to which MPTP and iron bind. But we know that divalent ions, including iron, do displace MPP^+ and MPTP from melanin. In our experiments we have confirmed that, but the reverse is not true. Neither MPTP nor MPP^+ displaces iron from melanin.

Fariello: I cannot speak about the iron binding, but, this year, there was a publication in Neuroscience Letters showing that MPTP does alter the electron-microscopy constitution of neuromelanin prior to insulting mitochondria. What I'm saying, and I don't know if it does relate to the iron story, is that we do have some evidence that MPTP damages neuromelanin in some way prior to insulting neurons, prior to the occurrence of mitochondrial damage.

Youdim: I really have no argument against that. I am not trying to deny that MPTP or MPP^+ doesn't bind to melanin. I was just saying that the "iron story" seems to fit in with the recent clinical and biochemical findings on Parkinson's disease. So, as you said, iron and MPTP produce the same end effect, but probably from a different route.

Riederer: Prof. Fariello, would you like to comment on the therapeutic possibilities of antioxidants or radical scavengers?

Fariello: I really don't think that at this stage we can make any logical hypothesis, we can just make some hopeful thinking. In our hands the antioxidants that we have tried with MPTP, they all have failed. The thing that we are pursuing is the ubiquinone story. We are thinking that, at least in the MPTP model, the damage is related to the electron transfer. It is a mitochondrial damage and there is a blockade of the electron transfer prior to ubiquinone.

That is what Thomas Singer thinks of the story, and our data are very much in keeping with that. We are hoping that if we can correct that particular link and if we can equilibrate the balance between oxidized and reduced ubiquinone, then we might prevent the damage which is caused by at least one parkinsongenic toxin.

Riederer: Thank you for your comment. Prof. Olanow, do you want to make a comment on that?

Olanow: What is exciting about all this is that bodies of information are coming in from different directions, suggesting that oxidation reactions and free radicals may contribute to the pathogenesis process.

I think at this point it is important to recognize that Parkinson's disease may be multifactorial, if in fact it's true. One need not depend entirely on the iron story per se for each and every patient. In that regard, I think it is important to note that chelation of central iron in animals has put the animals into coma. The problem is that iron is important for numerous enzymes and metabolic components within the dopamine system specifically, and throughout the brain. If we are going to chelate out iron as an effect of therapy for Parkinson's disease, somehow we are going to have to learn how to be selective about it. I am not really optimistic that just a broad chelation of iron in brain is going to necessarily be useful.

Dostert: Coming back to the 21-aminosteroids, these molecules have been shown not only to bind with iron, but also to act as radical chain blocking agents; it may be that the two mechanisms participate in the global effect of the drug.

Youdim: Maybe, as in fact the 21-aminosteroids inhibit radical formation and lipid peroxidation very effectively. I think it is interesting to mention that techniques have been developed to measure lipid peroxidation in vivo. These techniques are using drugs which can be transformed metabolically into specific metabolites by the action of the cytotoxic hydroxyl radical. To mention one or two of them: for example, when acetetylsalicylate, aspirin, interacts with hydroxyl radicals, a specific metabolite, 2,3-dihydroxybenzoic acid, is produced, which is normally not formed. This has been used as a technique for measuring free radical formation in vivo. The other one is uric acid, which is selectively transformed to allantoin by the hydroxyl radical. These are the techniques we could use, maybe in urine, in serum or CSF. They should have probably to be adapted but very sensitive HPLC systems have already been developed.

Calne: I just wanted to raise some questions: Prof. Youdim, you really did an extremely elegant and important presentation. But I would like to suggest a modification to your conclusion. I would like to suggest that the mechanism that you are discussing is responsible for the selective vulnerability of these regions to the change that occurs with aging, which may be of crucial importance in terms of the evolution and progression of the symptomatology of parkinsonism. But there must be some other factors. For example, I think that when you are looking for the etiology of a disease, traditionally the most productive approach has been epidemiological and if you look epidemiologically, then iron miners are not particularly vulnerable so far as we know to Parkinson's disease; there are no foci in regions of high iron content that we are aware of. So, I think that iron could be an extremely important general factor, but maybe not the sole explanation. I would like to suggest that you have two types of mechanism interplaying: the age-related change, in which iron and oxidation could well play a critical role, a necessary role, for the disease, but, in addition, some environmental factor, I would suggest unrelated to iron, because of the epidemiology. The kinds of models we have to think about here are indeed the MPTP situation, where it can be shown that those people who were exposed to MPTP but who had no clinical parkinsonism, who far exceeded the number of patients who were clinically affected, those patients have abnormalities in the striatum from PET scanning. Another model, that is equally cogent and relevant to the hypothesis I'm trying to formulate, is the Guamenian disease, where an environmental factor in the Guamenian lifestyle contributes to the disease. We know from the epidemiology, from the Guamenians who emigrated to the States, that something in their lifestyle damaged them in such a way that they had removed themselves from Guam for 20 or 30 years before the appearance and evolution of the disease. And the disease did indeed appear and did indeed evolve in the absence of that initial damaging factor. So, I would propose that both these mechanisms are of crucial importance and that both are engaged in the production of Parkinson's disease. An environmental factor or a series of environmental factors which may be quite diverse and operate in quite different ways, superimposed upon selective vulnerability to attrition with aging. In that attrition process, oxidation damage induced by iron may be critical.

Youdim: I think you may very well be right. Concerning transition metals, we know that manganese can be transported into the brain by transferrin. Manganese also changes its redox state between 2 and 3, giving rise to radicals. So, we could explain the cases of parkinsonism in manganese miners. I forgot to mention that in my talk. It also seems to be the case in paraquat toxicity. Paraquat liberates iron from ferritin and it's the free iron that produces the toxin. In fact, you can take mice, treat them with desferrioxamine and give paraquat; they are completely protected from toxicity.

Strolin Benedetti: Coming back to the in vivo lipid peroxidation discussed by Prof. Youdim, I think that a very nice in vivo method is the measurement of ethane and pentane in expired air; but, again, the problem is to have an in vivo method which is indicative of brain damage and not of whole in vivo lipid peroxidation.

LeWitt: I would like to raise a question that starts with Prof. Fariello's comments on MPTP as compared to Parkinson's disease in humans, and to come back to the question of the role of melanin. One thing about human parkinsonism induced by MPTP is that, in contrast with typical Parkinson's disease where the monoamine systems are diffusely damaged, human MPTP-induced parkinsonism has normally a slightly increased MHPG in the spinal fluid, showing that the noradrenergic system is spared. Now, in Parkinson's disease we know that the locus ceruleus is damaged, and that there are some melanin-containing neurons. Is there something different about the melanin in the locus ceruleus that might explain this difference, that is to say, potential wedge for understanding the toxic mechanism of MPTP as compared to the putative environmental or autotoxins for Parkinson's disease? Any thoughts on that?

Riederer: Dr. Lieberman, you would like to comment on that?

Lieberman: Yes. I was speaking to Bill Langston last week in New York, and he showed evidence that if you give MPTP slowly you can damage the locus ceruleus. So, the locus ceruleus damage in MPTP-induced parkinsonism depends on the way you give MPTP.

LeWitt: That may be, but it is not necessarily damaging the noradrenergic pathways, which recover even with chronic multiple administration. I think there is something much more selective about MPTP for dopaminergic nigrostriatal neurons than we hear of, whatever mechanisms lead to the loss of the serotonin, noradrenergic and dopaminergic systems. I think that's an important part of consideration of markers for the disease, which is a multineuronal disease. We have a model of something that is far more selective than the disease is, and in looking for markers and mechanisms, we should really go beyond what MPTP can do when it attacks a variety of neuronal systems and yet not every dopaminergic system in the brain.

Gotti: Concerning the selectivity of MPTP it can be said that there may be a pharmacokinetic reason; for instance, and Prof. Da Prada may comment on this, MPP^+ has been shown to utilize the dopamine transport system. So, it reaches, probably, much higher concentrations in the dopaminergic neurons than in other neurons. That might be a reason for selectivity.

Da Prada: I can make a general comment on the discussion between Prof. Fariello and Prof. Youdim about MPTP toxicity. When you take the platelets as model, they are not a good model because, though several mechanisms of uptake and storage of the monoamines are present, they don't reflect the neuron. In each case you take a toxin like 5,6-dihydroxytryptamine or MPTP, there is accumulation but you don't have toxicity for the platelets. In the platelets mitochondrial monoamine oxidase-B and other types of amine-oxidases are present. We don't have iron or neuromelanin, but this points to the fact that the peripheral tissue, for some reason, is much more resistant. In the nerves there should be a peculiar situation in which a toxin acts more selectively.

Riederer: Thank you for your comment. Are there any other comments? Please, Dr. Bianchine?

Bianchine: I would like to share the views of Prof. Olanow and Prof. Calne, who both speak to the issue that it is the oxidative stress and lipid peroxidation which are probably important in the genesis of deletions in the CNS. The case of reperfusion injury is another example where iron is thought to be important, but it is not alone because xanthine oxidase inhibition treatment is useful, as is superoxide dismutase treatment. So, this speaks to other issues. There is also evidence that suggests that some of the myocardial damage associated with anthracyclines is related to iron in this instance, and an iron chelator, ADR 529, which is now under clinical investigation, is unequivocally beneficial in that setting. I remind you that there is an old story about retroventricular fibroplasia, where it's the oxidative stress in the premature neonate which seems to be improved by vitamin E, and I don't know of evidence that suggests that iron is involved. And then the final example I can think of is acetaminophen toxicity. This is an example where overload of acetaminophen clearly decreases glutathione concentrations.

Poewe: Just to divert attention to another area of selective damage with respect to heavy metals. Taking the case of copper in Wilson's disease, there is a predilection for the striatum and the globus pallidus. That seems also to be the case with manganese. Are there any ideas why copper, for example, should have this selective affinity towards those areas of the brain, as opposed to iron in the nigra?

Riederer: Would you like to comment, Prof. Youdim?

Youdim: Copper can substitute for iron but I don't know what the selectivity is. I would like to come back to a previous comment. We're learning more about iron from the systemic organ lesions as a result of increase in iron. In rheumatoid arthritis there is an accumulation of iron and desferrioxamine is now being used for the treatment. We know that the cardiovascular lesions that occur in iron therapy are related to iron.

Fariello: I'd like to remind that we haven't talked at all about the possibility that calcium entrance, which is related to exitotoxicity, might have something to do with all the types of selective neurodegenerative disorders which have been mentioned here. We have to keep in mind, for instance, that MPTP is a powerful epileptogenic agent. It does have very powerful electrical and behavioural epileptogenicity, as we have observed in our laboratory; how this relates to the fact of lipid peroxidation, we don't know. Another fact that makes us think of the possibility that, if not exitotoxicity at least neuronal activity, in terms of neuronal firing, might have something to do with neurodegeneration, is the fact that ferric ion injections have been used in the cortex of rats to create a model of epilepsy. Therefore, I think we should not lose the sight of the possibility that the oxidative-reducing system may be reflected at the level of the neuronal channels, with excitability changes of the membrane which, in turn, may create a vicious circle.

Riederer: I think we should switch over to the next topic, namely early diagnostic test batteries in Parkinson's disease. Would you like to make your comments on that, Prof. Garattini?

Garattini: I would first like to make some comments on what we have heard today about early markers for Parkinson's disease. If a neurologist had been here today, he would acknowledge the lack of substantial elements enabling an early diagnosis of Parkinson. This is, in a way, the principal merit of this meeting, because it shows that it is timely. To find early markers, we need a proper methodology and this methodology should probably include longitudinal studies. I don't think that by looking at patients who have already developed Parkinson's disease we can find something that can be useful in terms of early markers. Therefore, there is a need to start longitudinal studies probably with populations which are at the very beginning of the disease, and, on this population, it would be necessary to use all the various tools that have been reported today. I hope that some of these determinations will be sensitive enough to allow the detection of the disease in the early stages, when the symptoms are at the very beginning. Therefore these early markers may, on the one hand, contribute to an early diagnosis, and, on the other hand, hopefully they could also contribute to design therapy. I think there is also need to have proper controls about the specificity of the marker. We have heard very little about what is common with other degenerative neurological disorders. It is absolutely necessary, to be sure that what is seen is specific enough. Finally, I would like to warn about the fashionable things: free radicals are very fashionable and they are now utilized to explain almost every disease of degenerative nature. It may not be so simple. They do certainly contribute, but the reason why they may determine different diseases probably lies in the fact that there are many other cofactors working at the same time. Free radicals may exert pathology only when a number of conditions appear, but I am very doubtful that, by themselves, they can explain so many different diseases.

Calne: I would just like to take up that point about the longitudinal studies. We have recently been repeating the fluorodopa PET scans that we undertook some years ago in a number of parkinsonian patients. This provides, to some extent, an in vivo assay of what is happening neurochemically in the brain. Over a three-year period, repeated fluorodopa scan shows a small change in parkinsonian patients, indicating that this is a slow process. It is a little bit faster than simple age-related attrition but it is not an aggressive, dramatic deterioration, as has been suggested from some morphological observations. One possible mechanism that would fit nicely with such findings has been alluded to in the previous discussions. Once you have a damaged nigra, as you obviously have in clinically overt disease, then the surviving neurons are going to be compensating and manifesting increased activity. Such a powerful mechanism has been shown in rodents with partial lesions due to 6-hydroxydopamine. So, I think the kind of changes that we have seen with longitudinal PET scans would be quite compatible with damage induced by overactivity of the surviving neurons undertaken in an effort to compensate functionally for the deficits.

Garattini: An early marker of Parkinson's disease should be something that allows the detection of the disease before it appears. So the question is: before the appearance of the disease, can you detect any kind of damage? That would be what I consider an early marker.

Calne: The problem is that PET scanning is an expensive procedure and you can't screen the large numbers that are required. But, in the process of looking at a reasonable number of control subjects we have found one dramatic outlier who is clinically normal, but way below the normal range. I think there is also such an observation from Hammersmith that David Marsden mentioned to me last year. As experience accumulates with PET scanning, such individuals will be identified, and it will be most important to study and observe them. But, unless you have a special at-risk group, you can't go screening large numbers of people with PET scanning.

Rinne: I understand the need for early markers for Parkinson's disease if we talk about drugs preventing the disease. But we have to realize that, today, we have only symptomatic dopaminergic treatment. Therefore, I'm wondering whether we really need more than an help to clinical diagnosis, before initiating that kind of powerful treatment which we have in our hands. We have followed the same patient population, initiating dopaminergic treatment in the midsixties at different stages of the disease. Treatment was initiated at stage 1, 2, 3, 4 or even 5 of Hoehn and Yahr, because it was the first time that levodopa was available. We have been following these patients for more than 20 years, studying such simple factors as survival and mortality. We studied how many extra years you are able to offer your patients initiating this kind of symptomatic dopaminergic treatment and we came to very interesting findings. When we initiated treatment in stages 1 and 2, patients behaved very close to normal up to 20 years. Survival was like in age-matched controls. But when initiation of treatment was postponed to stages 3, 4 and 5, the running was different from controls. This point is very critical: if you postpone the initiation of dopamine treatment to stage 3 even, then you are losing 5 − 6 extra years for your patients. So, I think it is not so urgent to have very early markers for Parkinson's disease but it is very important to initiate dopaminergic treatment in the early stages of the disease.

Leenders: I would like to mention the case of a patient having clinically Parkinson's disease, but with a fluorodopa study which was on all accounts completely normal. So it was the reverse of what Prof. Calne has seen.

Calne: It's not idiopathic parkinsonism, we know how difficult it is to make that clinical distinction.

Leenders: Clinically, this patient has a mild to moderate idiopathic-looking Parkinson's disease, with very long time course and responding to L-dopa. But that's the only exception so far; all the parkinsonians have in the putamen a very severe fluorodopa deficiency. Furthermore, I completely agree with Prof. Rinne when he says: do we need early markers and what is early? I think a premorbid marker is maybe completely unnecessary. An early Parkinson's disease is not a bad disease and you can treat it, provided L-dopa doesn't accelerate the disease. If you want to detect patients before they get clinical signs, then you need to screen the whole population with a very easy and simple screening method. I find it highly unlikely that such a thing does exist. So, I would agree with Prof. Rinne that probably there's no need for that.

Riederer: Are there any comments to that? Prof. Garattini?

Garattini: I think that early markers would be very important because they could give rise to new forms of therapy. We are not talking about what we know today, but, if we had some markers, they could constitute a tool on which we might build up new forms of therapies. It's not because we know what to do today we don't need early markers.

Heiss: I've been struggling with the analysis of PET data for 7 years now, and I wonder how you can use a non-quantitative model as is the fluorodopa injection, to make interindividual and intraindividual comparisons in the follow-up of the patients. You have to rely on the original takeup of the tracer and you have no information on the metabolites and on the amino acids which are also important for the transport of your tracer into the brain. So, I wonder if we can really use fluorodopa for really quantitating the progress of the disease.

Calne: I didn't use the word quantitative incidentally, but I would say semi-quantitative. But, whatever method of analysis you use, you can get substantial separation between the groups.

Da Prada: I agree completely with Prof. Garattini that we need a very early marker, because the problem is also how to prevent the progress of the disease. If we don't start something now, we will find nothing. Concerning the use of L-dopa, we have now a COMT inhibitor, which inhibits the methylation of dopa peripherally and centrally. We can eliminate the formation of 3-methyldopa and improve the penetration of dopa.

Rinne: As I said, if we had drugs preventing the disease, then it should be, of course, important to have early markers. But we have to realize that the markers presented here today are not specific at all.

Caraceni: What I would like to remind is that we don't know what is the etiology of Parkinson's disease. Therefore, I think that it's very important to give a better definition of the early diagnoses of idiopathic Parkinson's disease. In this way, I think that the characterization of Parkinson's disease with NMR and PET scan, would be of great interest, as also the characterization of all neurodegenerative diseases which have features similar to Parkinson's disease before initiating any kind of early markers. Perhaps it's excessive to say this, but I think that we don't know what Parkinson's disease really is.

Lieberman: One point is whether there's an incubation period for Parkinson's disease. We recognize the disease clinically, and we assume that there is some sort of preclinical incubation period, in conjunction with other neurological diseases. But there may not be. Except for a few cases of patients that I've seen, whom I've diagnosed as having postural tremors, and who a few years later had Parkinson's disease, there may not be an incubation period. The second question goes about as to how we examined patients for extrapyramidal diseases. We assumed that the examination we do monitors the state of the substantia nigra, of the putamen and of the caudate nucleus. But we really do not know what we are really doing when we are measuring the function of the pyramidal tract as modified by the extrapyramidal system. So, really another question may be: do we really understand what it is that the basal ganglia do. The basal ganglia probably have something to do with subconscious traces of postures and things that we do that we would never examine. It would be very nice here to have an early marker, but a more central question would be: is there an incubation period for the disease? and a second question is to re-examine what it is that we do when we examine patients with Parkinson's disease. We just apply a specialized program of measuring strength, speed, agility and assume that that's the measurement of the basal ganglia. It may not be.

Przuntek: As Prof. Garattini said, it is necessary to find early markers in Parkinson's disease, although none of us knows when we'll find the substance able to slow down the process of the disease. I would suggest a preselection be made with very simple and cheap methods because you have to investigate a very large population. I believe that the motory test could be suited for this preselection because this method is very simple, although not so specific. And then we could use more specific methods if we have made this preselection, like for example examination of the cerebrospinal fluid or PET scan.

Calne: There is obviously a debate which is of crucial significance to this meeting, which is whether we should be looking for early markers. I think there are two issues, and Prof. Caraceni said exactly the same thing. We are not just looking for a diagnostic test for a drug that we don't have. We are trying to identify the pathogenesis of the disease, and if we don't know the longitudinal time course of the disease, we can't begin to think about what kind of pathogenesis is taking place. As Prof. Lieberman said, does the disease start one month or 30 years before we see the symptoms? It is of critical importance to know this in terms of understanding what is going on, and also in terms of looking for the actual risk factor. If you're trying to get a drug that is going to have some impact on the progression of the disease and, possibly, some impact on preventing the disease, or if you're looking for some way of reducing the prevalence of the disease, clearly the crucial first step is to find the cause, or the causes. Then you can develop a rational approach towards preventive therapy, or towards therapy to slow down the progress of the illness.

Carvey: In terms of the criterion that you established as to what might be an early marker, I would like to offer the dopamine-neuron antibody. As far as we hope to, the test would be very easy to implement, very cheap and rapid. If the hypothesis is correct, we should be able to identify these patients prior to actual symptom expression. I know that MPTP-exposed non-symptomatic parkinsonian patients are a difficult groups to study, but I am really surprised that we haven't taken more advantage of that selective group which may tell us a lot about the progression of the disease in a presymptomatic population.

Riederer: I think we'll close this part now. My personal opinion is that probably a single early marker will not give us the results that we want. Probably, several test batteries will have to be used as a better diagnostic criterion. Now, finally we have some questions and comments to therapeutic strategies?

Rinne: I think that today, when we are still dealing with the initiation of symptomatic treatment, we have three main questions. First: when; second: what; and third: how. And we have to realize that still dopaminergic treatment is the most powerful antiparkinsonian treatment available today. In my opinion, we must realize today that to give L-dopa as we have been treating patients in the past few years by giving high doses of levodopa alone, is not good at all. As I said, we can even increase survival and decrease mortality of the patients. But, as we all know, very many patients are dealing with many late complications of this treatment, especially different kinds of motor fluctuations. I believe that, in trying to prevent these complications, it is better not to use levodopa but, vice versa, to initiate with dopamine agonists and to try to keep your patients on dopamine agonists as long as possible. When it is no longer possible, then it is time to add levodopa. I think that, with that kind of combination of dopamine agonists and levodopa, you can have a better treatment than by using levodopa alone. Thus you can offer extra years to your patients and also a better quality of life.

Da Prada: When you speak about dopaminergic agonists, it's important to know if the compound should be a D_2, or a D_1 or a D_1 plus D_2 agonist. Probably it should be a D_1 plus D_2 agonist, because dopamine stimulates both receptors. Since dopamine is the endogenous agonist, probably the best way should be to ameliorate the pharmacokinetics of L-dopa, rather than introducing a new agonist. As you know, pharmacokinetics of L-dopa is just terrible. So, I think it is important to maintain the level of L-dopa by slow-release formulations and to reduce the dose, maybe using COMT inhibitors.

Rinne: I agree with Prof. Da Prada that D_2 stimulation is not enough. We know from animal experiments that stimulation of both D_2 and D_1 receptors, and maybe a good balance between the stimulation of both receptors, is important; I think that by giving levodopa you cannot have a stimulation of presynaptic dopamine autoreceptors as good as by using dopamine agonists, which have higher affinity for the presynaptic dopamine autoreceptors which control the release of endogenous dopamine.

Poewe: I think that future treatment of Parkinson's disease might be simply to improve the delivery of our dopaminergic treatment, so as to more closely mimic tonic nigrostriatal impulse flow rather than having peaks and troughs as we do have. So I think Prof. Da Prada is right in saying that we should concentrate on improving pharmacokinetic variables.

Riederer: Before closing this session, I wish to express my thanks to all the participants in the round table and to the auditorium for their contribution to the discussions.

Early markers in Alzheimer's disease

The need for early markers in Alzheimer's disease

D. L. Murphy and T. Sunderland

Laboratory of Clinical Science, National Institute of Mental Health,
Bethesda, MD, USA

Summary

Currently, there are no definitive early markers for Alzheimer's disease (AD) or senile dementia of the Alzheimer type (SDAT). Nonetheless, there are some interesting marker candidates under evaluation. Several examples of candidate markers are considered in this review, including some biochemical, genetic, pharmacological, neuroimaging, and clinical measures. These examples are chosen to illustrate some aspects of an ideal marker, as well as the needs for such markers in different areas of AD and SDAT research and treatment.

Introduction: early markers, the need and strategies in SDAT

Some of the most apparent needs for early markers of Alzheimer's disease (AD) or senile dementia of the Alzheimer type (SDAT) have been widely discussed. Early differential diagnosis continues to be a problem in this disorder which, until recently, was often regarded as a diagnosis only made by exclusion of other types of dementia (Hachinski et al., 1974; McKhann et al., 1984; Barclay, 1988; Thal, 1988).

Treatable types of dementia obviously need to be differentiated from SDAT. While there continues to be debate about whether SDAT is distinct from normal aging or can be seen as a continuum including age-associated memory impairment or benign senescent forgetfulness (Brody, 1982; Berg, 1985; Brayne and Calloway, 1988), the prevalent view is that SDAT is a more severe distinct entity with a more rapid course of decline in functional capacities. Therefore, a marker for SDAT would be useful in establishing an accurate diagnosis or estimating prognosis.

Diagnostic uncertainty, which is currently estimated to range from 10% to 30% when clinical diagnoses are compared to neuropathological criteria (Ron et al., 1979; Wade et al., 1987), presents a number of problems for research and treatment of SDAT. A recent review pointed out the severe limitations of genetic linkage analysis in apparent familial forms of SDAT, which not only include diagnostic uncertainty regarding probands, but also the late onset of symptoms and subsequent short life span of patients, thereby limiting pedigrees which can be directly investigated to a few living affected individuals who rarely represent different generations (Tanzi et al., 1989). These factors, which would be obviated by a classic genetic marker, currently render potential identification of gene(s) for familial SDAT by linkage analysis alone an extremely difficult proposition.

In regard to treatment, the inclusion of inappropriate patients in treatment trials because of diagnostic uncertainty limits the validity of these trials and is currently viewed as a significant factor in slowing the identification of useful drugs or other potential treatments. In addition, should a treatment be found, for example, among neuronal growth-affecting factors under investigation, the availability of an early marker could be of great importance in slowing or even preventing neuronal damage and loss in SDAT.

Characteristics of an ideal marker have been frequently described for other disorders. These include two main characteristics: (1) specificity, i.e., the marker is linked to a distinct disorder, and identifies the disorder in every case with no overlap or misidentification of clinically similar disorders; (2) sensitivity, i.e., the marker detects the disorder in its early stages. Other desirable characteristics include the following: a definable relationship to the stage or progression of the disorder, relatively noninvasive (e.g., brain biopsies are not practical), technically feasible, and relatively inexpressive, to permit large scale screening.

Currently, there is no valid, specific and sensitive marker for SDAT. As the etiology of SDAT remains unknown, potential markers may be of different types. For instance, if SDAT proves to be a genetically based disorder, then a marker may be an altered gene product or DNA fragment. If SDAT is of infectious or toxic origin, then a marker may be an agent or product of infection or toxicity. While etiology remains uncertain or if SDAT proves to be a heterogeneous group of disorders, a nonspecific marker that reflects pathological severity may still be useful. In any case, a successful marker will eventually need to be validated in an autopsy- or biopsy-verified patient sample.

The pessimistic view towards finding a valid marker for SDAT begins with the evidence that SDAT may not be a distinct entity distinguishable from normal aging (Brayne and Calloway, 1988) or is, at least, a heterogeneous disorder (Friedland et al., 1988). In the latter instance, multiple markers would be required if the markers were etiologically based, although this need not be so if the marker were of the type that reflected severity of SDAT as a syndrome. However, as the clinical diagnosis of SDAT is not completely reliable, a marker related nonspecifically to SDAT severity may not be useful in distinguishing milder demented patients from controls. Finding a marker for SDAT is currently confounded by an additional issue: "controls" used to define a marker will include a certain proportion of individuals with a nonsymptomatic SDAT trait, especially given the lifetime prevalence of approximately 10% for SDAT (Thal, 1988).

A more optimistic view holds that even if an ideal marker as described above may not be found for SDAT as a single distinct entity, different specific markers may be found and used in specific families, or even in individuals followed over time. It should be noted that some clinical characteristics of SDAT are currently being evaluated or used as they potentially provide some marker-like information. These include the following: (1) age of onset (presenile vs. senile); (2) current symptoms and other characteristics of clinical picture (e.g., history of head trauma, agraphia); (3) course of illness (rapid vs. slow progression); (4) motor deficits (extrapyramidal syndrome, myoclonus); and (5) behavioural characteristics (depression, aggression, psychosis). Simply administered clinical tests [e.g., clock drawing (Sunderland et al., 1989)] can provide severity measures of visuospatial ability or other features of SDAT patients. Some examples of candidate neurobiological markers currently under investigations are worth discussion to illustrate some features of the strategies being used to identify potential biological markers.

Some illustrative examples of candidate markers in SDAT

Biochemical/neuropathological

Some of the earliest and also most intensive efforts to identify a biological marker for SDAT focused on the well studied brain cholinergic system damage believed to account for the characteristic cognitive and related dysfunctions in patients with SDAT (Davies and Maloney, 1976; Gottfries, 1985; Perry, 1988). As reviewed elsewhere, candidate markers include cerebrospinal (CSF) acetylcholine, acetylcholinesterase, and choline acetyltransferase concentrations (Thienhaus et al., 1985; Hollander et al., 1986; Cutler, 1988 a, b); these enzymes as well as choline uptake have also been measured in erythrocytes. Most studies did not find differences between SDAT patients and controls, or, when differences were found, similar changes were also found in patients with depression or with non-SDAT neurological disorders. In addition, differences in CSF somatostatin between SDAT patients and controls were also subsequently found to be equivalent to those in patients with multi-infarct dementia and other psychiatric patients vs. controls (Beal et al., 1986; Sunderland et al., 1987; Davis et al., 1988; Cutler, 1988 a, b).

An interesting example of a neuropathology-based candidate marker is the protein A 68 which can be detected in Alzheimer brains by Western blot analysis (Wolozin et al., 1986; Wolozin and Davies, 1987; Tabaton et al., 1988). Extensive studies have revealed abnormalities in the protein composition of brain tissues from SDAT patients represented histologically in the neuritic plaques, neurofibrillary tangles, and other abnormalities (Tanzi et al., 1989; Davies, 1988). A 68 is a 68,000 dalton protein which was originally identified by a monoclonal antibody, Alz 50, reactive to SDAT forebrain material. A 68 is found in low to nonexistent amounts in normal brains, but is elevated 15- to 30-fold in SDAT patient brains. Only trace amounts of A 68 are found in brains from patients with the Guam Parkinson dementia complex and Pick's disease. However, patients with progressive supranuclear palsy (PSP) have 10% of the A 68 protein that SDAT patients have, a change suggested to reflect the quantitatively more marked fibrillary pathology in SDAT than PSP (Tabaton et al., 1988). Studies of CSF A 68 in SDAT patients, age-matched controls, and neuropsychiatric patient comparison groups are now underway.

An example of a systemic abnormality that provides an apparent genetic marker for a subgroup of approximately 50% of SDAT patients is an abnormality in platelet membrane fluidity studied in some detail by Zubenko et al. (1987, 1988) (Chakravarti et al., 1989). Increased platelet membrane fluidity (PMF, defined as fluorescence ansiotropy of 1,6-diphenyl-1,3,5-hexatriene < 0.1920 at 37 °C, which is a cut-off point at the 90th percentile for healthy elderly controls) segregates SDAT patients into two subgroups. The subgroup with increased PMF has been suggested to have an earlier symptom onset and a more rapidly progressing course. Increased PMF is not found in higher incidence in patients with multi-infarct dementia or depression. The incidence of increased PMF in first-degree asymptomatic relatives of SDAT patients with increased PMF was 57% (13/23) vs. 10% (2/20) in SDAT patients without increased PMF (p = 0.0015). A 14 pedigree study showed increased PMF to be controlled by a single gene locus with two alleles that have additive effects. Other more preliminary studies of systemic abnormalities include the finding of elevated levels of high affinity antibody to human nerve growth factor in the blood of Alzheimer patients vs. normal controls (Roy et al., in press) and moderate differences in platelet monoamine oxidase (MAO) activity in behaviourally disturbed Alzheimer patients compared to control populations (Schneider et al., 1988; Funsch et al., 1989).

Pharmacological responsivity to cholinergic and serotonergic agents

Treatment of SDAT patients with cholinergic agents is under intensive study. Of interest as another possible strategy for identifying a marker are a series of studies comparing the biological and behavioural responses of SDAT patients, controls, and elderly depressed patients to cholinergic agents such as scopolamine, arecoline, and nicotine, and to the serotonergic agonist, m-chlorophenylpiperazine (Sunderland et al., 1987, 1988; Tariot et al., 1988; Newhouse et al., 1988; Lawlor et al., 1989 a, b). Of special interest was the finding that SDAT patients, when compared to elderly normal controls and elderly depressed patients, demonstrated a markedly greater, dose-dependent sensitivity to the cholinergic antagonist, scopolamine, on cognitive tests of semantic memory and free recall as well on behavioural measures, such as the Brief Psychiatric Rating Scale (Sunderland et al., 1987, 1988). The differentiation of SDAT from depressed patients was of note, because depressed patients can manifest concurrent cognitive impairments (i.e., pseudodementia) (Post, 1975; Wells, 1979; Reifler et al., 1982; Clark et al., 1985). Cognitive impairing effects of scopolamine have previously been characterized in young normal controls (Drachman and Leavitt, 1974). Supersensitivity to scopolamine has also been found in monkeys with experimentally induced lesions of the cholinergic cell bodies in the nucleus basalis of Meynert (Aigner et al., 1987). Our preliminary study of scopolamine testing in nondemented parkinsonian patients revealed some increased scopolamine sensitivity (Dubois et al., 1987), so the specificity of this test must be questioned. Testing other at-risk groups such as those with dementia associated with Parkinson's disease and Down's syndrome, and testing individuals with a strongly positive family history of Alzheimer's disease will be of interest.

Brain imaging and cerebral blood flow measurements

There have been extensive evaluations of the potential use of brain imaging techniques to yield an SDAT marker (Thienhaus et al., 1985; Hollander et al., 1986; Cutler, 1988 a, b). Computed tomography scans alone cannot distinguish among the various etiologies of cerebral atrophy. Preliminary results suggest that positron emission tomography (PET) may have some capability to provide early identification of SDAT patients, and to differentiate SDAT from multi-infarct dementia (Friedland et al., 1985 a, b; Kuhl et al., 1987). Single photon emission computed tomography (SPECT) offers the interesting possibility of quantitatively visualizing muscarinic cholinergic receptors using the radiolabeled cholinergic antagonist, ^{123}I-3-quinuclidinyl-4-iodobenzilate (QNB), as a ligand (Holman et al., 1985). Preliminary studies with QNB binding in SPECT scans of Alzheimer patients indicate a decrease in parietal cortex binding compared to controls (Weinberger et al., 1989). Potential techniques such as the use of radiolabeled antibodies reactive to elements of plaques or neurofibrillary tangles to locate semi-quantitatively actual neuropathology have yet to be evaluated.

One illustrative example of an evaluation of the use of cerebral blood flow measurement as a possible diagnostic marker is a study by Prohovnik et al. (1988) of 36 patients with AD and 12 healthy controls. Discriminant function analysis of ^{133}xenon regional cerebral blood flow values correctly identified > 90% of both patients and controls. Flow was most notably reduced in parietotemporal regions in the AD patients. However, when only mean global perfusion instead of the combined information from regional flow measurements obtained from 10 detectors were used in the analysis, 50% of the controls were misclassified. An important feature of this study was that a subgroup of patients with relatively short illness duration (< 2 years) and with relatively mild cognitive deficits (modified Mini-Mental Status scores > 30) were equally (> 90%) as well identified as were those patients with more severe AD. A prior study also found that ^{133}xenon blood flow measurements provided good diagnostic discrimination between patients with neuropathologically confirmed AD and those with major depressive disorder, multi-infarct disorder, diffuse parietotemporal degeneration, or Pick's disease and normal aging; however, it should be noted that the AD blood flow results used for the diagnostic comparisons in this study were those from the last year of life of the patients (Risberg and Gustafson, 1983). The AD patients in the study by Prohovnik et al. (1988) were carefully clinically classified, but postmortem neuropathological confirmation was not available. Further studies in more AD patients, other neuropsychiatric patients, and controls of the utility of regional blood flow as a valid marker are required because discriminant function analysis is a *post hoc* procedure that is designed to maximize differences between groups, and hence replication studies are necessary for validation and refinement of the elements used in this statistical procedure.

Conclusions

Currently, there is no definitive early marker for SDAT. While various biochemical, pharmacological, clinical, and genetic factors have been evaluated as potential contributors to the early detection of AD, none is as yet established as a completely valid, fully reliable measure, either diagnostically or prognostically. There is a great need for such a marker. During drug development, for instance, it is important that accurate diagnosis be established early in the course of illness so that homogeneous diagnostic groups can be assured and unnecessary variance avoided. Today, there is approximately a 20% error rate in clinically based diagnostic procedures compared to standard neuropathological criteria for SDAT; this error rate is found even in the best neuropsychiatric centers.

With families, there is tremendous strain and pressure surrounding the uncertainty of phenotypic penetrance of the proposed "Alzheimer gene". An early diagnostic marker would perhaps help in the counselling of at-risk individuals. Eventually, when more effective pharmacological agents are available, an early diagnostic marker might also allow for prophylactic treatment of individuals even before the development of clinical symptoms, or at least at earlier, perhaps even reversible stages of the disorder. Candidates for early detection of SDAT include brain specific proteins, pharmacological challenges with anticholinergic agents, neuroimaging studies with brain mapping or regional blood flow measuring techniques, and peripheral cell biochemical markers, just to name a few. If found to be reliable, early markers would enable researchers to profile the phenomenological and biological changes associated with AD longitudinally. In so doing, early markers would contribute to the development of new strategies for the treatment of this and other related conditions of the aging brain.

References

Aigner TG, Mitchell SJ, Aggleton JP, DeLong MR, Struble RG, Price DL, Wenk GL, Mishkin M (1987) Effects of scopolamine and physostigmine on recognition memory in monkeys with ibotenic-acid lesions of the nucleus basalis of Meynert. Psychopharmacology 92: 292–300

Barclay L (1988) Differential diagnosis of dementing diseases. Age 11: 19–22

Beal MF, Growdon JH, Mazurek MF, Martin JB (1986) CSF somatostatin-like immunoreactivity in dementia. Neurology 36: 294–297

Berg L (1985) Does Alzheimer's disease represent an exaggeration of normal aging? Arch Neurol 42: 737–739

Brayne C, Calloway P (1988) Normal ageing, impaired cognitive function, and senile dementia of the Alzheimer's type: a continuum? Lancet i: 1265–1267

Brody JA (1982) An epidemiologist views senile dementia: facts and fragments. Am J Epidemiol 115: 155–162

Chakravarti A, Slaugenhaupt SA, Zubenko GS (1989) Inheritance pattern of platelet membrane fluidity in Alzheimer's disease. Am J Hum Genet 44: 799–805

Clark DC, Clayton PS, Andreasen NC, Lewis C, Fawcett J, Scheftner WAK (1985) Intellectual functioning and abstraction ability in major affective disorders. Compr Psychiatry 26: 313–325

Cutler NR (1988 a) Utility of biologic markers in the evaluation and diagnosis of Alzheimer's disease. Brain Dysfunction 1: 12–31

Cutler NR (1988 b) Recent advances in the development of ante-mortem diagnostic markers for Alzheimer's disease. Curr Opin Psychiatry 1: 462–467

Davies P (1988) Neurochemical studies: an update on Alzheimer's disease. J Clin Psychiatry 49 [Suppl 5]: 23–28

Davies P, Maloney AJR (1976) Selective loss of central cholinergic neurons in Alzheimer's disease. Lancet ii: 1403–1405

Davis KL, Davidson M, Yang RK, Davis BM, Siever LJ, Mohs RC, Ryan T, Coccaro E, Bierer L, Targum SD (1988) CSF somatostatin in Alzheimer's disease, depressed patients, and control subjects. Biol Psychiatry 24: 710–712

Drachman DA, Leavitt J (1974) Human memory and the cholinergic system. Arch Neurol 30: 113–121

Dubois B, Danze F, Pillon B, Cusimano G, Lhermitte F, Agid Y (1987) Cholinergic-dependent cognitive deficits in Parkinson's disease. Ann Neurol 22: 26–30

Friedland RP, Brun A, Budinger TF (1985 a) Pathological and positron emission tomographic correlations in Alzheimer's disease. Lancet i: 228

Friedland RP, Budinger TF, Koss E, Ober BA (1985 b) Alzheimer's disease: anterio-posterior and lateral hemispheric alterations in cortical glucose utilization. Neurosci Lett 53: 235–240

Friedland RP, Koss E, Haxby JV, Grady CL, Luxenberg J, Schapiro MB, Kaye J (1988) Alzheimer's disease: clinical and biological heterogeneity. Ann Intern Med 109: 298–311

Funsch D, Sunderland T, Lawlor BA, Molchan SE, Mellow AM, Tariot PN, Hill JL, Murphy DL (1989) Platelet MAO activity in Alzheimer patients, older depressives, and age-matched controls. Society of Biological Psychiatry, San Francisco, CA (Abstract)

Gottfries CG (1985) Alzheimer's disease and senile dementia: biochemical characteristics and aspects of treatment. Psychopharmacology 86: 245–252

Hachinski VC, Lassen NA, Marshall J (1974) Multi-infarct dementia: a cause of mental deterioration in the elderly. Lancet ii: 207–210

Hollander E, Mohs RC, Davis KL (1986) Antemortem markers of Alzheimer's disease. Neurobiol Aging 7: 367–387

Holman BL, Gibson RE, Hill TC, Eckelman WC, Albert M, Reba RC (1985) Muscarinic acetylcholine receptors in Alzheimer's disease: in vivo imaging with iodine 123-labeled 3-quinuclidinyl-4-iodobenzilate and emission tomography. JAMA 254: 3063–3066

Kuhl DE, Small GW, Riege WH, Fujikawa DG, Metter EJ, Benson DF, Ashford JW, Mazziotta JC, Maltese A, Dorsey DA (1987) Abnormal PET-FDG scans in early Alzheimer's disease. J Nucl Med 28: 645 (Abstract)

Lawlor BA, Sunderland T, Mellow AM, Hill JL, Molchan SE, Murphy DL (1989 a) Hyperresponsivity to the serotonin agonist m-chlorophenylpiperazine in Alzheimer's disease: a controlled study. Arch Gen Psychiatry 46: 542–549

Lawlor BA, Sunderland T, Mellow AM, Hill JL, Newhouse PA, Murphy DL (1989 b) A preliminary study of the effects of intravenous m-chlorophenylpiperazine, a serotonin agonist, in elderly subjects. Biol Psychiatry 25: 679–686

McKhann G, Drachman D, Folstein M, Katzman R, Price D, Stadlan EM (1984) Clinical diagnosis of Alzheimer's disease: report of the NINCDS-ADRDA work group. Neurology 34: 939–944

Newhouse PA, Sunderland T, Tariot PN, Blumhardt CL, Weingartner H, Mellow A, Murphy DL (1988) Intravenous nicotine in Alzheimer's disease: a pilot study. Psychopharmacology 95: 171–75

Perry E (1988) Acetylcholine and Alzheimer's disease. Br J Psychiatry 152: 737–740

Post F (1975) Dementia, depression, and pseudo-dementia. In: Benson DF, Blumer D (eds) Psychiatric aspects of neurologic disease. Grune and Stratton, New York, pp 99–120

Prohovnik I, Mayeux R, Sackeim HA, Smith G, Stern Y, Alderson PO (1988) Cerebral perfusion as a diagnostic marker of early Alzheimer's disease. Neurology 38: 931–937

Reifler BV, Larson E, Hanley R (1982) Coexistence of cognitive impairment and depression in geriatric outpatients. Am J Psychiatry 139: 623–626

Risberg J, Gustafson L (1983) Xe cerebral blood flow in dementia and in neuropsychiatry research. In: Magistretti P (ed) Functional radionuclide imaging of the brain. Raven Press, New York

Ron MA, Toone BK, Garralda ME, Lishman WA (1979) Diagnostic accuracy in presenile dementia. Br J Psychiatry 134: 161–168

Roy BF, Sunderland T, Dauphin MM, Goodman A, Frazier JS, Murphy DL, Morihisa JM (in press) Human antibody to nerve growth factor: a preliminary study with implications for Alzheimer's disease. Ann Neurol

Schneider LS, Severson JA, Chui HC, Pollock VE, Sloane RB, Fredrickson ER (1988) Platelet tritiated imipramine binding and MAO activity in Alzheimer's disease patients with agitation and delusions. Psychiatry Res 25: 311–322

Sunderland T, Rubinow DR, Tariot PN, Cohen RM, Newhouse PA, Mellow AM, Mueller EA, Murphy DL (1987) CSF somatostatin in patients with Alzheimer's disease, older depressed patients, and age-matched control subjects. Am J Psychiatry 144: 1313–1316

Sunderland T, Tariot PN, Newhouse PA (1988) Differential responsivity of mood, behaviour, and cognition to cholinergic agents in elderly neuropsychiatric populations. Brain Res Rev 13: 371–389

Sunderland T, Hill JL, Mellow AM, Lawlor BA, Gundersheimer J, Newhouse PA, Grafman JH (1989) Clock drawing in Alzheimer's disease: a novel measure of dementia severity. J Am Geriatr Soc 37: 725–729

Tabaton M, Whitehouse PJ, Perry G, Davies P, Autilio-Gambetti L, Gambetti P (1988) Alz 50 recognizes abnormal filaments in Alzheimer's disease and progressive supranuclear palsy. Ann Neurol 24: 407–413

Tanzi RE, St George-Hyslop PH, Gusella JF (1989) Molecular genetic approaches to Alzheimer's disease. Trends Neurosci 12: 152–158

Tariot PN, Cohen RM, Welkowitz JA, Sunderland T, Newhouse PA, Murphy DL, Weingartner H (1988) Multiple-dose arecoline infusions in Alzheimer's disease. Arch Gen Psychiatry 45: 901–905

Thal LJ (1988) Dementia update: diagnosis and neuropsychiatric aspects. J Clin Psychiatry 49 [Suppl 5]: 5–7

Thienhaus OJ, Hartford JT, Skelly MF, Bosmann HB (1985) Biologic markers in Alzheimer's disease. J Am Geriatr Soc 33: 715–726

Wade JPH, Mirsen TR, Hachinski VC, Fisman M, Lau C, Merskey H (1987) The clinical diagnosis of Alzheimer's disease. Arch Neurol 44: 24–29

Wells CE (1979) Pseudodementia. Am J Psychiatry 136: 895–900

Weinberger DR, Gibson RE, Coppola R, Jones DW, Berman KF, Braun AR, Zeeberg BR, Sunderland T (1989) [123]IodoQNB SPECT in Alzheimer's and Pick's diseases. American Psychiatric Association, San Francisco, CA (Abstract)

Wolozin B, Davies P (1987) Alzheimer-related neuronal protein A 68: specificity and distribution. Ann Neurol 22: 521–526

Wolozin BL, Pruchnicki A, Dickson DW, Davies P (1986) A neuronal antigen in the brains of Alzheimer patients. Science 232: 648–650

Zubenko GS, Cohen BM, Boller F, Malinakova I, Keefe N, Chojnacki B (1987) Platelet membrane abnormality in Alzheimer's disease. Ann Neurol 22: 237–244

Zubenko GS, Huff FJ, Beyer J, Auerbach J, Teply I (1988) Familial risk of dementia associated with a biologic subtype of Alzheimer's disease. Arch Gen Psychiatry 45: 889–893

Descriptive and analytic epidemiology of Alzheimer's disease

L. Amaducci[1] and A. Lippi[2]

[1] Department of Neurology and Psychiatry, University of Florence, and
[2] S.M.I.D. (Italian Multicentre Study on Dementia) Center, Florence, Italy

Summary

The world population aged 60 years and over was about 415 million in 1985 and is expected to reach 1.1 billion by the year 2025. Therefore, the number of subjects at risk for age-associated disorder of the CNS, including dementia, is rapidly increasing.

In this article, the problems encountered in descriptive and analytic epidemiology of Alzheimer's disease in various countries are highlighted in an attempt to better define the selection of subjects for assessing the potential efficacy of new therapies.

Introduction

The world population aged 60 years and over was approximately 371 million in 1980. By 1985, its number had risen to 415 million and is expected to reach 1.1 billion by the year 2025 (United Nations, 1986). This phenomenon, due to reduction is fertility, infant mortality and deaths from infectious diseases is becoming of great importance in both developed and developing countries. In 1980, 6% of the world's total population aged 60 years and over, lived in developing countries. In 1985, this figure had risen to 6.3%. By the year 2025, people aged 60 years and over, living in developing regions, are expected to account for 11.9% of the world's total population (United Nations, 1986).

The growing number of elderly people will determine an increased number of subjects at risk for age-associated disorders of the nervous system, including dementia. Dementia is a clinical syndrome characterized by loss of memory associated with impairment of abstract thinking and judgement, disturbances of higher cortical functions and personality changes (American Psychiatric Association, 1987). This clinical syndrome may have several causes. Dementing disorders have been classified as primary and secondary according to their etiology. More recently, the National Institute of Health (U.S.A.) (1988), in the report of the Consensus Conference on the Differential Diagnosis of Dementing Diseases, held in July 1987, proposed a distinction between diseases that appear to be primary in the brain and those which are outside the brain and affect it secondarily; a useful distinction for clinical purposes was also made between progressive of fixed dementing pathological states, and arrestable or reversable causes of dementia.

Arrestable or reversible causes of dementia include intoxications, infections, metabolic disorders, nutritional disorders, vascular and space-occupying lesions, normal pressure hydrocephalus and affective disorders.

Arrestable or reversible dementias account for $10-15\%$ of all causes of dementia and efforts must be made to recognize them, as they are potentially reversible if appropriate therapy is applied (Marsden and Harrison, 1972).

However, the most frequent causes of dementia, as documented by clinical studies (Wells, 1979), post-mortem examinations (Tomlinson et al., 1970) and population-based surveys (Lippi et al., 1989) are Alzheimer's disease (AD) and vascular dementia, followed by secondary dementias. Jorm (1987) observed that the relative prevalence of AD and vascular dementia differs from country to country, e.g. vascular dementia is more common in Japanese and Russian studies, AD in West-European countries, while no significant differences were found in Finnish and American studies. In Italy, AD seems to be only slightly more frequent than vascular dementia (Lippi et al., 1989).

Descriptive epidemiology of Alzheimer's disease

Major problems in the descriptive epidemiology of Alzheimer's disease are the choice of the population study and the type of sample, the methodology for case ascertainment and the accuracy of clinical diagnosis.

The study population may be obtained from the overall population by randomization or by complete enumeration of the subjects. Otherwise, in community-based surveys, data will be collected from hospitals, nursing homes and Health Authorities. The latter approach was principally used in North-European surveys (Akesson, 1969; Molsa et al., 1982). However, the risk of this methodology is to underestimate the real frequency of the disease.

Two general strategies have been proposed for case ascertainment in population studies: the one- and the two-phase approach. The former is less expensive, but the latter affords a more accurate diagnosis. In the two-phase approach, a brief, highly sensitive cognitive test (a screening test) is performed on all the people to be investigated. Subjects scoring under the cut-off level at the screening test are extensively studied from a clinical point of view to confirm the presence of mental impairment, to diagnose dementia and classify dementias by type. However, for a definite diagnosis of dementing disorders, especially AD, a pathological examination of the brain is required.

Several sets of clinical diagnostic criteria have been proposed. Recent population-based surveys utilized the NINCDS-ADRDA criteria (McKhann et al., 1984). CAMDEX by Roth et al. (1986) provided a useful set of diagnostic criteria for differential diagnosis of mental disturbances in the elderly, and was successfully used in some recent British surveys (O'Connor et al., 1989).

Prevalence in Alzheimer's disease

Available data show that the prevalence of Alzheimer's disease ranges between 1.9 and 5.8% of the total population aged 65 and over (Rocca et al., 1986). This variability of the prevalence ratios can be influenced by different definitions of the disease under study and the different case ascertainment procedures adopted. Some investigations (Akesson, 1969; Molsa et al., 1982; Sulkawa et al., 1985; Schoenberg et al., 1985) considered only severe dementia, whereas in other studies also mild dementia was included (Kay et al., 1964; Broe et al., 1976). Akesson (1969), who found very low prevalence ratios, adopted very restrictive diagnostic criteria: constant disorientation as to time and place was required to make a diagnosis of dementia. Case ascertainment procedures also differ in various studies. Molsa et al. (1982) in Finland and Akesson (1969) in Sweden, who found lower prevalence ratios than some of their North-European colleagues, collected data only from health and social services. Sulkawa et al. (1985), in a random sample representative of the whole Finnish population, utilizing the door-to-door approach, found higher prevalence rations that were found in previous community-based surveys.

All the studies carried out until now consistently showed an exponential increase in the ratios with age. In the Finnish study by Molsa et al. (1982) the prevalence ratio increased from 0.025% in the age group 45 − 54 years to 6.3% in the age group 85 years and over. In an Italian survey, the prevalence ratio was 6.2% for subjects aged 60 and over rising to 8.4% for those aged 65 and over (Lippi et al., 1989).

In many studies, age-specific prevalence ratios were consistently higher for women than for men, especially in the older age groups (Broe et al., 1976; Molsa et al., 1982; Sulkawa et al., 1985). Molsa et al. (1982) found age-specific prevalence ratios to be higher for men in the age group 45 — 54 years and for females in the other age groups. Kaneko in Japan (1975) reported prevalence ratios consistently higher for males, while a more recent Japanese survey showed a pattern consistent with those reported in European and American studies (Karasawa et al., 1982). Finally, Schoenberg et al. (1985) in a population-based survey carried out in Copiah County, Mississippi, found consistently higher prevalence ratios for females and blacks.

Incidence of Alzheimer's disease

Similarly to the prevalence pattern, the incidence rate of AD rises exponentially with age. As summarized by Mas et al. (1987), the incidence rate can be estimated to be about 0.01‰ in the age group 40 — 60 years, rising to 1‰ in the age group 65 — 74 years and reaching 10‰ in subjects aged 75 years and over.

Akesson (1969) in Sweden, in a community-based survey, observed that incidence rates rapidly increased with age, from 0.29‰ in the age group 60 — 69 years to 4.58‰ in people over 80 years of age.

Rates were consistently higher for females in all age groups. Molsa et al. (1982) in Finland, in a community-based survey, found that age-specific annual incidence rates for AD were 0.06‰ in the age range 45 — 54 years and 11.44‰ in people aged 85 and over. The incidence rates were higher in females, except for the age group 45 — 54 years. Nilsson (1984) in Sweden, in a sample of 70 to 79 year-old people found a rapid age-related increase in the incidence rates, from 3.6‰ in subjects aged 70 — 74 to 13.3‰ in those aged 75 — 79, and higher rates in men. Treves et al. (1986) in Israel, using the Israel National Neurological Register as source of cases, found lower incidence rates, though increasing with age: from 0.01‰ in the age range 45 — 49 years, to 0.87‰ in subjects aged 60 years and over. Higher incidence rates were observed for females after the age of 59. Moreover, in this study, age-specific incidence rates were higher in the European-American born citizens than in the Afro-Asian-born ones.

No significant variations seem to have been found in AD incidence rates across the years. Kokmen (1988) in Rochester, Minnesota, had determined the age- and sex-specific annual incidence rates for dementing disorders and AD between 1960 and 1974, showing that the incidence of dementia as a whole or dementia of the Alzheimer type had not particularly increased in this 15-year period.

Analytic epidemiology of Alzheimer's diseases

Our knowledge of these risk factors for AD is mainly based on findings derived from case-control studies. In these retrospective studies, one group of AD patients (cases) is compared with one or more groups of subjects without AD (controls), with respect to their past exposure to one or more potential risk factors. Because demented people suffer from memory impairment, information was usually recorded from a surrogate respondent and the same procedure should be adopted for controls for comparability reasons. Rocca et al. (1986) assessed the reliability of surrogate respondents to provide data for the specific items of a case-control study of AD conducted by Amaducci et al. (1986) in Italy. Agreement was over 80% for the majority of questions, lower for alcohol consumption and number of cigarettes smoked per day and very poor for information about the use of antacid drugs.

The case-control studies carried out until now pointed out a positive family history of dementia and head trauma in the subject's history as risk factors for AD.

In the case-control study by Amaducci et al. (1986) the presence of dementia in first-degree relatives yielded odds ratios equal to 5 in the comparison with hospital controls and over 2 in the comparison with population controls, whereas the presence of dementia in siblings yielded odds ratios equal to 11 in the comparison with hospital controls and over 5 in the comparison with population controls. Other case-control (Heyman et al., 1984; Graves et al., 1987; Shalat et al., 1987) and comparative studies (Heston et al., 1981; Whalley et al., 1982) provided similar findings.

Mortimer (1990), using the Mantel-Haenszel statistical method (Schlesselman, 1982) carried out a meta-analysis of case-control studies of AD and found an odds ratio (95% C.I.) for the risk factor "family history of dementia" equal to 3.96 (2.84 − 5.52).

Indications as to the role of genetic factors in the etiology of AD were derived also form pedigree, twin and genetic studies. Rocca and Amaducci (1988) reviewed epidemiologically all the contributors to the problem of familial aggregation of AD. The authors suggested that at least 3 types should be considered in terms of occurrence of the disease: autosomal dominant, familial, and sporadic. The autosomal dominant type is linked to a genetic defect located on chromosome 21, at least in the families studied (St. George-Hyslop et al., 1987). The familial type identifies AD cases with familial aggregation but without a clear relationship with inheritance. These cases could be of polygenic origin. Finally, sporadic cases could be due to environmental risk factors or to a combination of genetic and environmental risk factors.

Among the potential environmental risk factors for AD, a previous head trauma in the history of the subject was found to be significantly associated with the disease in case-control studies (Heyman et al., 1984; Mortimer et al., 1985). In some studies (Amaducci et al., 1986; Chandra et al., 1987 a) head trauma was more frequent in AD patients than in controls, but the difference was not considered statistically significant. On the other hand, several studies could find no association between head trauma and AD. Mortimer (1990) observed that the association with head trauma is highly significant ($p < 0.001$) when considering the studies as a whole, despite the fact that most of them reported no significant associations with this risk factor.

The other factors found to be associated with AD in case-control studies are smoking (Shalat et al., 1987) and exposure to aluminium (Graves et al., 1987). These associations were negative in all the other case-control studies (Bharucha et al., 1983; Heyman et al., 1984; French et al., 1985; Amaducci et al., 1986; Chandra et al., 1987 b).

References

Amaducci LA, Fratiglioni L, Rocca WA, Fieschi C, Livrea P, Pedone D, Bracco L, Lippi A, Gandolfo C, Bino G, Prencipe M, Bonatti ML, Girotti F, Carella F, Tavolato B, Ferla S, Lenzi GL, Carolei A, Gambi A, Grigoletto F, Schoenberg BS (1986) Risk factors for clinically diagnosed Alzheimer's disease: a case-control study of an Italian population. Neurology 36: 922–931

American Psychiatric Association (1987) Diagnostic and statistical manual of mental disorders, 3rd edn (Revised). American Psychiatric Association, Washington, p 103

Akesson HO (1969) A population study of senile and arteriosclerotic psychoses. Hum Hered 19: 546–566

Bharucha NE, Schoenberg BS, Kokmen E (1983) Dementia of Alzheimer's type (DAT): a case-control study of association with medical conditions and surgical procedures. Neurology 33 [Suppl 2]: 85

Broe GA, Akhtar AJ, Andrews GR, Caird FI, Gilmore AJJ, McLennan WJ (1976) Neurological disorders in the elderly at home. J Neurol Neurosurg Psychiatry 39: 362–366

Chandra V, Kokmen E, Schoenberg BS (1987a) Head trauma with loss of consciousness as a risk factor for Alzheimer's disease using prospectively collected data. Neurology 37 [Suppl 1]: 152

Chandra V, Philipose V, Bell PA, Lazaroff A, Schoenberg BS (1987 b) Case-control study of late onset "probable Alzheimer's disease". Neurology 37: 1295–1300

French LR, Schuman LM, Mortimer JA, Hutton JT, Boatman RA, Christians B (1985) A case-control study of dementia of the Alzheimer type. Am J Epidemiol 121: 414–421

Graves AB, White E, Koepsell T, Reifler B (1987) A case-control study of Alzheimer's disease. Am J Epidemiol 126: 754

Heyman A, Wilkinson WE, Stafford JA, Helms MJ, Sigmon AH, Weinberg T (1984) Alzheimer's disease: a study of epidemiological aspects. Ann Neurol 15: 335–341

Heston LL, Mastri AR, Anderson VE, White J (1981) Dementia of the Alzheimer type: clinical genetics, natural history, and associated conditions. Arch Gen Psychiatry 38: 1085–1090

Jorm AF, Korten AE, Henderson AS (1987) The prevalence of dementia: a quantitative integration of the literature. Acta Psychiatr Scand 76: 465–479

Kaneko Z (1975) Care in Japan. In: Howells JG (ed) Modern perspectives in the psychiatry of old age. Brunner/Mazel, New York, pp 519–530

Karasawa A, Kawashima K, Kasahara H (1982) Epidemiological study of senile dementia in Tokyo's metropolitan area. In: Ohashi (ed) Proceedings of the World Psychiatric Association Regional Symposium, Tokyo 1982. The Japanese Society of Psychiatry and Neurology, pp 285–289

Kay DWK, Beamish P, Roth M (1964) Old age mental disorders in Newcastle upon Tyne, part 1. A study of pevalence. Br J Psychiatry 110: 146–158

Kokmen E (1988) Trends in incidence of dementing illness in Rochester, Minnesota, in three quinquennial periods, 1960–1974. Neurology 38: 975–979

Lippi A, Rocca WA, Bonaiuto S, Luciani P, Turtu F, Cavarzeran F, Amaducci L (1989) Prevalence of Alzheimer's disease (AD) and other dementing disorders: a door-to-door survey in Appignano, Macerata province, Italy. Neurology 39 [Suppl 1]: 180

Marsden CD, Harrison MJG (1972) Outcome of investigation of patients with presenile dementia. Br Med J 2: 249

Mas JL, Alperovitch A, Derouesne C (1987) Epidémiologie de la démence de type Alzheimer. Rev Neurol 3: 161–171

Mckhann G, Drachman D, Folstein M, Katzman R, Price D, Stadlan EM (1984) Clinical diagnosis of Alzheimer's disease: report of the NINCDS-ADRDA Work Group under the auspices of the Department of Health and Human Service Task Force on Alzheimer's disease. Neurology 34: 939–944

Molsa PK, Marttila RJ, Rinne UK (1982) Epidemiology of dementia in a Finnish population. Acta Neurol Scand 65: 541–552

Mortimer JA, French LR, Hutton JT, Schuman LM (1985) Head injury as a risk factor in Alzheimer's disease. Neurology 35: 264–267

Mortimer JA (1990) Epidemiology of dementia: cross-cultural comparisons. Adv Neurol 91: 27–34

National Institute of Health Consensus Development Conference Statement (1988) Differential diagnosis of dementing diseases. Alz Dis Assoc Disord 2: 4–28

Nilsson LV (1984) Incidence of severe dementia in an urban sample followed from 70 to 79 years of age. Acta Psychiatr Scand 70: 478–486

O'Connor DW (1989) The prevalence of dementia as measured by the CAMDEX. Acta Psychiatr Scand 79: 190–198

Rocca WA, Amaducci L (1988) The familial aggregation of Alzheimer's disease: an epidemiological review. Psychiatr Dev 1: 23–26

Rocca WA, Amaducci LA, Schoenberg BS (1986) Epidemiology of clinically diagnosed Alzheimer's disease. Ann Neurol 19: 415–424

Rocca WA, Fratiglioni L, Bracco L, Pedone D, Groppi C, Schoenberg BS (1986) The use of surrogate respondents to obtain questionnaire data in case-control studies of neurological diseases. J Chron Dis 39: 907–912

Roth M, Tym E, Mountjoy CQ, Huppert FA, Hendrie H, Verma S, Goddard R (1986) CAMDEX, a standardized instrument for the diagnosis of mental disorders in the elderly with special reference to the early detection of dementia. Br J Psychiatry 149: 698–709

Schlesselman JJ (1982) Case-control studies: design, conduct, analysis. New York, Oxford University Press

Schoenberg BS, Anderson DW, Haerer AF (1985) Severe dementia. Prevalence and clinical features in a biracial US population. Arch Neurol 42: 740–743

Shalat SL, Seltzer B, Pidcock C, Baker EL Jr (1987) Risk factors for Alzheimer's disease: a case-control study. Neurology 37: 1630–1633

St George-Hyslop PH, Tanzi RE, Polinsky RJ, Haines JL, Nee L, Watkins PC, Myers RH, Feldman RG, Pollen D, Drachman D, Growdon J, Bruni A, Foncin JF, Salmon D, Frommelt P, Amaducci L, Sorbi S, Piacentini S, Stewart GD, Hobbs WJ, Conneally PM, Gusella JF (1987) The genetic defect causing familial Alzheimer's disease maps on chromosome 21. Science 235: 885–890

Sulkava R, Wikstrom J, Aromaa A, Raitasalo R, Lehtinen V, Lahtela K, Palo J (1985) Prevalence of severe dementia in Finland. Neurology 35: 1025–1029

Tomlinson BE, Blessed G, Roth M (1970) Observations on the brains of demented old people. J Neurol Sci 11: 205

Treves T, Korczyn A, Zilber N, Kahana E, Leibowitz Y, Alter M, Schoenberg BS (1986) Presenile dementia in Israel. Arch Neurol 43: 26–29

United Nations (1986) Report of the interregional seminar to promote the implementation of the international plan of action on aging. United Nations Publication Scales No. E.86.IV.5

Whalley LJ, Carothers AD, Colleyr S, De Mey R, Frackiewicz (1982) A study of familial factors in Alzheimer's disease. Br J Psychiatry 140: 249–256

Differential diagnosis of early Alzheimer's disease

C. G. Gottfries

Department of Psychiatry and Neurochemistry, Gothenburg University, Sweden

Summary

The concept of Alzheimer type dementia, as used today, does not imply a homogeneous group of dementias. There is no scientific basis for sampling the presenile form, Alzheimer's disease (AD), with the senile dementia of the Alzheimer type (SDAT). The rare form of AD seems to be fairly homogeneous. SDAT is much more common than AD and is, evidently, a heterogeneous group. The differentiation of the subgroups is made by thorough clinical investigation, brain imaging techniques and analysis of cerebrospinal fluid and blood. Data indicate that AD is a cortical disorder, while SDAT also involves subcortical areas. The differentiation between SDAT and secondary dementias, such as vitamin B 12 deficiency and leukoaraiosis, is difficult. The SDAT group is assumed to include also forms of normal aging with cognitive impairment.

Introduction

Originally, Alzheimer's disease was considered a relatively rare presenile, dementing illness. Today, however, Alzheimer's disease is regarded as the fourth or fifth most common cause of death in the United States and a major public health trap.

One reason for this change in the prevalence and the diagnosis of Alzheimer's disease is that the presenile form of Alzheimer's disease (AD) has been combined with the senile dementia of the Alzheimer type (SDAT) (Katzman, 1976). One reason for combining the two disorders was that the presenile and the senile forms of the dementia disorder had very similar clinical phenomenology. The main reason, however, was that the Alzheimer lesions, senile plaques and neurofibrillary tangles, were found in both forms at postmortem histological investigations of the brain. Investigations by Blessed et al. (1968) also suggest that there is a relationship between the amount of structural changes in the brain and behavioural disturbances.

However, from a scientific point of view, the above-mentioned factors cannot be considered sufficient for the sampling of the two disorders. As shown below, there are differences in the clinical picture of the two disorders. The specificity of the Alzheimer lesions must also be questioned. These structural changes are found in several dementing disorders and are a form of Alzheimer encephalopathy that can be the common pathway for several disorders. Even though there is a significant correlation between the level of dementia and the frequency of Alzheimer lesions, this does not indicate a causal relationship between the two phenomena.

Although the DSM-III-R is based on specific criteria for the diagnosis of AD/SDAT, it is obvious that the most important criterion is that other diseases which may produce mental impairment must be excluded. This means that, in fact, AD/SDAT is an exclusion diagnosis.

When discussing differential diagnosis it must be kept in mind that there is no scientific basis for sampling AD with SDAT. AD is a rare form of presenile dementia, while SDAT is a very common form of a dementia that may be heterogeneous. The aim of the present paper is to discuss differential diagnosis between subgroups of the Alzheimer type dementia and between the Alzheimer type dementia and other forms of dementia disorders.

Differential diagnosis between subgroups of the Alzheimer type dementias

Early versus late onset of Alzheimer dementia

Early onset AD was originally separated from SDAT. Several factors indicate that this separation is still valid. Population studies (Sjögren et al., 1952; Heston et al., 1981) have shown that there is an increased risk of dementia disorder among relatives. The highest risk was found among siblings of autopsied Alzheimer patients who had a comparatively early onset of the disorder. In this group the cumulative risk was 50 : 50. The lowest risk was found in siblings of patients who developed a dementia disorder after the age of 70. In the Swedish population study by Sjögren et al. (1952) the familial appearance of the two disorders indicated that they must be distinguished.

One reason for sampling the two Alzheimer dementia disorders was a similar phenomenology. This is not quite correct, however. As pointed out by Lauter (1970), Gustafson (1985) and Bråne et al. (1989), there are differences in the clinical picture of the two disorders. This was also confirmed in a prospective investigation at our institute (unpublished results). AD is a cortical dementia with a relatively typical parieto-temporal "syndrome". This focalization is far less prominent in the SDAT group, in which the manifestation of a "subcortical dementia" is more evident. These patients show a psychomotor slowing, forgetfulness and an apathetic, depressed affect, while aphasia, amnesia, apraxia and agnosia are less evident.

Ratings (Bråne, 1989) and psychometric investigations have limited value in the differentiation between AD and SDAT. The most sensitive investigation is a thorough neurological and psychiatric investigation of the patient performed by an experienced neuropsychiatrist.

Brain imaging technique are now greatly improved, and the localization and degree of morphological changes in the brain, when present, can be described relatively clearly. In an investigation at our institute (Wallin et al., 1989 a), morphological changes are presented according to the CT scans. The analysis of the data show that the frequency of white matter low attenuation (WMLA) was significantly lower in the AD group than in the SDAT group. The frequency of central atrophy in the AD group was also significantly lower than in the SDAT group. There were no significant differences between the AD and SDAT groups with regard to cortical atrophy. The lack of subcortical findings on CT, as well as WMLA in AD patients, suggests that AD is mainly a cortical disorder, while in SDAT there is also an involvement of subcortical regions in the degenerative process. It is obvious that not only CT scans but also nuclear magnetic resonance (NMR), positron emission tomography (PET), single photon emission tomography (SPECT), and rCBF should be used in further studies of the AD and SDAT subgroups.

The use of EEG to differentiate between AD and SDAT may also be of value. It has been shown that the late positive component of the evoked cortical potential is changed in neurophysiological investigations. Recently it has been shown that cortical and subcortical dementias can be distinguished by sensory evoked responses on EEG. Patients with subcortical dementias have normal EEGs, while only 20% of patients with AD have normal EEGs. Sensory evoked potentials are most often abnormal in subcortical dementias, but normal in cortical dementias (Verma et al., 1987).

Although the structural changes found in AD and SDAT are similar, there are differences in the neurochemical findings in the two disorders. As pointed out by Rossor et al. (1984) and Gottfries et al. (1985), there are more severe neurochemical changes in brains from patients with AD than from those with SDAT. Brun and Gustafson (1976) described structural white matter changes in brains from patients with AD/SDAT. These findings were confirmed in neurochemical investigations of white matter (Gottfries et al., 1985; Svennerholm et al., 1988). However, it was evident that the white matter changes in both types of investigation were more pronounced in SDAT than in AD.

In the cerebrospinal fluid (CSF), monoamine metabolites and activity of acetylcholine esterase (AChE) can be investigated. Studies in patients with AD and SDAT have been performed, and cumulative data indicate that the concentration of homovanillic acid (HVA), of 5-hydroxyindoleacetic acid (5-HIAA), and the activity of AChE are reduced in AD. However, the SDAT group does not differ from age-matched control groups to the same extent (Gottfries, 1983; Bråne et al., 1989).

When peripheral markers were studied, some differences also emerged in the diagnosis of AD and SDAT. Regland et al. (1988) studied a group of patients including AD and SDAT diagnoses. In this clinical sample, the concentration of vitamin B 12 in serum and the activity of monoamine oxidase (MAO) in platelets were determined. In the AD group, the vitamin B 12 levels were normal, while in the SDAT group 23% (13 of 56) were found to have significantly reduced vitamin B 12 concentrations (below 130 pmol/l). A group of demented patients with low vitamin B 12 levels had also significantly increased concentrations of MAO in platelets. The latter finding indicated the presence of immature platelets, possibly due to a mild peripheral deficiency of vitamin B 12. None of the demented patients with pathologically low vitamin B 12 levels had pernicious anaemia. According to further studies by Regland (unpublished results), the patients with low vitamin B 12 levels also had reduced pepsinogen I in serum, indicating atrophic gastritis.

When the above-mentioned data are taken together, it is obvious that AD should be distinguished from SDAT. AD appears to be mainly a cortical disease in which the degeneration, at least at the beginning of the disorder, is restricted to the parieto-temporal lobes. It can be assumed that it is a rare form of dementia and a rather homogeneous group. Mayeux et al. (1985), however, suggested heterogeneity also in AD and described some subgroups according to prospective investigations. This is also partly supported by Kaye et al. (1988) who found a special pattern of monoamine metabolite changes in the CSF of AD patients with extrapyramidal features.

SDAT is a more common and evidently also more heterogeneous group in which the brain disturbance is localized not only to the cortex but also to the subcortical areas. It is evident that within the SDAT group there is a subgroup of leukoaraiosis, and the pathogenetic importance of these white matter changes must be further evaluated. Within the SDAT group there also seems to be a subgroup with reduced concentrations of vitamin B 12. As changes in the mucous membrane of the gut were found in these patients, it might be assumed that in a subgroup of SDAT there are degenerative changes in the gut giving rise to brain deficiencies of essential nutrients. This group should, of course, be named secondary dementias.

From population studies (Adolfsson et al., 1981; Jorm et al., 1987) it is evident that there is a strong relationship between the frequency of dementia and age. This is most evident at a very old age. Neurochemical data also indicate that the activity in some neurotransmitter systems decreases with the normal aging process. Thus, it can be assumed that, in the age groups above 85 − 90 years, the normal aging process may cause a mental impairment that is impossible to differentiate from that caused by pathological processes in SDAT. The concepts of age-associated memory impairment (AAMI, Crook et al., 1986) and benign senescent forgetfulness (Kral, 1962) are indicative of different forms or more benign forms of dementing disorders in the later part of life.

Differential diagnosis between AD/SDAT and vascular dementia

A differential diagnosis between AD/SDAT and multi-infarct dementia (MID) is not difficult to make. The slowly progressing disorder of AD/SDAT can easily be separated from the MID disorder with acute onset and a stepwise course. However, according to Wallin et al. (1989 b), there are other forms of vascular dementia than MID. Probable vascular dementias (PVD) are forms in which evident risk factors are indicative of a vascular dementia. Heart vessel disorders, diabetes mellitus, etc. make a vascular disorder probable, although the onset is insidious and the stroke attacks are not seen or seen only in the final stages of the disorder. In these cases, haemodynamic disorders and small vessel disorders are assumed to be of pathogenetic importance. Wallin et al. (1989 b) showed that the neurochemical changes in the brains of these patients were similar to those found in patients with SDAT. In investigations using CT scans, Wallin et al. (1989 a) found that WMLA and atrophy of the brain were similar in SDAT and in PVD. Thus, it is obvious that the differentiation between SDAT and non-MID vascular dementias may create difficulties.

Postmortem human brain investigations have shown relatively severe disturbances of serotonin metabolism in brains from patients with PVD. It is then of interest to find decreased concentrations of 5-HIAA in CSF from patients with VD, as done at our institute (unpublished data).

The use of rating scales is of importance in the differentiation between vascular dementia and AD/SDAT. The well-known ischemic score of Hachinski et al. (1975), the modified version of Portera Sanchez et al. (1982), and the modification by Loeb and Gandolfo (1983) can be recommended.

AD/SDAT versus Pick's disease

Pick's disease belongs to the dementias affecting the frontal and the temporal lobes. The clinical picture in this disorder is dominated by a destruction of personality, behaviour, and speech, while cognitive functions are less affected. Hyperactive, disinhibited, stereotyped behaviour may be seen in the early forms. It is of interest that, as reported by Gustafson (1985), a frontal lobe dementia is not as uncommon as usually assumed. The name has therefore been changed into frontal lobe dementia (FLD), as not all of these dementias have the significant changes seen in Pick's disease. In FLD, EEG is usually reported to be normal even at a late stage of the dementia. The pattern of symptoms can be recognized using a rating scale for the diagnosis of Pick's disease (Gustafson and Nilsson, 1982). The differentiation between AD and FLD is further strenghtened by EEG, rCBF, and psychometric data (Gustafson et al., 1981; Johanson et al., 1986).

AD/SDAT versus other forms of primary degenerative dementia

The differentiation between AD/SDAT and Huntington's chorea usually poses no differential diagnostic problems. Parkinsonism may sometimes be combined with dementia. It is obvious that there is overlap between AD/SDAT and parkinsonism. 10% of patients with AD are sometimes assumed to have parkinsonistic symptoms. On the other hand, a subgroup of patients with Parkinson's disease is assumed to have dementia symptoms.

AD/SDAT versus pseudodementia

Depression in old age may be comined with cognitive impairment, and differential diagnosis of SDAT may be difficult. Usually the two disorders can be differentiated by neuropsychological tests. EEG may also be of help. According to rCBF studies this method can also be used to differentiate the two disorders. The differentiation between dementia and depression is important, as the latter can be successfully treated using electroshock therapy or psychotropic drugs.

In DSM-III-R is stated that a diagnosis of dementia must be made in the absence of a major confusion state. A differential diagnosis between dementia and delirium attacks with acute onset is usually not difficult. Sometimes a dementia syndrome may be hidden behind an acute delirium attack. Then dementia cannot be diagnosed until the confusion symptoms have disappeared.

In studies at our institute we have found that in SDAT, according to ratings using the GBS scale (Bråne, 1989), the presence of mild confusion is frequent. A careful anamnesis reveals that some of the SDAT patients have long shown symptoms of a fluctuating mild confusion (sundown syndrome), which has been slowly progressing over many years. The pathogenetic importance for the dementia syndrome of this mild confusion must be further investigated. It cannot be excluded that a group of dementias is contaminated with a group of patients with mild and slowly progressing confusion.

References

Adolfsson R, Gottfries CG, Nyström L, Winblad B (1981) Prevalence of dementia disorders in institutionalized Swedish old people. The work load imposed by caring for these patients. Acta Psychiatr Scand 63: 245–252

Blessed G, Tomlinson BE, Roth M (1968) The association between quantitative measures of dementia and of senile change in the cerebral matter of elderly subjects. Br J Psychiatry 114: 797–811

Bråne (1989) The GBS scale — a geriatric rating scale — and its clinical application. Dissertation, Gothenburg University, Sweden

Bråne G, Gottfries CG, Blennow K, Karlsson I, Lekman A, Parnetti L, Svennerholm L, Wallin A (1989) Monoamine metabolites in cerebrospinal fluid and behavioural ratings in patients with early and late onset of Alzheimer's dementia. Alz Dis Assoc Disord 3: 148–156

Brun A, Gustafson L (1976) Distribution of cerebral degeneration in Alzheimer's disease. A clinicopathological study. Arch Psychiatr Nervenkr 223: 15–33

Crook T, Bartus R, Ferris S, Gershon S (1986) Treatment development strategies for Alzheimer's disease. Mark Powley, Connecticut, pp 385–450

Gottfries CG (1983) Biochemical changes in blood and cerebrospinal fluid. In: Reisberg (ed) Alzheimer's disease. The standard reference. The Free Press, Collier MacMillan, London, pp 122–130

Gottfries CG, Karlsson I, Svennerholm L (1985) Senile dementia — a "white matter" disease? In: Gottfries CG (ed) Normal aging. Alzheimer's disease and senile dementia. Aspects on etiology, pathogenesis, diagnosis and treatment. Proceedings of two symposia held at the C.I.N.P. Congress, June 22–23, 1984, Florence, Italy. L'Université de Bruxelles, Brussels, pp 11–118

Gustafson L (1985) Differential diagnosis with special reference to treatable dementias and pseudodementia conditions. Dan Med Bull 32 [Suppl 1]: 55–60

Gustafson L, Nilsson L (1982) Differential diagnosis of presenile dementia on clinical grounds. Acta Psychiatr Scand 65: 194–209

Gustafson L, Risberg J, Silverskiöld P (1981) Regional cerebral blood flow in organic dementia and affective disorders. In: Mendlewicz J, van Prag HM (eds) Advances in biological psychiatry, vol 6. Karger, Basel, pp 109–116

Hachinski VC, Iliff LB, Phil M, Zilhka E, Du Boulai GH, McAllister VL, Marshall J, Russell RW, Symon L (1975) Cerebral blood flow in dementia. Arch Neurol 32: 632–637

Heston LL, Mastri AR, Anderson E, White J (1981) Dementia of the Alzheimer type: clinical genetics, natural history and associated conditions. Arch Gen Psychiatry 38: 1085–1090

Johanson AM, Gustafson L, Risberg J (1986) Behavioural observations during performance of the WAIS Block Design Test related to abnormalities of regional cerebral blood flow in organic dementia. J Clin Exp Neuropsychol 8: 201–209

Jorm AF, Korten AE, Henderson AS (1987) The prevalence of dementia: a quantitative integration of the literature. Acta Psychiatr Scand 76: 465–479

Katzman R (1976) The prevalence and malignancy of Alzheimer's disease. Arch Neurol 33: 217–218

Kaye JA, May C, Daly E, Atack JR, Sweeney DJ, Luxenberg JS, Kay AD, Kaufman S, Milstien S, Friedland RP, Rapoport SI (1988) Cerebrospinal fluid monoamine markers are decreased in dementia of the Alzheimer type with extrapyramidal feastures. Neurology 38: 554–557

Kral VA (1962) Senescent forgetfulness: benign and malignant. J Can Med Assoc 86: 257–260

Lauter H (1970) Über Spätformen der Alzheimerschen Krankheit und ihre Beziehung zur senilen Demenz. Psychiatr Clin 3: 169–189

Loeb C, Gandolfo C (1983) Diagnostic evaluation of degenerative and vascular dementia. Stroke 14: 399–401

Mayeux R, Stern Y, Spanton S (1985) Heterogeneity in dementia of the Alzheimer type. Evidence of subgroups. Neurology 35: 453–461

Portera-Sanchez A, del Ser T, Bermejo (1982) Clinical diagnosis of senile dementia of Alzheimer type and vascular dementia. In: Perry RD, Bolis SL, Toffano F (eds) Neural aging and its implications in human neurological pathology. Raven Press, New York, pp 169–188 (Aging, vol 18)

Rossor MN, Iversen LL, Reynolds GP, Mountjoy CQ, Roth M (1984) Neurochemical characteristics of early and late onset types of Alzheimer's disease. Br Med J 288: 961–964

Sjögren T, Sjögren H, Lindgren AGH (1952) Morbus Alzheimer and morbus Pick. A genetic, clinical and pathoanatomical study. Acta Psychiatr Neurol Scand [Suppl 82]

Svennerholm L, Gottfries CG, Karlsson I (1988) Neurochemical changes in white matter of patients with Alzheimer's disease. In: Serlupi Crescenzi G (ed) A multi-disciplinary approach to myelin disease. Plenum Publishing Corporation, pp 319–328

Verma MP, Greiffenstein MF, Verma N, King SD, Caldwell DL (1987) Electrophysiologic validation of two categories of dementias — cortical and subcortical. Clin Electroencephalogr 18: 26–33

Wallin A, Blennow K, Uhleman C, Långström G, Gottfries CG (1989 a) White matter low attenuation on computed tomography in Alzheimer's disease and vascular dementia — diagnostic and pathogenetic aspects. Acta Neurol Scand 80: 518–523

Wallin A, Alafuzoff I, Carlsson A, Eckernäs SÅ, Gottfries CG, Karlsson I, Svennerholm L, Winblad B (1989 b) Neurotransmitter deficits in a non-multi-infarct category of vascular dementia. Acta Neurol Scand 79: 397–406

Down syndrome: a model for the study of Alzheimer's disease and aging

J. M. Delabar[1], J. L. Blouin[2], Z. Rahmani[1], N. Créau-Goldberg[2],
Z. Chettouh[1], A. Nicole[1], A. Bruel[1], M. C. de Blois[3], and
P. M. Sinet[1]

[1] URA CNRS 1335, Laboratoire de Biochimie Génétique,
[2] INSERM U 173, and
[3] Service de Cytogénétique, Hôpital Necker, Paris, France

Summary

Down syndrome is defined by the association of a number of features usually observed in patients with trisomy of chromosome 21. Different forms of aneuploidy are encountered: free trisomy 21 (the most frequent), partial trisomy 21, unbalanced translocation, mosaicism, and microduplications of a short chromosomic fragment. Comparison of genotype and phenotype permitted us to define a critical region for the pathogenesis of Down syndrome, on the proximal part ar 21 q 22.3. Down syndrome is also associated with early manifestation of aging and Alzheimer like neuropathology; this pathology results from the presence of an extra copy of one or few genes on chromosome 21. We hypothesize that Alzheimer's disease might, at least in certain patients, be secondary to a chromosome 21 defect susceptible to induce the overexpression of this or these genes. Experimental approaches for testing this hypothesis have been recently developed, both in familial and sporadic Alzheimer's disease.

Introduction

The characteristics of Down syndrome are early manifestation of aging and Alzheimer like neuropathology. This syndrome, the most frequent among birth defects (1 out of 700 newborns), results from the presence of an extra chromosome 21 in all cells of affected individuals (Lejeune, 1959), except for the rare cases that we will discuss later. Down syndrome is defined by an association of clinical features, the most typical i.e. the most commonly observed of which are: 1) unusual morphological traits of the face (flat nasal bridge, oblique palpebral fissures, epicanthic eye fold, brachycephaly, highly arched and narrow palate), of the hands (broad and short hands, short and incurved 5th finger, transverse palmar crease, and unusual dermatoglyphic patterns), and of the feet (broad with gap between first and second toes), 2) hypotonia and joint hyperflexibility, 3) visceral malformations (mostly heart defect), and 4) susceptibility to infections and increased risk of leukemia.

None of these signs is either constant or pathognomonic but this is the association of a number of them in an individual which evokes trisomy 21. Thus, Jackson et al. (1976) proposed a diagnostic index based on a cheklist of 25 physical signs of Down syndrome. Comparison between karyotypcially normal and trisomy 21 individuals indicates that subjects with 13 signs or more can be confidently diagnosed as having Down syndrome. Between 5 and 12 signs there is an overlapping between normal and trisomy 21 populations. These data support the well-known notion that the clinical expression of trisomy 21 is variable from one individual to another. This variability in Down syndrome expression has three noteworthy exceptions, i.e. features considered as constant in trisomy 21: growth retardation, mental delay, and Alzheimer like neuropathology. This last aspect of Down syndrome is specially interesting in the context of this meeting.

Dementia is observed in more than 25% of Down patients over the age of 30 (Wisniewsky et al., 1985) and its incidence increases progressively with age (Lai et al., 1987). Besides symptoms related to Alzheimer pathology, patients with Down syndrome show a gradual loss of their intellectual functions with age (Melyn, 1973). Brain lesions similar to those observed in Alzheimer, i.e. senile plaques and paired helical filaments (PHF) are present in all Down patients aged over 30 − 40 years (Wisniewski et al., 1985), although PHFs seem to appear years after the first diffuse plaques (Mann, 1988). The amyloid plaque core protein in Down syndrome shows the same amino acid composition as in Alzheimer's disease (Glenner and Wong, 1984). Antibodies against PHF from Alzheimer patients recognize epitopes of the PHFs found in Down syndrome (Anderson et al., 1982). Alz-50 protein antigen in brain, specific of Alzheimer's disease, is high in Down syndrome (Wolozin et al., 1988). The same neurochemical changes are also observed in both conditions, i.e. decreased choline acetyl transferase and acetyl cholinesterase activities (Yates et al., 1980). Thus, Down syndrome patients develop neuropathologic features apparently identical to those observed in Alzheimer's disease. This data has two implications:

i) one or few genes on chromosome 21, when in 3 copies instead of two, can induce Alzheimer like pathology.

ii) the hypothesis of a genetic defect on chromosome 21 or/and of a dysregulation in the expression of chromosome 21 genes as to be tested in Alzheimer's disease.

We will successively consider recent results related to these two lines of research.

1. Chromosome 21 and Down syndrome

One or few genes on chromosome 21, when duplicated, are able to induce, with 100% likelihood, Alzheimer like neuropathology. It can be expected that the discovery of this gene orgenes will be a crucial step in our understanding of the pathogenesis of Alzheimer's disease. Both the identification of new genes on chromosome 21 and the study of chromosome 21 rearrangements in relation to their phenotypic expression contribute to advances in this domain.

1.1 Genes on chromosome 21 and hypotheses on their role in the pathogenesis of Alzheimer like neuropathology in Down syndrome

Among the genes located on chromosome 21 two have been particularly studied: one coding for the Cu-Zn superoxide dismutase SOD 1, and the other coding for the precursor of the amyloid peptide, APP.

Increased activity of SOD 1 in trisomy 21 was the first gene dosage effect to be described – i.e. a correlation between a gene copy number and the amount of its product (Sinet et al., 1976). Many hypotheses (Sinet, 1982) have been formulated on the potential role of an excess of SOD 1 in the rapid aging observed in Down patients. It was also assumed that excess brain SOD 1 might participate in the pathogenesis of Alzheimer's disease through increased production of H_2O_2. Recent experimental observations have reinforced these hypotheses: i) large pyramidal neurons which are potentially susceptible of degenerative processes in Alzheimer's disease have been shown by immunostaining (Delacourte et al., 1988) to contain higher amounts of Cu-Zn SOD than other brain cells. ii) In cellular models of SOD 1 overexpression (murine L-cells and neuroblastoma transfected with the SOD 1 gene) glutathione peroxidase is increased as is also observed in trisomy 21. iii) Transgenic mice with high levels of human Cu-Zn superoxide dismutase show an abnormal neuromuscular junction in the tongue as was observed on the tongue muscles of Down syndrome patients (Avraham et al., 1988).

The APP gene located proximally to SOD 1 gene (Blanquet et al., 1987), but distally to the locus for familial Alzheimer's disease (St. Georg Hyslop et al., 1987; Van Broeckhoven et al., 1988) encodes a precursor of the amyloid peptide which constitutes both in Down and Alzheimer patients the amyloid fibril protein of plaques, cerebral vessels, and perhaps enters in the composition of paired helical filaments. Two alternative forms of the precursor have been identified with insertion of either a 56 amino acid or a 75 amino acid additional region, similar in both cases to Kunitz protease inhibitor sites. It is not yet possible to understand the implications of the different precursor forms in amyloidogenesis. In situ hybridization studies have shown that the gene is expressed in large neurons from the hippocampus and cortex (Higgins et al., 1988); an overexpression of the gene has been described in the Meynert nucleus basalis and locus coeruleus from AD patients (Cohen et al., 1988; Palmert et al., 1988). Further studies are needed to know whether the increased expression seen in the brain cells of Down syndrome patients is linked to the deposition of A 4 protein in the plaques and whether there is a similar mechanism in specific cells of Alzheimer patients.

1.2 Chromosome 21 imbalance and Down syndrome pathogenesis

As described before, Down syndrome is usually the clinical expression of a trisomy for the whole chromosome 21 in all the cells of affected individuals. This definition applies to the majority of Down syndrome patients. Exceptions are nevertheless observed and, although very rare, some are of considerable interest with regard to the definition of one or several regions on chromosome 21, the genetic content of which could be crucial for the pathogenesis of specific features of Down syndrome.

1.2.1 Trisomy 21 mosaicism

Trisomy 21 mosaicism has been observed in lymphocytes and/or fibroblasts of Down syndrome patients.

Although not exceptional $(2-3\%)$ (Grouchy and Turleau, 1982) the exact proportion of these cases among the Down population is difficult to assess, since some may have a very small number of cells with 46 chromosomes, which may not be detected by usual karyotypic analysis. Also, mosaicism varies among tissues and with time, at least in lymphocytes where there is a tendency to a decreased proportion of trisomic 21 cells with age (Taylor, 1968). Therefore, the phenotypic expression of Down syndrome in mosaic individuals probably depends on the ratio between normal and trisomic 21 cells in certain tissues at a certain time of the foetal development. Similarly, mosaicism for trisomy 21 may not have a pathological expression, depending on the tissue affected by mosaicism.

Indeed, phenotypically normal individuals have been shown to have trisomy 21 mosaicism in lymphocytes and/or fibroblasts (Uchida and Freeman, 1985; M. Prieur, personal communication) and, in some cases, in gonads (Uchida and Freeman, 1985). One might then consider the possibility that a somatic trisomy 21 mosaicism affecting specific populations of neurons could be a potential etiopathogenic factor in some cases of Alzheimer's disease.

1.2.2 Partial trisomies 21 and definition of a "Down syndrome region" on chromosome 21

The duplication of a specific region of chromosome 21 could be responsible for the main features of Down syndrome. Indeed, in very rare cases (less than 1/1000), Down syndrome is associated to a partial trisomy 21. Karyotypic analysis of such cases has shown that only the distal part of chromosome 21, band 21 q 22, is involved in the pathogenesis of the syndrome (Aula et al., 1973). In addition to individuals with partial trisomy for the entire band 21 q 22, cytogenetic studies have identified two other groups of patients characterized by either a duplication including the proximal part of the band 21 q 22 (Poisonnier et al., 1976; Cantu et al., 1980; Jenkins et al., 1983) or a duplication of only the distal part of 21 q 22 (Mattei et al., 1981; Habedank and Rodewald, 1982). These observations could be explained by a duplication of only a portion of the band 21 q 22, adjacent to sub-band 21 q 22.2, critical for the expression of Down syndrome (Park et al., 1987), which is present in both groups. We tested this hypothesis by studying, at the molecular level, one patient from each group. Both patients (Poissonier, 1978; Mattei et al., 1981) had many features of Down syndrome associated with partial duplication of distinct regions of chromosome 21, respectively q 11.205 − q 22.300 and q 22.300 − qter (Fig. 1). The number of copies of DNA sequences located on chromosome 21, i.e. SOD-1, D21S17, D21S55, ETS 2, and D21S15 (Fig. 1) was assessed by using a "slot blot" method that we designed as an alternative to other methods of gene dosage, which does not require DNA digestion or Southern blotting (Rahmani et al., 1989; Blouin et al., 1989). Results indicated that the D21S55 sequence was duplicated in both patients (Fig. 1). By means of pulsed field gel analysis and knowing the regional mapping of the probes D21S17, D21S55 and ETS 2, it was possible to estimate the size of the common duplicated region between 400 Kb and 3000 Kb. This region, located in the proximal part of 21 q 22.3 (Fig. 1) is suspected to contain genes, the overexpression of which is crucial in the pathogenesis of at least the following features of trisomy 21: Down facies, hypotonia, short and incurved 5th finger, gap between

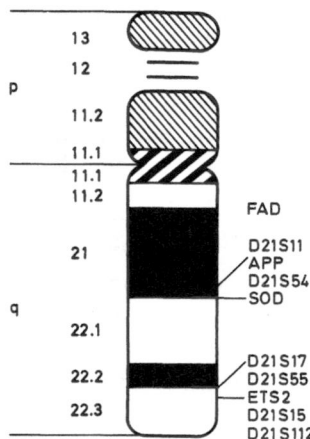

Fig. 1. Location of human chromosome 21 sequences

the 1st and 2nd toe and mental retardation. These data are consistent with preliminary reports (Korenberg et al., 1988; Mc Kirmink et al., 1988) on the molecular analysis of patients with partial trisomy 21 suggesting that the "Down syndrome region" includes 21 q 22.3 and extends proximally to distal 21 q 22.1. Keeping in mind that Down syndrome is a heterogeneous and complex clinical disorder, it is likely that other genes, located outside the D21S55 region, may also play a role.

An obvious and crucial question, which has yet to be answered, is to know whether both patients or only one or even none of them will develop Alzheimer like neuropathology when they get older. The first eventuality would occur if the gene(s) involved in this process is (are) located within the D21S55 region. The second eventuality would correspond to the gene(s) located either proximally or distally to the D21S55 region. The third possibility cannot be excluded if duplication of several genes, located on both sides of D21S55, is required for the pathogenesis of Alzheimer like neuropathology in Down syndrome. Among known genes or loci on chromosome 21, APP and SOD 1 genes (see chapter 1.1) and FAD locus (see chapter 2.1), both proximal to D21S55 (Fig. 1) are good candidates for playing a role in this process.

1.2.3 Down syndrome without visible karyotypic anomaly

Two observations of patients with features of trisomy 21 but no apparent modification of the karyotype have been recently reported (Huret et al., 1987; Delabar et al., 1987). No mosaicism was detected in the lymphocytes and fibroblasts of these patients. In both cases, SOD 1 activity in red cells was within the range of the trisomy 21 values. Molecular analysis of the DNA suggested a duplication of the SOD 1 gene. For one of these patients, SP (Huret et al., 1987), human/CHO hybrid cells were obtained by fusion of CHO cells with SP's fibroblasts (D. Patterson, Eleanor Roosevelt Institute, Denver, CO, USA). Two clones, PRC 13 and PRC 20, were selected, each containing only one of the two SP's chromosome 21. Gene dosage indicated that PRC 13 had two SOD 1 gene copies whereas PRC 20 had only one. Another sequence, D21S54, very closely linked to SOD 1 on chromosome 21, was also found to be duplicated in PRC 13. As previously suggested (Huret et al., 1987), these data (Rahmani et al., 1989 b) indicate that one of the chromosomes 21 in this patient carries a microduplication of a segment of chromosome 21 containing SOD 1 and D21S54.

Analysis of the DNA of other patients with Down syndrome features and normal karyotype (4 unpublished cases under study in our laboratory, and Mc Kormik et al., 1988) has failed to show any duplication among 14 tested chromosome 21 probes.

These observations suggest at least two remarks: i) As discussed previously, Down syndrome is defined by the association of a number of features usually observed in trisomy 21. In the absence of an extra chromosome 21, a detailed clinical examination is required to describe precisely the phenotype. As suggested (Rahmani et al., 1989 c), the use of the Jackson's checklist plus additional signs (dermatoglyphic patterns, mental retardation, etc.) is the best way to fulfill this task. As different genes are likely to be involved in different features of Down syndrome, on detailed phenotypic descriptions of the patients will allow to establish possible links between phenotype and genotype. ii) Karyotypic analysis by high resolution banding techniques must be carried out in lymphocytes and fibroblasts of the patients and, when possible, their parents in order to avoid two potential pitfalls: − the presence in the patient of an unbalanced translocation on another chromosome of a sub-band of chromosome 21, the size of which represents the limit of detection of a chromosome rearrangement by cytogenetic analysis. − the presence of a trisomy 21 mosaicism. Thus, two siblings, brother and sister, have been reported to have typical Down syndrome and chromosome analyses gave the following results: normal karyotype in the lymphocytes (100 counted cells) and fibroblasts (150 counted cells) in the boy; normal karyotype in 118 cells and 47, XX, + 21 in 3 cells in the lymphocytes of the girl (Parloir et al., 1979). The finding of a limited number of trisomy 21 cells in the sister supports the hypothesis that, in both patients, Down syndrome is due to a normal/trisomy 21 mosaicism, not easily detectable in tissues accessible to karyotypic analysis.

Thus, some of the Down patients with normal karyotype may have either an unbalanced translocation of a small part of chromosome 21 or an undetectable trisomy 21 mosaicism. In addition, the observation of patient SP suggests that microduplications on chromosome 21 may be associated with a Down phenotype.

Conversely, microduplication of chromosome 21 sequences might be present in people without either Down syndrome features or any other apparent pathology. Preliminary results (J. L. Blouin et al., unpublished) on DNA analysis of the mother of one of the Down syndrome patients with normal karyotype, under study in our laboratory, support this hypothesis, showing a duplication for a chromosome 21 gene both in blood and in fibroblasts.

2. Chromosome 21 and Alzheimer's disease

Alzheimer like neuropathology in Down syndrome results from the presence of an extracopy of one or few genes on chromosome 21, the nature of which is still unknown. As was extensively documented for many genes on chromosome 21, gene overdosage is associated with gene overexpression in trisomy 21. One can then speculate that Alzheimer's disease might, at least in certain patients, be secondary to a chromosome 21 defect susceptible of inducing the overexpression of this gene or genes. Experimental approaches for testing this hypothesis have been recently developed, both in familial and sporadic Alzheimer's disease.

2.1 Genetic defect predisposing to familial Alzheimer's disease (FAD) on chromosome 21

A link between Down syndrome and Alzheimer's disease has directed the search for a genetic defect in familial Alzheimer's disease towards chromosome 21. Using linkage analysis with DNA restriction fragment length polymorphism in families in which early onset Alzheimer's disease segregates as a dominant character, a genetic defect has been localized on chromosome 21. The locus of this defect, or FAD locus, is close to the centromere and distant from the APP gene and the Down syndrome critical region (Fig. 1). First reported by St George-Hyslop et al. in 1987, this data has been confirmed by two other groups (Van Broeckhoven et al., 1988; Goate et al., 1989) in different early onset families. Meanwhile, two separate studies on late onset families were unable to demonstrate a link to chromosome 21 (Pericak-Vance et al., 1988; Schellenberg et al., 1988). These last results

led to the suggestion that familial Alzheimer's disease is a heterogeneous disorder, with different chromosome loci in its late and early
onset forms. However, no evidence of a second FAD locus has been
reported yet.

Families in which Alzheimer's disease segregates as autosomal
dominant character are rare. Many of the Alzheimer cases appear
sporadically, that is with no evidence of heritability of the disease.
This suggests at least two hypotheses. Alzheimer's disease might not
be the result of a modification of the genome, but could be simply
due to the action of environmental factors, such as toxins or virus
infection. Although this cannot be excluded, there is so far no real
evidence for a predominant role of environment in the pathogenesis
of the disease. The other possibility is that there is a modification of
the genome, which, however, may be not expressed as a fully dominant
trait in all individuals or not transmissible, i.e. restricted to somatic
cell lineages. As discussed previously, the hypothesis of a genetic defect
implying chromosome 21 or/and of a dysregulation in the expression
of chromosome 21 genes has to be tested in sporadic Alzheimer's
disease.

2.2 Chromosome 21 imbalance in sporadic Alzheimer's disease

As a microduplication of a small segment of chromosome 21 with
overexpression of the duplicated gene, i.e. SOD 1, has been described
in patients with Down syndrome with normal karyotype (Huret et al.,
1987; Delabar et al., 1987), the hypothesis of similar genetic defects in
Alzheimer's disease was tested by several groups. Preliminary results
in a limited number of sporadic cases suggested duplications of SOD 1
(Schweber et al., 1987), APP (Delabar et al., 1987) or ETS 2 (Delabar
et al., 1987). Subsequent reports (St George-Hyslop et al., 1987; Tanzi
et al., 1987; Podlisny et al., 1987; Sacchi et al., 1988) showed no evidence
for such duplications in a large number of sporadic and familial Alzheimer patients. With the use of our new "slot blot" method for gene
dosage (Rahmani et al., 1989 a; Blouin et al., 1989), we analysed blood
DNAs from 5 patients with sporadic Alzheimer's disease, two age-
matched controls and one trisomy 21 subject. The clinical diagnosis
of Alzheimer's disease was based on criteria defined by the NINCDS.
Evaluation of the copy number of four chromosome 21 sequences,
APP, SOD 1, D21S17, and ETS 2 (Table 1, Fig. 1) gave no evidence
of gene duplication. Similar gene dosage experiments on fibroblasts
DNA from the same patients, control and trisomy 21 individuals are
in progress. Preliminary results on one patient are intriguing. Two
skin biopsies were performed and the samples were cultured separately.

Table 1. Quantification of the copy number of 4 CHR 21 sequences in blood DNA from 5 AD patients (AD), 2 normal controls (N), 1 trisomy 21 patient (T 21)

Subjects	CHR 21 Probes			
	APP	SOD 1	D 21 S 17	ETS 2
1 T21	nd	nd	3	3
2 N	ni	2	2	2
3 AD	2	nd	i	i
4 AD	2	2	2	2
5 N	2	2	2	2
6 AD	2	2	2	2
7 AD	2	2	2	2
8 AD	nd	2	2	2

The quantification of each sequence was assessed on two different membranes with two different reference probes either COL 1 A 1 or COL 1 A 2. All the experiments were run on coded samples. *Nd* not determined; *i* intermediate values for the slope of the (X) regression line between the (N) and (T 21) regression lines. *ni* not informative

In culture 1, gene dosage was normal for all tested chromosome 21 probes. In culture 2, all tested chromosome 21 sequences were found in 1 copy instead of two. Restriction fragment length polymorphysm with D21S11/EcoRI and D21S112/Rsa I showed an almost complete loss of one of the two alleles, confirming that a monosomy 21 clone was predominant in this culture. Further studies are required to know whether this monosomic clone resulted from the occurrence of an aneuploidy during the culture or pre-existed in the skin of the patient. Studies of other patients will be necessary to assess the significance of these data with regard to a possible genetic instability of fibroblasts in culture or to the presence of aneuploidies for chromosome 21 (monsomy and/or trisomy) in skin fibroblasts, two events which could be linked to the aging process itself or more specifically to Alzheimer's disease.

2.3 Expression of chromosome 21 genes in fibroblasts of patients with sporadic Alzheimer's disease: application to SOD 1

As previously discussed, Alzheimer's disease might be associated, at least in certain cases, with overexpression of chromosome 21 genes. Recent publications reporting an increase in APP mRNA levels in hippocampus (Higgins et al., 1988) nucleus basalis (Cohen et al., 1988;

Palmert et al., 1988) and locus coeruleus (Palmert et al., 1988) neurons of Alzheimer patients support this hypothesis. If changes in the expression of chromosome 21 genes in Alzheimer's disease are due to a genetic defect, they should be observed in other tissues besides brain. To investigate this hypothesis, we have undertaken studies on the expression at the mRNA level of several chromosome 21 genes in fibroblasts of patients suffering from sporadic Alzheimer's disease compared to age-matched control and Down syndrome subjects. Results on the quantification of SOD 1 mRNAs have already been obtained. In a first study (Delabar et al., 1988 a, b), including 5 patients, 3 controls and 2 trisomy 21 subjects, Northern blot analysis showed that the ratio between SOD 1 mRNA and actin mRNA was significantly increased in the trisomy 21 subjects and in 4 of the 5 patients as compared to controls. Another series of experiments has been recently performed with the same cultures, using a slot blot method for the quantification of mRNAs (Table 2). Based on glyceraldehyde-3-phosphate dehydrogenase (gene on chromosome 12) mRNA levels, SOD 1 mRNA amounts were found tube significantly increased in Down syndrome patients, comprised between control and Down levels in 4 patients and above the Down levels in one patient. Thus, our results indicate that fibroblastic SOD 1 mRNA amounts are higher in Alzheimer patients than in controls. Several mechanisms are possible for explaining such a difference: i) an increase in SOD 1 gene dosage resulting from either a trisomy 21 mosaicism or a duplication limited to the SOD 1 locus. This last possibility seems unlikely since SOD 1 gene dosage in blood DNA of these patients was found normal. ii) a change in the regulation of mRNA synthesis or stability. In any case, these data strongly suggest the presence of a genetic defect in the

Table 2. CuZn-SOD mRNA quantification: slot blots of mRNA prepared from fibroblasts of controls (3), AD patients (5) and trisomy 21 individuals (3) were successively hybridized with a CuZn-SOD cDNA and a GAPD cDNA probes. The table indicates the slopes of the regression lines obtained after the graphical and the statistical analysis of the results

Subjects	C (3)	X	D (3)
I	0.26 + 0.006	0.92 + 0.042	0.40 + 0.03
II	0.15 + 0.005	0.36 + 0.014	1.01 + 0.034
III	0.07 + 0.005	0.34 + 0.016	0.57 + 0.019
IV	0.08 + 0.006	0.19 + 0.02	0.85 + 0.038
V	0.23 + 0.012	0.44 + 0.035	1.7 + 0.15

fibroblasts of these Alzheimer patients, the identification of which would lead to a significant improvement in our understanding of the pathogenesis of Alzheimer's disease.

Acknowledgements

We are indebted to Centre National de la Recherche Scientifique, Ministère de la Recherche, Ministère de la Défense, Faculté de Necker and Bayer Pharma France for financial support. We thank Dr. Prieur for helpful discussions, Dr. M. Roudier, Dr. Y. Lamour and Dr. P. Davous for providing us with blood samples and skin biopsies from Alzheimer patients, Dr. J. Lejeune and Dr. M. O. Rethoré for providing us with blood samples of trisomy 21 patients. We thank respectively Y. Groner, Stehelin, D. Goldgaber, G. Stewart, F. Ramirez and J. L. Mandel for the gift of their SOD1, ETS2, APP, D21S17, Collagen and GAPD probes.

References

Anderson BH, Breinburg D, Downes MJ, Green PJ, Tomlinson BE, Ulrich J, Wood JN, Kahn J (1982) Monoclonal antibodies show that neurofibrillary tangles and neurofilaments share antigenic determinants. Nature 298: 84–86

Aula P, Leisti J, von Koskull H (1973) Partial trisomy 21. Clin Genet 4: 241–251

Avraham KB, Schickler M, Sapoznikov D, Yarom R, Groner Y (1988) Down's Syndrome: abnormal neuromuscular junction in tongue of transgenic mice with elevated levels of human Cu-Zn superoxide dismutase. Cell 54: 823–829

Blanquet V, Goldgaber D, Turleau C, Creau-Goldberg N, Delabar JM, Sinet PM, Roudier M, de Grouchy J (1987) The B amyloid protein cDNA hybridizes in normal and Alzheimer individuals near the interface of 21 q 21 and 21 q 22. 1. Ann Genet 30: 68–69

Blouin JL, Rahmani Z, Chettouh Z, Prieur M, Fermania J, Poissonnier M, Leonard C, Nicole A, Mattei JF, Sinet PM, Delabar JM (1990) Slot blot method for the quantification of DNA sequences and mapping of chromosome rearrangements: application to chromosome 21. Am J Hum Genet 46: 518–526

Cantu JM, Hernandez A, Plascencia L, Vaca G, Moller M, Rivera H (1980) Partial trisomy and monosomy 21 in an infant with an unusual de novo 21/21 translocation. Ann Genet 23: 183–186

Ceballos I, Delabar JM, Nicole A, Kamoun P, Sinet PM (1988) Expression of transfected human superoxide dismutase in mouse L-cells and NS20Y neuroblastoma cells induces enhancement of glutathione peroxidase activity. Biochim Biophys Acta 949: 58–64

Cohen ML, Golde TE, Usiak MF, Younkin LH, Younkin SG (1988) In situ hybrization of nucleus basalis neurons shows increased B-amyloid mRNA in Alzheimer disease. Proc Natl Acad Sci USA 85: 1227–1231

Delacourte A, Defossez A, Ceballos I, Nicole A, Sinet PM (1988) Preferential localization of copper zinc superoxide dismutase in the vulnerable cortical neurons in Alzheimer's disease. Neurosci Lett 92: 247–253

Delabar JM, Sinet PM, Chadefaux B, Nicole A, Gegonne A, Stehelin D, Fridlansky F, Creau-Goldberg N, Turleau C, Grouchy J de (1987 a) Submicroscopic duplication of chromosome 21 and trisomy 21 phenotype (Down syndrome). Hum Genet 76: 225–229

Delabar JM, Goldgaber D, Lamour Y, Nicole A, Huret JL, Grouchy J de, Brown P, Gajdusek DC, Sinet PM (1987 b) Amyloid gene duplication in Alzheimer's disease and karyotypically normal Down syndrome. Science 235: 1390–1392

Delabar JM, Lamour Y, Gegonne A, Davous P, Roudier M, Ceballos I, Amouyel P, Stehelin D, Sinet PM (1986) Rearrangement of chromosome 21 in Alzheimer's disease. Ann Genet 29: 226–228

Delabar JM, Rahmani Z, Blouin JL, Nicole A, Creau-Goldberg N, Ceballos I, Grouchy J de, Huret JL, Lamour Y, Roudier M, Davous P, Sinet PM (1988 a) Gene dosage in Down syndrome and Alzheimer's disease. In: Brown P, Bolis CL, Gajdusek D (eds) Discussions in neurosciences. Magistrati PJ, V: 3: 97–101

Delabar JM, Rahmani Z, Blouin JL, Nicole A, Creau-Goldberg M, Ceballos I, Lamour Y, Roudier M, Davous P, Sinet PM (1988 b) Molecular analysis of patients with Alzheimer's disease. In: Sinet PM, Lamour Y, Christen Y (eds) Research and perspectives in Alzheimer's disease. Springer, New York, pp 151–156

Glenner GG, Wong MD (1984) Alzheimer's disease and Down syndrome: sharing of a unique cerebrovascular amyloid fibril protein. Biochem Biophys Res Comm 122: 1131–1135

Goate AM, Haynes AR, Owen MJ (1989) Predisposing locus for AD on chromosome 21. Lancet i: 352–355

Grouche J de, Turleau C (1982) Atlas des maladies chromosomiques. Expension Scientifique Française, Paris

Habedank M, Rodewald A (1982) Moderate Down's syndrome in three siblings having partial trisomy 21 q 22.2 − qter and therefore no SOD-1 excess. Hum Genet 60: 74–77

Higgins GA, Lewis DA, Bahmanyar S, Goldgaber D, Gajdusek DC, Young WG, Morrison JH, Wilson MC (1988) Differential regulation of amyloid protein mRNA expression within hippocampal neuronal subpopulations in Alzheimer's disease. Proc Natl Acad Sci USA 85: 1297–1301

Huret JL, Delabar JM, Marlhens F, Aurias A, Nicole A, Berthier M, Tanzer J, Sinet P (1987) Down syndrome with duplication of a region of chromosome 21 containing the CuZn superoxide dismutase gene without detectable karyotypic abnormality. Hum Genet 75: 251–257

Jackson JF, North ER III, Thomas JG (1976) Clinical diagnosis of Down's syndrome. Clin Genet 9: 483–487

Jenkins EC, Duncan J, Wright CE, Gordano FM, Wilbur L, Wisniewski K, Sklower SL, French JH, Jones C, Brown WT (1983) A typical Down syndrome and partial trisomy 21. Clin Genet 24: 97–102

Korenberg JR, Pulst SM, Kawashima H, Ikeuchi T, Yamamoto K, Ogasawara
    N, Schonberg SA, West R, Kojis T, Epstein CJ (1988) Familial Down
    syndrome with normal karyotype: molecular definition of the region.
    Am J Hum Genet 43: A0439
Lai F, Williams R, Waltham MA (1987) Alzheimer's dementia in Down's
    syndrome. Neurology 37: 332
Lejeune J, Gautier M, Turpin R (1959) Etude des chromosomes somatiques
    de neuf enfants mongoliens. C R Acad Sci Paris 248: 1721–1722
McCormik MK, Schinzel A, Petersen MB, Mikkelsen M, Driscoll D, Cantu
    E, Stetten G, Watkins PC, Antonorakis SE (1988) Molecular genetic
    characterization of the "Down syndrome region" of chromosome 21.
    Am J Hum Genet 43: A 035
Mann DM (1988) Neuropathology of Alzheimer's disease: towards an un-
    derstanding of the pathogenesis. Biochem Soc Trans 17: 73–75
Mattei JF, Mattei MG, Beateman MA, Giraud F (1981) Trisomy 21 for the
    region 21 q 22.3: identification by high resolution R banding patterns.
    Hum Genet 56: 409–411
Palmert MR, Golde TE, Cohen ML, Kovacs DM, Tanzi RE, Gusella JF,
    Usiak MF, Younkin LH, Younkin SG (1988) Amyloid protein precursor
    messenger RNAs differential expression in Alzheimer's disease. Science
    241: 1080–1084
Park JP, Wurster-Hill DH, Andrews PA, Cooley WC, Graham JM Jr (1987)
    Free proximal trisomy 21 without the Down syndrome. Clin Genet 32:
    342–348
Parloir C, Fryns JP, Van den Berghe (1979) Down's syndrome in brother
    and sister without evident trisomy 21. Hum Genet 51: 227–230
Pericak-Vance MA, Yamaoka LN, Haynes CS (1988) Genetic linkage studies
    in Alzheimer disease families. Exp Neurol 102: 271–279
Podlisny MR, Lee G, Selkoe DJ (1987) Gene dosage of the amyloid B
    precursor protein in Alzheimer's disease. Science 238: 669–671
Poissonier M, Saint-Paul B, Dutrillaux B, Chassaigne M, Gruyer P, Blignières-
    Strouk G de (1976) Trisomie 21 partielle (21 q 21–21 q 22.2). Ann Genet
    19: 69–73
Rahmani Z, Blouin JL, Creau-Goldberg N, Wahkins PC, Mattei JF, Pois-
    sonnier M, Prieur M, Chettouh Z, Nicole A, Aurias A, Sinet PM, Delabar
    JM (1989 a) Critical role of the D21S55 region on chromosome 21 in the
    pathogenesis of Down syndrome. Proc Natl Acad Sci USA 86: 5958–5962
Rahmani Z, Barton J, Sinet PM, Huret JL, Bertier M, Patterson D, Delabar
    JM (1989 b) Submicroscopic duplication of SOD 1 on individual chro-
    mosome 21 cloned in a human/CHO hybrid established from a patient
    with mild Down syndrome phenotype. Cytogenet Cell Genet 51: 1063
Sacchi N, Nalbantoglu J, Sergovich FP, Papas TS (1988) Human ETS 2 gene
    is not rearranged in Alzheimer disease. Proc Natl Acad Sci USA 85:
    7675–7679
St Georg-Hyslop T, Tanzi R, Polinsky R, Haines JL, Nee L, Watkins PC,
    Myers RM, Feldman RG, Pollen D, et al (1987) The genetic defect causing
    familial Alzheimer disease maps on chromosome 21. Science 235: 885–890

Schellenberg GD, Bird TD, Wijsman EM, Moore DK (1988) Absence of linkage of chromosome 21 q 21 markers to familial Alheimer's disease. Science 241: 1507–1510

Schweber M, Tuson C, Shiloh R, Ben-Neriah Z (1987) Triplication of chromosome 21 material in Alzheimer's disease. Neurology 37 [Suppl 1]: 222

Sinet PM, Couturier J, Dutrillaux B, Poissonnier M, Raoul O, Rethoré MO, Allard D, Lejeune J, Jerome H (1976) Trisomie 21 et superoxide dismutase (IPOA). Exp Cell Res 97: 47–55

Sinet PM (1982) Metabolism of oxygen derivatives in Down syndrome. Ann NY Acad Sci 396: 83–94

Tanzi RE, Bird ED, Latt SA, Neve RL (1987) The amyloid protein gene is not duplicated in brains from patients with Alzheimer's disease. Science 238: 666–669

Taylor A (1968) Cell selection in vivo in normal/G trisomic mosaics. Nature 219: 1028–1031

Uchida IA, Freeman VCP (1985) Trisomy 21 Down syndrome — Parental mosaicism. Hum Genet 70: 246–248

Van Broeckhoven C, van Hul W, Backoven H, van Camp G, Vandergerghe A (1988) The familial Alzheimer disease gene is located close to the centromere of chromosome 21. Am J Hum Genet 43: A 205

Wisniewski KE, Dalton AJ, Crapper Mc Lachlan DL, Wen GY, Wisniewski HM (1985) Alzheimer's disease in Down's syndrome: clinicopathological studies. Neurology 35: 957–961

Wolozin B, Scicutella A, Davies P (1988) Reexpression of a developmentally regulated antigen in Down syndrome and Alzheimer's disease. Proc Natl Acad Sci USA 85: 6202–6206

Yates CM, Simpson J, Maloney AFJ, Gorden A, Reid AH (1980) Alzheimer like cholinergic deficiency in Down syndrome. Lancet ii: 979

PET scanning for the detection of Alzheimer's disease

W.-D. Heiss, B. Szelies, R. Adams, J. Kessler, G. Pawlik, and **K. Herholz**

Max-Planck-Institut für neurologische Forschung und Universitätsklinik für Neurologie, Köln (Lindenthal), Federal Republic of Germany

Summary

At present, PET is the only technology affording the quantitative, three-dimensional imaging of various aspects of brain function. Since function and metabolism are coupled, and since glucose is the dominant substrate of the brain's energy metabolism, studies of glucose metabolism by PET of $2(^{18}F)$-fluoro-2-deoxy-D-glucose (FDG) are widely applied to the investigation of the participation of various brain systems in simple or complex stimulations and tasks. In focal or diffuse disorders of the brain, functional impairment of affected or inactivated brain regions is a reproducible finding.

While glucose metabolism slightly decreases with age to a regionally different degree, in most types of dementia severe changes in glucose metabolism are observed. Degenerative dementia of the Alzheimer type is characterized by a metabolic disturbance most prominent in the parieto-occipito-temporal association cortex and later in the frontal lobe, while primary cortical areas, basal ganglia, thalamus, brainstem and cerebellum are not affected. Thanks to this typical pattern Alzheimer's disease can be differentiated from other dementia syndromes, such as Pick's disease (with the metabolic depression mostly prominent in the frontal and temporal lobe), multi-infarct dementia (with multiple focal metabolic defects), Huntington's chorea (with metabolic disturbances in the neostriatum) and other diseases leading to cognitive impairment with more or less typical metabolic patterns. A ratio calculated form CMRGl of affected (temporo-parieto-occipital and frontal association cortex) and non-affected brain regions (primary cortical areas, brainstem, cerebellum) enabled us to separate clearly AD patients from age-matched controls and to discriminate those patients suffering from cognitive impairment of other origin in 82% of the cases. The discrimination power can be further improved by specific activation studies. In demented patients PET can also be used to assess the effects of treatment on disturbed metabolism. Such studies demonstrated an equalization of metabolic heterogeneities in patients responding to muscarinic cholinergic agonists, as well as a diffuse increase of metabolism during treatment with piracetam and phosphatidylserine. The therapeutic relevance of such metabolic effects, however, remains to be proved in controlled clinical trials.

Introduction

Since disturbances of cerebral function are followed by changes in metabolism and blood flow, and pathological impairments of blood supply and energy metabolism themselves lead to functional deficits, in many diseases of the CNS these parameters are measurably altered without it being possible to draw conclusions with respect to the etiology. Dementias, which are clinically manifested primarily as non-localizable disturbances of cerebral function, can hardly be diagnosed by conventional supplementary neurological investigations, which detect mainly localized morphologic lesions. Although regional structural cerebral injuries can be demonstrated to occur in many forms of secondary dementia, the degree of dementia often depends on functional disorders of cerebral regions not primarily affected by the disease. The primary (degenerative) diseases leading to dementia are accompanied by atrophic changes in the brain visible at CT only in the late stages. Progressive cell loss and reduced cell and synaptic activity lead to a reduction of metabolism and blood flow which can be visualized with the aid of functional imaging techniques. Since glucose is the most important substrate of cerebral energy metabolism, studies of glucose metabolism are currently the best method of detecting and quantifying functional disturbances of the brain. The glucose metabolic rate can be determined regionally and three-dimensionally in the brain by means of positron emission tomography.

Glucose metabolism in healthy subjects

The different rates of glucose metabolism in various regions of the brain depending on their functional activity have been determined in a number of studies (review by Heiss et al., 1984). The overall metabolic level depends to a great degree on internal (anxiety, vigilance) and external (illumination, environmental noise) conditions (Mazziotta et al., 1982) so that the resting conditions for the investigations must be defined. In our studies, which were carried out on subjects with their eyes closed in a darkened room with low levels of noise from equipment and manipulation, the mean glucose turnover rate of 42 normal subjects (age 43 ± 19.1 years, 14 women, 28 men) was $34.6 \pm 3.83 \, \mu mol/100 \, g$ cerebral tissue/min. Highly significant regional differences (Fig. 1 a) with values between 40 and 50 $\mu mol/100 \, g/$

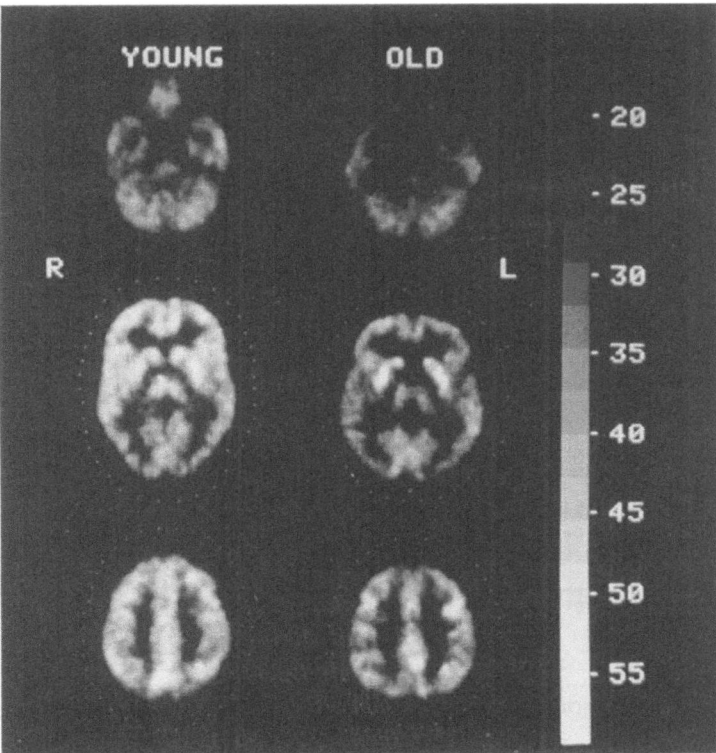

a

Fig. 1. a PET scans of glucose metabolism (μmol/100 g/min) according to gray scale in cerebral sections at the level of cerebellum, basal ganglia, thalamus and semioval centre in young (23 years) and old (67 years) healthy subjects. The individual brain structures can be differentiated according to different metabolic rates, metabolism decreases slightly in all regions in older patients

min were detected in the striatum, upper limbic system, insula, frontal cortex and primary visual cortex, between 35 and 40 μmol/100 g/min in the other gray structures of the hemispheres, between 30 and 35 μmol/100 g/min in the cerebellum and hippocampal structures and below 20 μmol/100 g/min in the medullary layer. Our studies confirm additionally a certain age dependency: the global rate of cerebral glucose metabolism showed a decline with advancing age which, although statistically significant (p < 0.05) represented less than 2% per decade (Fig. 1 b). A detailed analysis revealed that the individual brain regions were affected symmetrically but to rather different degrees

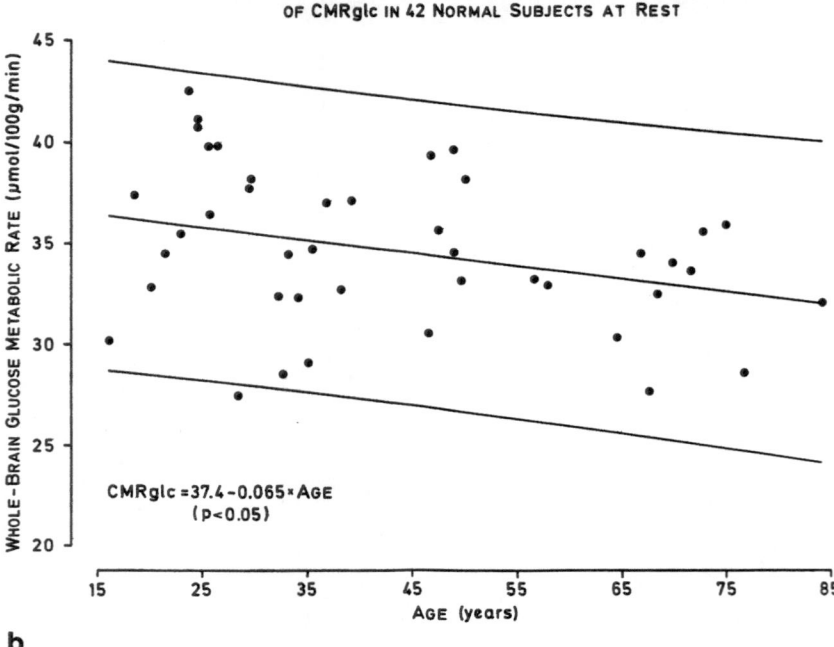

Fig. 1. b Decrease of mean global glucose metabolic rate in 42 healthy subjects with increasing age. The regression line shows a significant relationship despite the large range of variation

(p < 0.0001). After the subjects were divided into three age groups each comprising 14 normal subjects, the greatest age dependent changes were detected in the frontal cortex, the insula and the upper part of the limbic system, as well as in the parieto-temporal region and to a lesser extent also in the perirolandic region and medullary layer. No appreciable age-dependency could be demonstrated for the metabolism of the subcortical gray structures.

Metabolic disturbances in dementia syndromes

Dementias are a very heterogeneous group of diseases sharing some common features such as deterioration of intellectual function and memory and a degeneration of the personality, which are clearly differentiable from the usual age-related changes — slight forgetfulness, change in the intelligence structures — and greatly exceed them in scope. Clinical symptoms include impairment of learning ability and memory and reduction of attention, orientation, critical faculty and judgement. These are frequently associated with disturbances of visual-spatial orientation, speech functions and apraxia.

Primary degenerative dementias

Primary degenerative dementias of the Alzheimer type (AD) which are accompanied by a loss of cortical neurons for unknown reasons (Terry et al., 1981), disturbance of various transmitter systems (Rossor et al., 1982) but also by selective reduction of specific projection systems [especially the cholinergic system, (Coyle et al., 1983)] and also by typical pathological changes (plaques and fibrils), account for more than 50% of all dementia disorders. Patients with AD show a reduction of cerebral glucose metabolism [similar to oxygen utilization and blood flow, (Frackowiak et al., 1981)] proportional to the severity of the dementia, in which the reduced metabolism is detectable (Fig. 2) before the occurrence of atrophic changes in CT and shows significant regional differences: the bilateral local reductions are especially pronounced in the parieto-temporal and frontal cortex (Fig. 2) and do not affect the primary visual and sensorimotor cortex or the subcortical structures and the cerebellum (Kuhl et al., 1983; DeLeon et al., 1983; Friedland et al., 1983; Duara et al., 1986).

In studies which compared results of PET and neuropsychological tests, it was possible to demonstrate a relationship between leading symptoms and the localization of especially reduced glucose metabolism: when an aphasic disorder was predominant, glucose metabolic disturbance was more pronounced in the left than in the right parietal lobe, when apraxic symptoms predominated this parameter was disturbed more on the right than on the left, while on predominance of amnestic deficits no asymmetry was present (Foster et al., 1983).

In the second, but much rarer form of primary degenerative dementia, Pick's disease, the first and most marked metabolic changes — in analogy to the primary localization of pathological changes — are seen in the frontal and temporal lobe (Szelies and Karenberg, 1986). This distinctly different pattern of damage allows Pick's disease to be differentiated from AD; in moderate cases it is often not possible to make this distinction on the basis of clinical findings alone. The typical reduction of metabolism in the frequently asymmetrically affected atrophic frontal lobes and lower temporal lobes and in the much less changed parietal lobes as well as in the basal ganglia and the thalamus, is correlated with the degree of gliosis and cell depletion (Kamo et al., 1987).

This pattern of metabolic disturbance is also characteristic of Huntington's chorea which in addition to the extrapyramidal-hyperkinetic

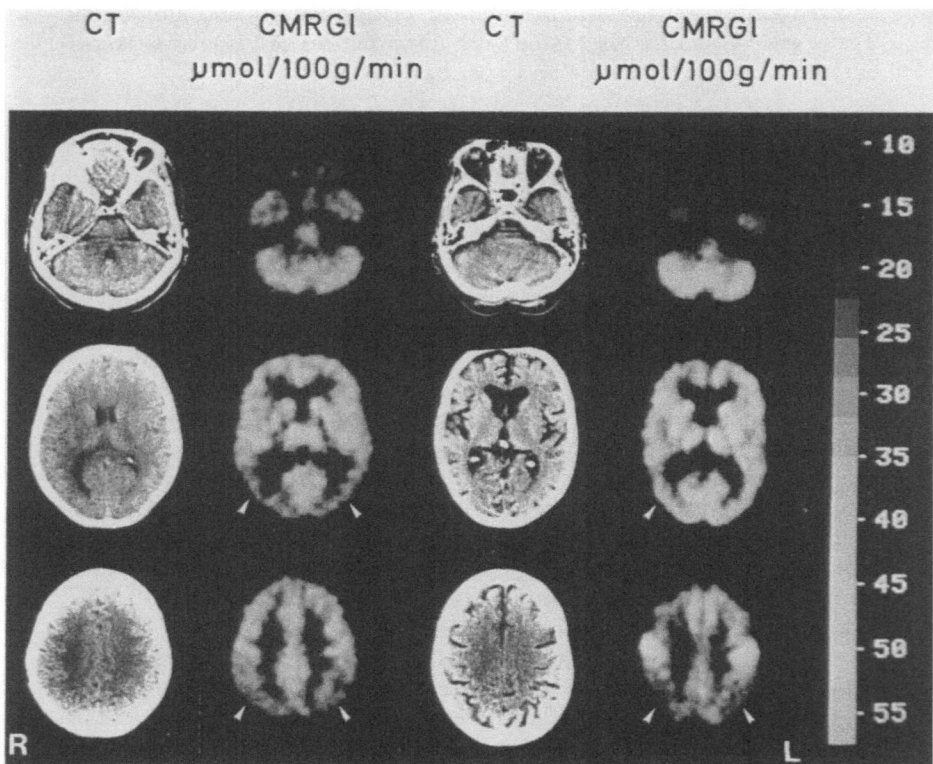

Fig. 2. CT and PET scans of glucose metabolism in patients with mild (left) and severe (right) Alzheimer's dementia. In mild AD metabolic rates in the normal CT are distinctly reduced in the parieto-occipito-temporal region, in severe dementia there is diffuse atrophy and pronounced reduction of cortical metabolism with recessing of primary somatosensory and visual areas and of basal ganglia/thalamus and cerebellum

syndrome is always accompanied by dementia disorders. The glucose turnover rate in the neostriatum is already significantly reduced in the early stages of this disease (Fig. 3) and as the severity and duration of the disease increase, metabolism is seen to be reduced in the nucleus caudatus and putamen, and later (according to the degree of severity of dementia) also in the cerebral cortex (Kuhl et al., 1984). Since the metabolic disturbances precede the clinical manifestations of the disease, PET studies may help to identify persons at risks in chorea families, and these studies as well as genetic investigations can be used

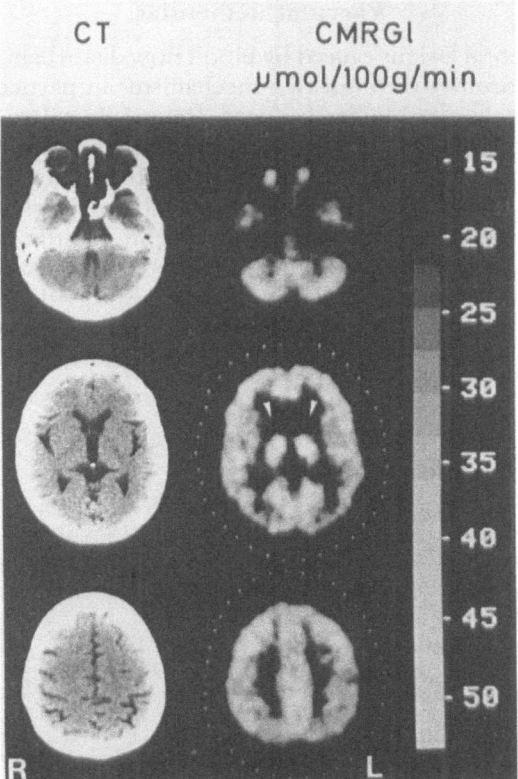

Fig. 3. CT and PET scans of glucose metabolism of patients with Huntington's chorea. In still non-pathological CT, metabolism is reduced in the neostriatum (caudate nucleus and putamen)

to establish a prognosis for the subsequent appearance of the disease (Hayden et al., 1987; Mazziotta et al., 1987).

In Parkinson's disease, a degeneration of the dopaminergic nigrostriatal system, glucose metabolism is usually not altered, in contrast to the reduction of the dopaminergic endings in the basal ganglia demonstrated by means of PET of ^{18}F-dopa (Nahmias et al., 1985). Only on development of a dementia, a frequent concomitant of Parkinson's disease, are the metabolic changes typical of AD also in evidence (Kuhl et al., 1985).

Vascular dementias

Focal cerebral lesions caused by blood flow disturbances can induce dementia syndromes through two mechanisms in particular: multiple lesions in mostly neurologically silent, frequently subcortical regions impair cerebral function in the form of dementia (multi-infarct dementia) when they exceed a total volume that cannot be precisely defined ($80-150\,\text{cm}^3$). In rare cases, relatively small infarcts of critical localization can cause a dementia syndrome in addition to the focally dependent neurological symptoms. Chronically inadequate blood flow in the cerebral tissue which leads to a persisting hypofunction and thus to a disturbance of intellectual function which cannot be precisely

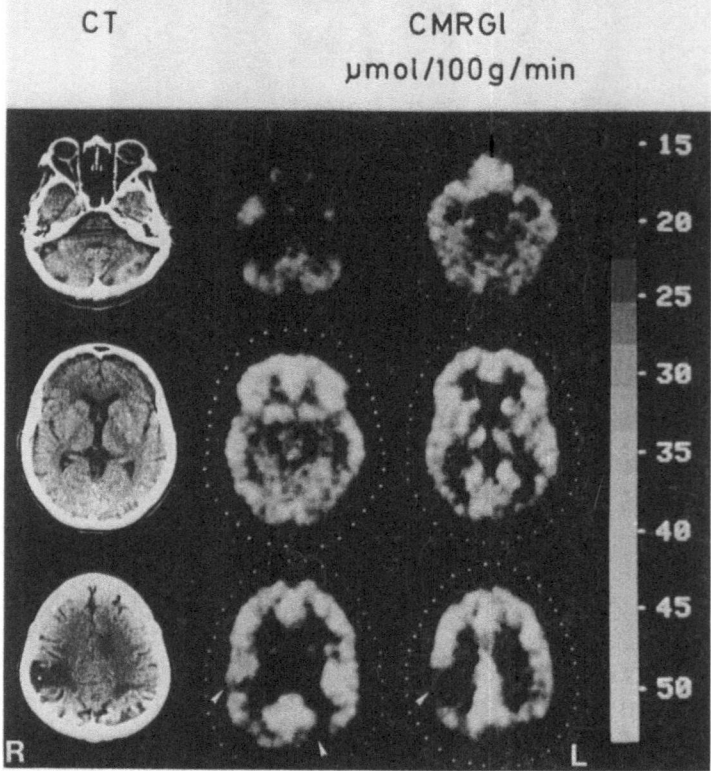

Fig. 4. CT and PET scans of glucose metabolism in patients with multiple infarcts: several morphological lesions (CT) cause regional metabolic disturbances in the infarcted area or in the superjacent deafferentated cortex

localized, is likely to be present only in exceptional cases or to be of temporary duration following transient blood flow disturbances. Such deficient perfusion syndromes probably are only very rarely a cause of dementia since in the usual forms no corresponding disproportions between blood flow and oxygen consumption or glucose metabolism could be demonstrated (Frackowiak et al., 1981; Gibbs et al., 1986).

Multi-infarct dementias (MID) together with the AD-MID mixed forms account for about 30% of all dementia syndromes. A clinical differentiation on the basis of rating scales (Hachinski et al., 1975) is often difficult, and diagnostic classification is often easier on the basis of morphological lesions demonstrated by CT or MRI. In MID patients, PET can clearly differentiate mostly multilocular metabolic reductions from the pattern typical of AD (Kuhl et al., 1983). Ischaemic lesions in the medullary layer in MID and Binswanger's disease can be detected with great sensitivity by means of T_2-weighted MRI (Heiss et al., 1986; Alavi et al., 1987), and the regions of reduced metabolism will then correspond to the superjacent deafferentated cortical areas (Fig. 4).

Small isolated infarctions in the cerebral regions which are particularly important for the integrity of the personality lead to disturbances of behaviour, affect, mood and intellectual performance. This is especially true of infarcts in the area supplied by the anterior cerebral artery, but also for small localized infarcts in strategically important regions, for example unilaterally in the anterior centre of the thalamus or bilaterally in the median thalamus: they also lead to permanent cognitive and amnestic losses.

Dementias of other etiology

Various other causes — inflammatory diseases, as herpes encephalitis, HIV encephalopathy, Creutzfeldt-Jakob disease; posttraumatic encephalopathy; toxic affections of the central nervous system; communicating hydrocephalus — can lead to severe dementias which then must be differentiated from those having more frequent etiologies. Their representation in PET does not usually follow a typical pattern — as AD or Pick's disease — or metabolic changes are not restricted to small areas — as in postinfarction states. Therefore, a differentiation from the more frequently occurring forms of dementia — e.g., AD and MID — can usually be achieved.

It is often difficult to differentiate the affective disorders with impairment of drive and psychomotor activity seen in depressions from similar symptoms in the early stages of dementia. The metabolic patterns observed in depressed patients are not comparable to the characteristic changes seen in dementias, particularly in AD. When the overall metabolic level is in relation to the mood (Baxter et al., 1987) there are sometimes regional differences of varying distribution; a pattern typical of depression or correlations between the metabolic values of certain regions with the severity of specific symptoms or function deficits have not so far been described.

Differentiation of AD from other dementias

In order to test the value of FDG-PET in the differential diagnosis of dementias, rCMRGl measurements of 19 patients with probable Alzheimer's disease according to the NINCDS-ARDA criteria (McKhann et al., 1984) were compared to those from 19 age-matched healthy subjects and 22 patients with cognitive impairment due to other diseases. In comparison to the 19 healthy controls the AD patients (age 49 to 71 years, cognitive deterioration for $1-5$ years, average mini-mental status 14.5 ± 7.3, average global deterioration scale 4.9 ± 0.9) were characterized by significantly lower global cerebral metabolic rates for glucose ($30.3 \pm 3.2 \mu mol/100 g/min$) and considerable metabolic asymmetries with the most remarkable decrease of regional CMRGl (in terms of Z-transforms) found in the supramarginal and angular gyrus, the adjacent parts of the superior temporal gyrus and the medial frontal gyrus. While abnormally low metabolism ($Z < -2$) in supramarginal/angular gyrus was observed in all AD patients and is a highly sensitive indicator of the disease, abnormal metabolism in temporal and frontal association areas supports the diagnosis, but is not mandatory.

However, despite the decrease of global CMRGl in AD, some regions maintained a strikingly normal metabolism even in most severely affected patients. Those regions were the cerebellum, brain stem, primary sensorimotor cortex and the occipital cortex including the cortex around the calcarine fissure. Notably, relative cerebellar CMRGl was normal (within 2 SD of normal range) in all 19 AD patients, and even absolute cerebellar rCMRGl was normal in 17 of the 19 AD patients.

Considering the contrast between typically affected and non-affected regions, a ratio R of the rCMRGl in those regions was calculated. Its average value in controls was 1.09 (SD 0.084, range 1.01 to 1.15) vs. 0.77 (SD 0.11, range 0.60 to 1.00), thus a complete separation of the two groups was achieved.

The diagnostic criteria of AD deduced from the metabolic pattern of FDG-PET-reduced temporo-parietal metabolism, normal cerebellar metabolism, reduced CMRGl ratio of temporo-parietal and frontal association areas to primary sensory areas, brainstem and cerebellum — were fulfilled only by 4 of the 22 patients with cognitive impairment of other origin than AD. One of those suffered from dementia in the course of Parkinson's disease, one had multiple infarcts and in two cognitive impairment was a sequel of diffuse hypoxic or hypoglycemic brain damage. The overall specificity of this diagnostic procedure in this sample was 82%. FDG-PET reaches a diagnostic sensitivity and specificity unsurpassed by other imaging modalities.

Since AD affects mainly and primarily parieto-temporo-occipital areas involved in the processing of visual information, a test procedure was developed to further improve the diagnostic sensitivity of PET measurements. When a continuous visual recognition task (Kessler et al., 1989) is performed during metabolic studies a significant activation is observed in normal controls averaging 24% for global CMRGl with the most prominent changes (38%) in occipital regions. In AD the global increase during the performance of an adapted visual recognition task was only 8%, and the changes mainly occurred in areas not primarily affected by the disease. This finding points to the impaired functional reserve capacity of the brain in degenerative dementia of the Alzheimer type. The pattern typical of AD — low CMRGl in parieto-temporo-occipital regions, high CMRGl in primary cortical areas, brainstem and cerebellum — becomes more obvious enhancing the diagnostic contrast between AD patients and normal age-matched controls or patients with non-AD dementia.

Evaluation of drug effects

The effect of therapeutic interventions in dementia syndromes is difficult to assess because degenerative diseases often progress slowly or intermittently or are interrupted by phases without appreciable deterioration. Because of the differences in the disease dynamics, it is also very difficult to make comparisons between individuals. Because metabolic disturbances show a characteristic distribution particularly in degenerative dementias of the Alzheimer type and furthermore are correlated with the severity and duration of the disease, and because the functional activity is reflected in metabolic values, measurements of this kind may be valuable in assessing the effects of drugs. Metabolic investigations could then also be useful in providing objective evidence of therapeutic results within a relatively short time, when clinical improvements or a slowing of the progression of the deficits would not yet be apparent.

In the last few years some principles have been elaborated for therapeutic strategies aimed at improving certain clinical deficits in

senile or presenile degenerative dementia of the Alzheimer type (AD).
These concepts include measures for substitution of cholinacetyltrans-
ferase deficiency that is presumed to be specific (Davies and Maloney,
1976) and the destruction of cholinergic neurons (Coyle et al., 1983).
This can be done by inducing a presynaptic increase in the synthesis
and release of acetylcholine, by inhibiting the breakdown of acetylcholine at the synapse and by postsynaptic stimulation of the acetylcholine receptors. All these therapeutic approaches centering on the
cholinergic activity have reportedly brought about improvements in
the memory disturbances typical of AD, with inhibition of cholinesterase with physostigmine (Davis et al., 1978) and the administration
of tetrahyodroaminoacridine in larger controlled clinical studies yield-
ing successful results in individual cases (Summers et al., 1986). Muscarinic agonists have also been able to improve the symptoms in cases
that are not so advanced (Szelies et al., 1986). In contrast, precursors
of acetylcholine were not effective if given as the sole form of therapy
(review by Kurz et al., 1986; Hollander et al., 1986) although specific
memory disturbances were improved by a combination with nootropic
substances which stimulate cerebral metabolism (Ferris et al., 1982;
Smith et al., 1984).

The use of PET to objectivize the effects of drugs is still rare and
has so far been limited to small groups of patients. Glucose metabolism
was monitored for six to twelve weeks in eight patients with AD of
different severity undergoing therapy with the muscarinic choline agonist (RS 86 Sandoz, 2.5 − 3 mg/d, Szelies et al., 1986). Over this period
the global metabolic rate decreased under therapy, but there was a
compensation of the heterogeneous metabolic pattern typical of AD
with a particular reduction of the slightly elevated values (sensorimotor
and visual cortex) measured before starting treatment and there was
only a slight influence on the typically lowered parieto-occipital to
temporal values. This effect was especially pronounced in patients who
became clinically stabilized using this therapy and showed improved
performance in several functions; this group which profited from the
therapy originally showed regional glucose metabolic rates diverging
relatively little from the norm and were also those with less severe
AD. This study therefore shows the importance of initiating therapy
at an early stage before severe cell destruction takes place and suggests
that a metabolic decoupling takes place between different regions of
the brain as the functional substrate of the specific symptoms. Another
study (Heiss et al., 1988) examined whether piracetam, which improves
memory performance when administered in combination with precursors of acetylcholine (Ferris et al., 1982; Smith et al., 1984) has
metabolic effects in AD. Of 16 patients with dementia syndrome (DSM-

III, American Psychiatric Association, 1980) nine fulfilled the criteria for AD (McKhann et al., 1984) and the remaining seven were graded as MID or unclassifiable and used as a control group. Between the PET investigations all patients received 6 g piracetam b.i.d. (Nootrop®, Chemie) for 14 days as a rapid infusion. The groups differed significantly from each other and from an age-matched control group with respect to regional rates of glucose metabolism, the reductions being particularly pronounced in the parieto-temporo-occipital regions for the AD groups. Under piracetam treatment the glucose metabolism values in the AD group increased in the frontal, central, parieto-occipital, visual, auditory and cingulate cortex, basal ganglia and thalamus whereas no significant changes were detected in the non-AD group. The differences in the effects of treatment between AD and non-AD groups were statistically significant (ANOVA $p < 0.02$ for interactions between regions, treatment and group); on the basis of the ANOVA, the increase in the individual regions was checked by paired t-test. The results were supported by improvements in five AD patients during the short therapy phase with respect to their clinical deficits and their performance in tests. Similar results were obtained in 8 AD patients under treatment with phosphatidylserine (FIDIA, 500 mg/d for 3 weeks) which is suggested to have an effect on membrane structure and cell function. In these patients rCMRGl increased during the treatment with the most significant effect in occipital areas ($+17\%$, $p < 0.05$). For all drugs shown to be effective on one aspect of AD — the regional disturbance of metabolism — controlled clinical studies will be needed in order to justify their clinical use.

References

Alavi A, Fazekas F, Chawluk J, Zimmerman R (1987) Magnetic resonance imaging of the brain in normal aging and dementia. In: Meyer JS, Lechner H, Reivich M, Ott EO (eds) Cerebral vascular disease 6. Excerpta Medica, Amsterdam New York Oxford, pp 191–195

American Psychiatric Association (1980) Diagnostic and statistical manual of mental disorders, 3rd edition (DSM-III). Washington DC, pp 124–126

Baxter LR, Phelps ME, Mazziotta JC, Guze BH, Schwartz JM, Selin CE, (1987) Local cerebral glucose metabolic rates in obsessive-compulsive disorder — a comparison with rates in unipolar depression and in normal controls. Arch Gen Psychiatry 44: 211–218

Coyle JT, Price DL, Delong MR (1983) Alzheimer's disease: a disorder of cortical cholinergic innervation. Science 219: 1184–1190

Davies P, Maloney AJF (1976) Selective loss of control cholinergic neurons in Alzheimer's disease. Lancet ii: 1403

Davis KL, Mohs RC, Tinklenberg JR, Pfefferbaum A, Hollister LE, Kopell
    BS (1978) Physostigmine: improvement of long-term memory processes
    in normal humans. Science 201: 272–274
DeLeon MJ, Ferris SH, George AE, Reisberg B, Christman DR, Kricheff
    II, Wolf AP (1983) Computed tomography and positron emission tran-
    saxial tomography evaluations of normal aging and Alzheimer's disease.
    J Cereb Blood Flow Metab 3: 391–394
Duara R, Grady C, Haxby J, Sundaram M, Cutler NR, Heston L, Moore A,
    Rapoport SI (1986) Positron emission tomography in Alzheimer's disease.
    Neurology 36: 879–887
Ferris SH, Reisberg B, Crook T, Friedman E, Schneck K, Mir P, Sherman
    KA, Corwin J, Gershon S, Bartus RT (1982) Pharmacologic treatment
    of senile dementia: choline, L-dopa, piracetam, and choline plus pira-
    cetam. Aging 19: 475–481
Foster NL, Chase TN, Fedio P, Patronas NJ, Brooks RA, Di Chiro G (1983)
    Alzheimer's disease: focal cortical changes shown by positron emission
    tomography. Neurology 33: 961–965
Frackowiak RSJ, Pozzilli C, Legg NJ, Du Boulay GM, Marshall J, Lenzi
    GL, Jones T (1981) Regional cerebral oxygen supply and utilization in
    dementia. A clinical and physiological study with oxygen-15 and positron
    tomography. Brain 104: 753–778
Friedland RP, Budinger TF, Ganz E, Yano Y, Mathis CA, Koss B, Ober
    BA, Huesman RH, Derenzo SE (1983) Regional cerebral metabolic al-
    terations in dementia of the Alzheimer type: positron emission tomog-
    raphy with (18F)fluorodeoxyglucose. J Comput Assist Tomogr 7:
    590–598
Gibbs JM, Frackowiak RSJ, Legg NJ (1986) Regional cerebral blood flow
    and oxygen metabolism in dementia due to vascular disease. Gerontology
    32 [Suppl 1]: 84–88
Hachinski VC, Iliff LD, Zilkha E, Du Boulay GH, Mc Allister VL, Marshall
    I, Ross Russell RW, Symon L (1975) Cerebral blood flow in dementia.
    Arch Neurol 32: 632–637
Hayden MR, Hewitt J, Stoessl AJ, Clark C, Ammann W, Martin WRW
    (1987) The combined use of positron emission tomography and DNA
    polymorphisms for preclinical detection of Huntington's disease. Neu-
    rology 37: 1441–1447
Heiss WD, Pawlik G, Herholz K, Wagner R, Göldner H, Wienhard K (1984)
    Regional kinetic constants and CMRGlu in normal human volunteers
    determined by dynamic positron emission tomography of (18F)-2-fluoro-
    2-deoxy-D-glucose. J Cereb Blood Flow Metab 4: 212–223
Heiss WD, Herholz K, Böcher-Schwarz HG, Pawlik G, Wienhard K, Stein-
    brich W, Friedmann Cr (1986) PET, CT, and MR imaging in cerebro-
    vascular disease. J Comput Assist Tomogr 10: 903–911
Heiss WD, Hebold I, Klinkhammer P, Ziffling P, Szelies B, Pawlik G,
    Herrholz K (1988) Effect of piracetam on cerebral glucose metabolism
    in Alzheimer's disease as measured by PET. J Cereb Blood Flow Metab
    8: 613–617

Hollander E, Mohs RC, Davis KL (1986) Cholinergic approaches to the treatment of Alzheimer's disease. Br Med Bull 42: 97–100

Kamo H, McGeer PL, Harrop R, McGeer EG, Calne DB, Martin WRW, Pate BD (1987) Positron emission tomography and histopathology in Pick's disease. Neurology 37: 439–445

Kessler J, Adams R, Herholz K, Szelies B, Heiss WD (1989) Impaired metabolic activation (FDG-PET) in patients with Alzheimer's disease under stimulation by continuous recognition. In: Aging of the brain and dementia: ten years later. Conf. World Federation Neurology Florenz, May 31–June 3, 1989

Kuhl DE, Metter EJ, Riege WH, Hawkins RA, Mazziotta JC, Phelps ME, Kling AS (1983) Local cerebral glucose utilization in elderly patients with depression, multiple infarct dementia, and Alzheimer's disease. J Cereb Blood Flow Metab 3 [Suppl 1]: S 494–S 495

Kuhl DE, Metter EJ, Riege WH, Markham CH (1984) Patterns of cerebral glucose utilization in Parkinson's disease and Huntington's disease. Ann Neurol 15 [Suppl]: S 119–S 125

Kuhl DR, Metter EJ, Benson DF, Ashford JW, Riege WH, Fujikawa DG, Markham CH, Mazziotta JC, Maltese A, Dorsey DA (1985) Similarities of cerebral glucose metabolism in Alzheimer's and Parkinsonian dementia. J Cereb Blood Flow Metab 5 [Suppl 1]: S 169–S 170

Kurz A, Rüster P, Romero B, Zimmer R (1986) Cholinerge Behandlungsstrategien bei der Alzheimer'schen Krankheit. Nervenarzt 57: 558–569

Mazziotta JC, Phelps ME, Carson RE, Kuhl DE (1982) Tomographic mapping of human cerebral metabolism: sensory deprivation. Ann Neurol 12: 435–444

Mazziotta JC, Phelps ME, Pahl JJ, Huang S-G, Baxter LR, Riege WH, Hoffman JM, Kuhl DE, Lanto AB (1987) Reduced cerebral glucose metabolism in asymptomatic subjects at risk for Huntington's disease. N Engl J Med 316: 357–362

McKhann G, Drachman D, Folstein M, Katzman R, Price D, Stadlan EM (1984) Clinical diagnosis of Alzheimer's disease. Neurology 34: 939–944

Nahmias C, Garnett ES, Firnau G, Lang A (1985) Striatal dopamine distribution in Parkinsonian patients during life. J Neurol Sci 69: 223–230

Rossor MN, Emson PC, Mountjoy CQ, Roth M, Iversen LL (1982) Neurotransmitters of the cerebral cortex in senile dementia of Alzheimer type. Exp Brain Res [Suppl 5]: 153–157

Smith RC, Vroulis G, Johnson R, Morgan R (1984) Pharmacologic treatment of Alzheimer's type dementia: new approaches. Psychopharmacol Bull 20: 542–545

Summers WK, Majovski LV, Marsh GM, Tachiki K, Kling A (1986) Oral tetrahydroaminoacridine in long-term treatment of senile dementia, Alzheimer-type. N Engl J Med 315: 1241–1245

Szelies B, Karenberg A (1986) Störungen des Glukosestoffwechsels bei Pick'scher Erkrankung. Fortschr Neurol Psychiat 54: 393–397

Szelies B, Herholz K, Pawlik G, Beil C, Wienhar K, Heiss WD (1986) Zerebraler Glukosestoffwechsel bei präseniler Demenz vom Alzheimer-Typ — Verlaufskontrolle unter Therapie mit muskarinergem Cholin-agonisten. Fortschr Neurol Psychiat 54: 364–373

Szelies B, Wullen T, Adams R, Grond M, Karbe H, Herholz K (1989) Comparison between cerebral glucose metabolism and late evoked potentials in patients with Alzheimer's disease. J Neural Transm (P-D Sect) 1: 141

Terry RD, Peck A, De Teresa R, Schechter R, Horoupian DS (1981) Some morphometric aspects of the brain in senile dementia of the Alzheimer type. Ann Neurol 10: 184–192

Reference-free evaluation of auditory evoked potentials-P 300 in aging and dementia

T. Dierks and **K. Maurer**

Department of Psychiatry, University of Würzburg, Federal Republic of Germany

Summary

Latency, field range (amplitude), and topography of AEP-P 300 were compared in a young and a geriatric control group vs a group of DAT patients, using reference-independent, spatial-related measurements. Latencies in the geriatric control group were significantly prolonged and field range was diminished in geriatric controls and in the DAT-group as compared to younger controls. The topography changed in DAT-patients.

Introduction

Auditory evoked potentials (AEP) have been used in research in an attempt to discover deficits of brain function due to various diseases of the nervous system such as dementia. The most commonly used parameters for evoked potentials are latency and amplitude. To determine these values peaks or troughs of the waveforms are selected. However, this is an ambiguous method for the description of physiological brain functions, since waveforms are perpetually depending on the chosen reference. A description of brain function using reference-dependent parameters is therefore not meaningful.

To describe P 300 parameters in normal and pathological aging in a physiological and unambiguous way, we used different reference-independent measurements first described by Lehmann and Skrandies (1980).

The aim of this study was to describe the results obtained with reference-independent measurements to determine latencies, amplitudes and topography, and their differences between healthy young and aged controls and DAT-patients.

Methods

Subjects

Three groups were studied, each consisting of eleven subjects (Table 1). All volunteers were free of psychiatric or physical illnesses and right-handed. Patients were 45 subjects suffering from dementia of Alzheimer type (DAT). They were moderately demented according to the Brief Cognitive Rating Scale (BCRS Reisberg et al., 1983) with scores between 24 and 32. To exclude multi-infarct dementia (MID) the modified Hachinski scores (Rosen et al., 1984) and computer tomographic findings were used. Only 11 exhibited evidence of a marked P 300 according to the parameters described later. In the DAT-group, all subjects were right-handed.

P 300 recordings

The AEP-300 as a late endogenous component of the auditory evoked potential was elicited using a bitonal stimulation mode, where a task was connected to one of the tones. A low-pitched tone (1000 Hz, 10 ms rise and fall time, 50 ms duration) was presented regularly with a frequency of 1 tone burst each two seconds. Another high-pitched tone (2000 Hz, 10 ms rise and fall time, 50 ms duration), to which a task was connected, was presented randomly with a probability of 1:5, compared to the low-pitched tone. Subjects were asked to pay attention to the high pitched tone, and to count the number of its occurrences. A short test trial was performed to enhance comprehension of the task. Subjects were regarded to have concluded the P 300 paradigm correctly, only if the recalled number of the task-connected tones was in the range of ± 20% compared to the real number of high pitched tones presented. If the subjects did not comprehend the task or forgot it during the test, the P 300 paradigm was regarded as not concluded. Recording were performed using 20 channels from the scalp. For electrode placement the 10 − 20 system was used (Jasper, 1958). Electrode locations were: Fp 1, Fp 2, F 7, F 3, Fz, F 4, F 8, T 3, C 3, Cz, C 4, T 4, T 5, P 3, Pz, P 4, T 6, O 1, Oz, O 2. Brain responses to low and high pitched tones were separately averaged using a Bio-logic Brain Atlas III (Bio-logic Systems Corporation). Approximately 30 samples of high pitched tones (targets) were averaged on line in each subject. The signals were sampled with 250 Hz and a bandpass of 0.3 Hz to 70 Hz. Analysis time was 1024 ms; 220 ms were used as predelay for baseline control. As reference linked mastoid electrodes were used.

AEP-P 300 identification

Multilead recordings create statistical problems (Oken et al., 1986; Kahn et al., 1988). To avoid statistical pitfalls methods for data-reduction are recommended. Lehmann and Skrandies (1980, 1984) proposed the global field-power (GFP) as reference-independent method of spatial data reduction. The global fieldpower is the spatial standard deviation at each recorded point in

Table 1. Summary of demographic and statistical data

Group	Age (years)		Latency (ms)			Field range (µV)		
	mean	std. d.	range	mean	std. d.	range	mean	std. d.
Young	29	6	290–395	339	32	13.5–33.4	21.6	5.4
Ger.	71	11	348–580	426	55	10.5–20.6	14.9	3.0
DAT	69	10	275–572	394	69	9.1–19.9	13.6	2.9

time and thus a measurement of the field variability in space. By envisioning the P 300 component as a model equivalent dipole, this dipole will display its maximum strength when GFP, the field variability, reaches its peak. The GFP is a reference-independent measurement since the relative differences between electrodes at certain points in time will not change whatever reference is used due to the fact that the electrical "landscape" generated by the brain and not influenced by electrode locations. The latency of P 300 is equivalent to the latency where the GFP reaches its maximum in a time window between 280 and 580 ms after stimulus onset. The spatial amplitude distribution at this time corresponds to the electrical field distribution on the scalp of the equivalent P 300 dipole. As measurement of the amplitude (strength) of P 300, the range (difference between maximal and minimal value in the field) was used. The direction of the P 300 was plotted so as to lie between the location of the smallest and that of the largest value in the field at the latency with the maximal GFP.

Statistical analysis

Latencies, as determined by GFP, and amplitudes were compared using a non-parametric one-way ANOVA (Kruskall-Wallis test). If significance was attained the groups were separated with the Mann-Whitney U-test. Topography was compared using the Mann-Whitney U-test.

Results

The mean GFP, disregarding peaks, was calculated for all three groups (Figs. 1 − 3). N 100 and P 300 were the two main components (Fig. 1). The individual latencies are plotted in Fig. 4. Results of the statistical computations are presented in Table 1 and Fig. 5. The one-way ANOVA indicated a significant difference (p = 0.0013) between the three groups (Fig. 5). Statistical comparisons with the Mann-Whitney U-test revealed a significantly (p = 0.004) prolonged latency in geriatric subjects compared to healthy young controls. The latencies of DAT patients did not significantly differ from those of geriatric controls (p = 0.42); however they diverged significantly from the

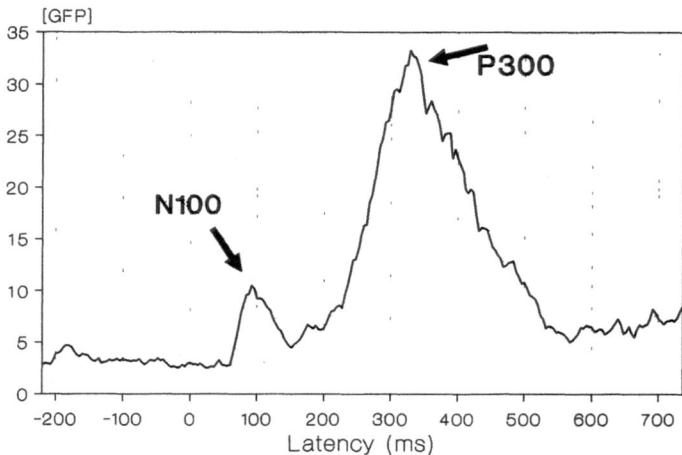

Fig. 1. Mean GFP for 11 young healthy subjects. GFP was calculated on the basis of a 20-channel recording of the AEP-300. N 100 and P 300 peaks are marked. The distinct P 300-peak of the mean GFP indicates low variance of P 300 latency

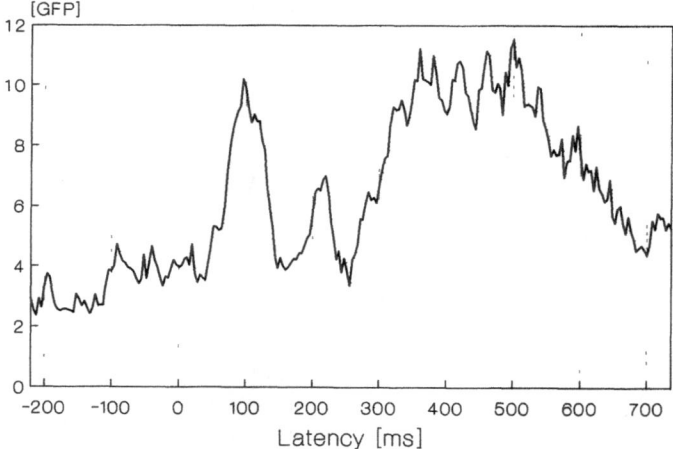

Fig. 2. Mean GFP for 11 geriatric healthy subjects. GFP was calculated on the basis of a 20-channel recording of the AEP-300. The broad-based P 300-peak of the mean GFP indicates high variance of the P 300 latency

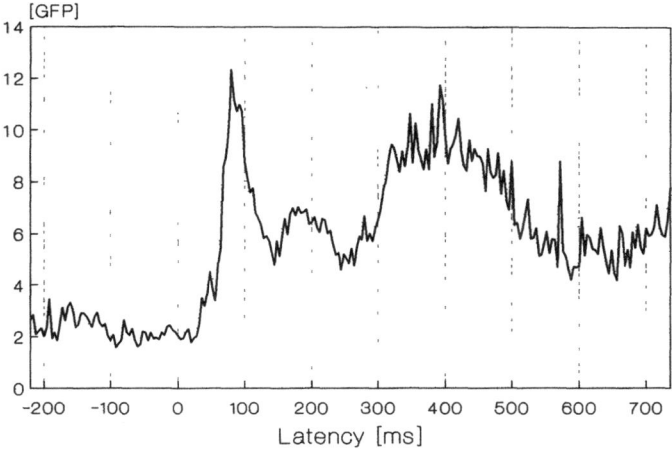

Fig. 3. Mean GFP for 11 DAT patients. GFP was calculated on the basis of a 20-channel recording of the AEP-300. The broad based P 300-peak of the mean GFP indicates a high variability of the P 300 latency. The N 100 peak of the mean GFP is distinct and appears at the same latency in all three groups, indicating low interindividual and intergroup variance (Figs. 1, 2 and 3)

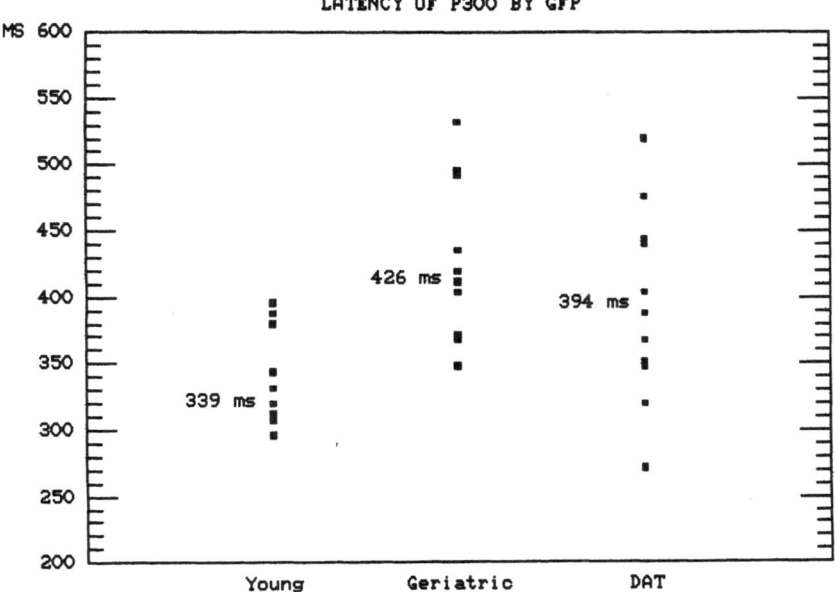

Fig. 4. Individual latencies, determined by GFP, plotted for each group

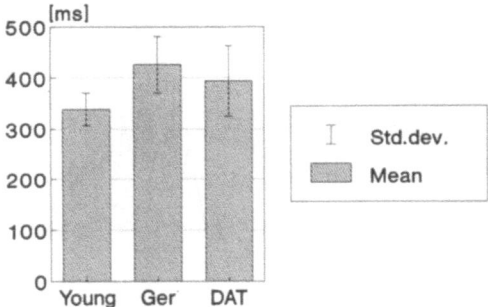

Fig. 5. Mean latency of the P 300 in the three groups (± standard deviation)

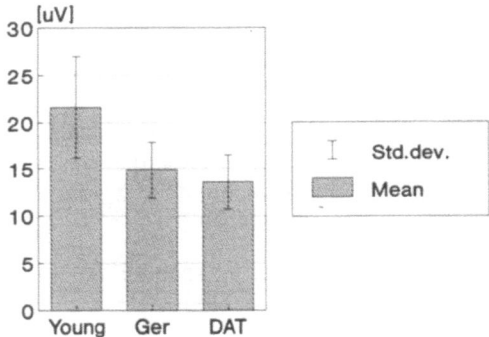

Fig. 6. Mean field-range of the P 300 in the three groups (± standard deviation)

young control group (p = 0.021). The field range, i.e. the difference between the smallest and largest value in the spatial field at the latency determined by the GFP, was then calculated for each subject in all three groups. Table 1 and Fig. 6 show the statistics of the field range calculations. A non-parametric one-way ANOVA (Kruskall-Wallis) was used to determine statistically significant differences between the groups (p = 0.0031). With the Mann-Whitney U-test group differences were delineated. Amplitudes in the young control group were significantly higher compared to those in the geriatric control (p = 0.023) and in the DAT-group (p = 0.009). However there was no significant amplitude difference (p = 0.56) between the geriatric control group and the group consisting of DAT-patients.

   To enable a topographic comparison between groups, the localization of individual maxima and minima in the spatial fields corre-

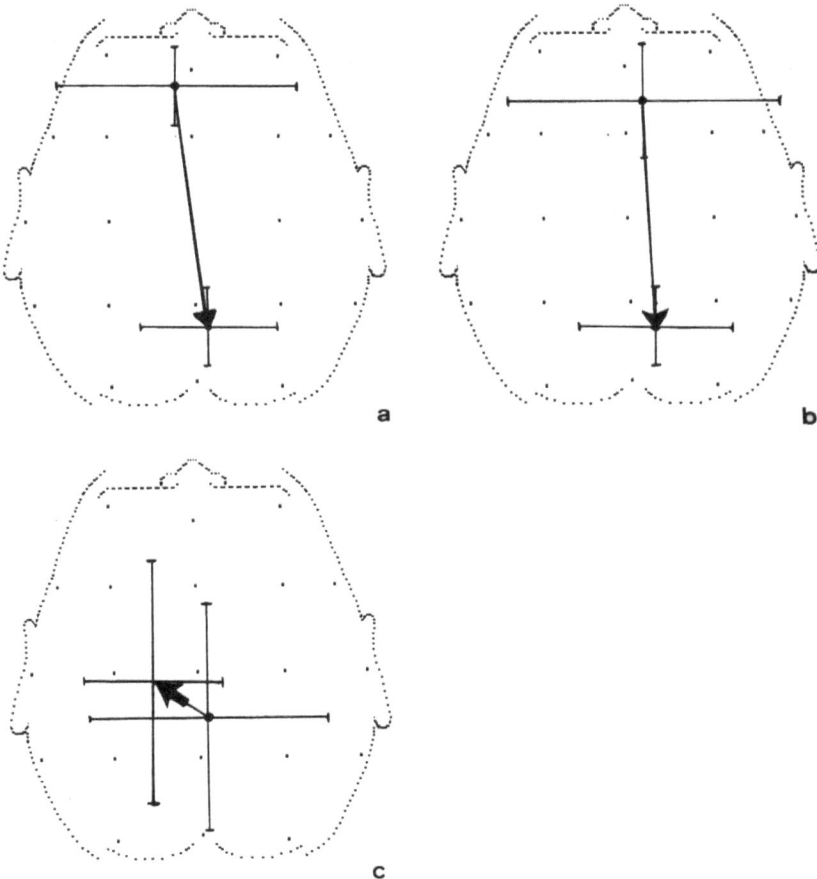

Fig. 7. Mean minimum and mean maximum with standard deviation in transversal and sagittal direction plotted for **a** the young healthy group, **b** the healthy geriatric group and **c** the DAT-group. Head seen from above. Arrow originates from the mean minimum and ends at the mean maximum giving an impression of direction of the model equivalent dipole

sponding to the latency with maximal GFP were evaluated and mean maxima and minima with standard deviation were calculated and plotted (Fig. 7 a − c). The young control group exhibited a clear frontal minimum of the P 300 field. The mean maximum was clearly parieto-occipitally localized with a slight preference to the right side. In geriatric controls a similar mean topography was found; the mean minima

were, however, in the right hemisphere compared to a left localization
in young controls, but due to a high variance this was not significant
(Mann-Whitney U-test). In DAT patients the topography revealed a
different pattern. The mean minimum was more occipitally localized
than the mean maximum. The mean minimum was localized almost
in the midline in the centro-parietal region (Fig. 7 c). The mean maximum was located in the left central region between electrodes Cz and
C 3. The high variability of the individual localization of minima and
maxima resulted in no significant difference between minimum and
maximum localization in the DAT group, which had been the case in
the young and geriatric control groups. The minima in the DAT group
were significantly (p = 0.006, Mann-Whitney U-test) more occipitally
localized compared to the young and geriatric control group. The
maxima in the DAT group were significantly (p = 0.025, Mann-Whitney U-test) more frontally localized compared to the young control
group, and almost significant (p = 0.058, Mann-Whitney U-test) compared to the geriatric control group. Thus the topography in DAT
patients altered in such a way that the model equivalent dipole shifted
from a marked frontal to parietal direction in both healthy groups to
a more parieto- to central direction in DAT patients.

Discussion

Using stereotaxically implanted electrodes in amygdala, hippocampus, and parahippocampal gyrus in patients with temporal lobe epilepsy, Halgren et al. (1986) demonstrated that the medial temporal lobe
(MTL) is involved in generating the scalp P 300. Magnetic encepha-
lographic (MEG) studies (Okada et al., 1983) also point to MTL as
the generator of P 300. This generator would equalize a current model
dipole which is directed toward the parieto-occipital regions. The P 300
component was negative at the nasopharyngeal wall (Perrault et al.,
1983) and the corresponding positive fields were found over the parietal
regions (Maurer and Dierks, 1987; Morstyn et al., 1983; McCarley et al.,
1986).

When trying to motivate demented patients to participate in a
study which makes demands on their cognitive function, problems
arise due to their limited comprehension of the task they are asked to
solve, the difficulty in keeping their mind focused on the paradigm
and to their forgetting the instructions given. Duffy et al. (1984) were
fully aware of this problem, and discussed the possibility that the
amplitude reduction in demented patients may be better explained by
differential adequacy of task performance. By interrupting the session
when necessary and reinforcing the previously given instructions, they

tried to keep such artifacts to a minimum. Polich et al. (1986) also tried to avoid this artifacts checking performance by asking the subjects to press a button each time an infrequent tone was heard; when not correctly done the subject was reminded verbally of the instructions.

We did not use a button to measure performance, since in our opinion this may contribute to time locked artifacts due to activation of cortical regions of motor function. The performance was controlled by the instruction to count the infrequent tones and only subjects with an error range of ± 20% of the number of tones presented were included. This strict procedure contributed partly to the relatively high (75%) drop-out in the demented group. Another reason was the decision to include only subjects with a clear GFP. The identification of the P 300 wave is an issue solved in various ways, one of which consists in using the most positive point in a defined time window as latency of P 300 (Squires et al., 1975; Pfefferbaum et al., 1984; Polich et al., 1986; Gordon et al., 1986). However, the time window used varies to some extent between different authors. Goodin et al. (1978) used the intersection of extrapolated lines from the ascending and descending slopes of each peak. Generally the peak was determined in one single electrode not taking into account the variable spatial distribution of the P 300. A more spatial related method was used by Morstyn et al. (1983) and Duffy et al. (1984), where the spatial evolution of the P 300 was described and features using Significance Probability Mapping (SPM) (Duffy et al., 1981) were delineated between groups, allowing integrals to be calculated in fixed time windows disregarding individual latency differences. The methods described above may be used in studies with a limited number of electrodes; however, when recording P 300 with a multilead arrangement more spatial related methods have to be applied. From n electrodes $n*(n-1)$ different waveshapes can be recorded simultaneously, and by using the common average reference (Offner, 1950) additional $2*n$ waveshapes are possible. With a 20 lead system this would come up to 420 different waveshapes. This demonstrates that the visual definition of the so-called components, e.g. (N 100, P 300) is a doubtful procedure, with results relying on the actual reference, which is pre-selected. For measuring the amplitude of the P 300, most authors used a prestimulus time which was averaged and the average set to zero (e.g. Gordon et al., 1986) to determine baseline. The amplitude was then determined to be the amplitude between the peak and the averaged baseline. A pre-stimulus mean baseline makes a physiological interpretation more difficult. Lehmann and Skrandies (1984) are of the opinion that the use of a prestimulus mean baseline may produce meaningless results in some cases where the brain electrical state at the time of stimulation is not meaningfully related to the evoked activity. It would therefore be ideal to use measurements that are independent of reference and of prestimulus manipulations and are related to spatial distribution. Global fieldpower (spatial standard deviation, Lehmann and Skrandies, 1980) is one such measurement, attributable to the fact that the relative field in time does not change, independent of the reference chosen. The reference solely imposes a spatial DC offset to the field, whereas the common average reference is a sort of spatial highpass filter. The GFP allows a reference-free estimation of latency. The field range, difference between minimal and maximal values in the field, is a measurement for the strength (amplitude) of the field and is likewise not connected to the choice of reference. By making the simplistic assumption that the P 300 is generated by a model equivalent dipole, this dipole can be described in its direction by the location of minima and maxima in the field and is thus relatively easy to compare between groups or individuals (Skrandies, 1984).

Our findings of prolonged latencies in aged persons confirm those described by many authors. Goodin et al. (1978) found a prolongation of the P 300 in the order of 1.8 ms/year; Pfefferbaum et al. (1984), Syndulko et al. (1982), Brown et al. (1983) and Picton et al. (1984) all found age/latency relationships between 0.94 and 1.36 ms/year. The study of latency in demented patients did not provide unanimous results. Syndulko et al. (1982) found certain latency prolongations in dementia; however, Canter et al. (1982) and Pfefferbaum et al. (1984) found only slightly prolonged values. Our results show that latencies of P 300 vary to a great extent in patients with DAT.

Our results cannot be compared to those of the multi-lead recordings made by Duffy et al. (1984), where a totally different way of describing P 300-data was used. With regard to the amplitude of P 300, Pfefferbaum et al. (1984) did not find any correlation between age and amplitude; however, this result was obtained when studying exclusively the lead Pz. The same authors, however, found that the spatial relief measures (difference in amplitude across the 3 used leads) were negatively correlated with age which is in accordance with our results of reduced field range in age. In dementia, amplitudes have been described by most authors as diminished compared to age-matched controls (Pfefferbaum et al., 1984). However, like in the study by Pfefferbaum, the spatial component was not taken into account. Comparing our data, using SPM (Duffy et al., 1981), between geriatric controls and demented patients, there were significant differences in amplitudes between the two groups. In contrast, when using the field range as a spatial related measurement for amplitude, the differences between the control and the DAT group were no longer observed.

The topographic change in DAT patients compared to that in the young and the geriatric control groups can be seen as a reflection of neuropathological findings (Brun and Gustafson, 1976) and biochemical and other studies using functional imaging methods (e.g. PET; Friedland et al., 1983) demonstrated the temporoparietal region was the most affected part of the brain. Thus, reference-independent and spatial-oriented methods of AEP evaluation show methodical advantages and a higher correlation to the results obtained by other functional imaging methods such as SPECT and PET.

References

Brown WS, Marsh JT, LaRue A (1983) Exponential electrophysiological aging: P 3 latency. Electroencephalogr Clin Neurophysiol 55: 277–285

Brun A, Gustafson L (1976) Distribution of cerebral degeneration in Alzheimer's disease. Arch Psychiatr Nervenkr 223: 15–33

Canter NL, Hallett M, Growdon JH (1982) Lecithin does not affect EEG spectral analysis or P 300 in Alzheimer's disease. Neurology 32: 1260–1266

Duffy FH, Bartels PH, Burchfiel JL (1981) Significance probability mapping: an aid in the topographic analysis of brain electrical activity. Electroencephalogr Clin Neurophysiol 51: 455–462

Duffy FH, Albert MS, McAnulty G, Garvey AJ (1984) Age-related differences in brain electrical activity of healthy subjects. Ann Neurol 16: 430–438

Duffy FH, Albert MS, McAnulty G (1984) Brain electrical activity in patients with presenile and senile dementia of the Alzheimer type. Ann Neurol 16: 439–448

Friedland RP, Budinger TF, Ganz E, Yukio Y, Mathis CA, Koss B, Ober B, Huesman RH, Derenzo SE (1983) Regional cerebral metabolic alterations in dementia of the Alzheimer type: positron emission tomography with [18F]-fluorodeoxyglucose. J Comput Assist Tomogr 7(4): 590–598

Goodin DS, Squires KC, Henderson BH, Starr A (1978) Age-related variations in evoked potentials to auditory stimuli in normal human subjects. Electroencephalogr Clin Neurophysiol 44: 447–458

Gordon E, Kraiuhin C, Harris A, Meares R, Howson A (1986) The differential diagnosis of dementia using P 300 latency. Biol Psychiatry 21: 1123–1132

Halgren E, Stapleton JM, Smith M, Altafullah I (1986) Generators of the human scalp P 3(s). In: Cracco RQ, Bodis Wollner I (eds) Evoked potentials. Alan R Liss, New York, pp 269–284

Jasper HH (1958) Report of committee on methods of clinical examination in electroencephalography. Electroencephalogr Clin Neurophysiol 10: 370–375

Kahn EM, Weiner RD, Brenner RP, Coppola R (1988) Topographic maps of brain electrical activity – pitfalls and precautions. Biol Psychiatry 23: 628–636

Lehmann D, Skrandies W (1980) Reference-free identification of components of checkerboard-evoked multichannel potential fields. Electroencephalogr Clin Neurophysiol 48: 609–621

Lehmann D, Skrandies W (1984) Spatial analysis of evoked potentials in man — a review. Prog Neurobiol 23: 227–250

Maurer K, Dierks T (1987) Brain-Mapping — topographische Darstellung des EEG und der evozierten Potentiale in Psychiatrie und Neurologie. Z EEG-EMG 18: 4–12

McCarley RW, Torello M, Shenton M, Duffy FH (1986) The topography of P 300 and spectral energy in schizophrenics and normals. In: Shagass et al (eds) Biological psychiatry. Elsevier, pp 389–391

Morstyn R, Duffy FH, McCarley RW (1983) Altered P 300 topography in schizophrenia. Arch Gen Psychiatry 40: 729–734

Offner FF (1950) The EEG as potential mapping: the value of the average monopolar reference. Electroencephalogr Clin Neurophysiol 2: 215–216

Okada YC, Kaufman L, Williamson SJ (1983) The hippocampal formation as a source of the slow endogenous potentials. Electroencephalogr Clin Neurophysiol 55: 417–426

Oken BS, Chiappa KH (1986) Statistical issues concerning computerized analysis of brainwave topography. Ann Neurol 19: 493–494

Perrault N, Wolfe R, Picton T (1983) Nasopharyngeal recordings of event related potentials. Ann NY Acad Sci 30: 205–211

Pfefferbaum A, Ford JM, Wenegrat B, Roth WT, Kopell BS (1984) Clinical application of the P 3-component of event-related potentials. I. Normal aging. Electroencephalogr Clin Neurophysiol 59: 85–103

Pfefferbaum A, Ford JM, Wenegrat B, Roth WT, Kopell BS (1984) Clinical application of the P 3-component of event-related potentials. II. Dementia, depression and schizophrenia. Electroencephalogr Clin Neurophysiol 59: 104–124

Picton TW, Stuss DT, Champagne SC, Nelson RF (1984) The effects of age on human event-related potentials. Psychophysiology 21: 312–326

Polich J, Ehlers CL, Otis S, Mandell AJ, Bloom FE (1986) P 300 latency reflects the degree of cognitive decline in dementing illness. Electroencephalogr Clin Neurophysiol 63: 138–144

Reisberg B, London E, Ferris SH, Borenstein J, Scheie L, deLeon M (1983) The brief cognitive rating scale: language, motoric and mood, in concomitants in primary degenerative dementia (PDD). Psychopharmacol Bull 19: 702–708

Rosen WG, Mohs RC, Davis KL (1984) A new rating scale for Alzheimer's disease. Am J Psychiatry 14: 1356–1364

Skrandies W (1984) Scalp potential fields evoked by grating stimuli: effects of spatial frequency and orientation. Electroencephalogr Clin Neurophysiol 58: 325–332

Squires NK, Squires KC, Hillyard SA (1975) Two varieties of long-latency positive waves evoked by unpredictable auditory stimuli in man. Electroencephalogr Clin Neurophysiol 38: 387–401

Syndulko K, Hansch E, Cohen SN (1982) Long latency related potentials in normal aging and dementia. Adv Neurol 32: 279–293

Enzymes and glial cells in brain damage and neurodegenerative diseases

G. J. McBean[1], E. B. Horner[1], I. Couée[1], J. P. Phillips[3], M. O'Brien[2], T. C. Lee[2], and K. F. Tipton[1]

Departments of [1] Biochemistry and [2] Anatomy, Trinity College, [3] Department of Neurosurgery, Beaumont Hospital, Dublin, Ireland

Summary

Neurodegenerative diseases may also involve damage to glial cells. An apparent glial cell involvement in the neurotoxicity of kainate is, for example, demonstrated by the protection afforded by D,L-α-aminoadipate at concentrations where it selectively affects glial cells. Immunological determination of the serum levels of creatine kinase-BB can be used to monitor astrocyte damage following head injury and may also be of value in degenerative conditions. The recessive adult-onset form of olivopontocerebellar atrophy has been reported to involve a deficiency of a specific membrane-bound isoenzyme of glutamate dehydrogenase (GDH). However attempts to distinguish this enzyme from the well-studied mitochondrial GDH using polyclonal antibodies were unsuccessful.

Introduction

The search for early markers for neurodegenerative diseases would be easier if the details of their aetiologies were more completely understood. In the cases of such conditions which result from vitamin deficiencies direct analysis of the tissue or urinary vitamin levels may be sufficient for identifying the causes. An example would be the subacute combined degeneration that results from a deficiency of the tissue levels of functional cobalamin (vitamin B_{12}). This is commonly due to a defective absorption of the vitamin due to the lack of the intrinsic factor although it may also result from the change in the oxidation state of the cobalt in cobalamin due to nitrous oxide inhalation (see Scott et al., 1986). Other degenerative processes which involve vitamin deficiencies include the Wernicke-Korsakoff syndrome that results from thiamine deficiency. The symptoms of the earlier stages of this disease may be completely reversed by the administration of thiamine, but irreversible changes develop as the disease progresses. The degeneration which may result from alcohol abuse resembles this condition and it has been argued that the alcohol-induced disorder may be due to malnutrition rather than to a direct effect of alcohol. However detailed studies on the development of the conditions in alcoholism and thiamine deficiency have suggested that direct neurodegenerative effects of ethanol are involved although these may be exacerbated by malnutrition (see Freund, 1985).

In neurodegenerative diseases where there is a strong genetic link, such as Huntington's disease (Gusella et al., 1985), direct genetic analysis may be the most appropriate means of assessing susceptibility. However in other cases, such as Parkinson's disease (Ward et al., 1984) any genetic predisposing factors appear to be weak or absent.

The changes in central enzyme and metabolite levels that occur during the normal processes of ageing and their possible relationship to Alzheimer's disease have been recently discussed in detail (see e.g. Strolin Benedetti and Dostert, 1989). In this contribution attention will be focused on physical brain damage as a possible indication of early markers for neurodegenerative diseases, the possible relationships between glial cell function and such diseases and, in particular, neurodegenerative conditions in which the excitatory neurotransmitter glutamate appears to be a contributory factor.

Materials and methods

Creatine kinase-BB and -MB were assayed by peroxidase-linked immunoassays with pre-coated microtite plates which were a kind gift from Celltech Diagnostics Ltd, Slough, U.K. Neurone-specific enolase was determined using a double-antibody immunoassay from Pharmacia, Uppsala, Sweden. The Student's T-test was used to test the significance of any differences between pre- and post-exercise groups. 17 amateur oarsmen and 8 amateur boxers (age range $18-28$ years) took part in the study. They had been training for at least 5 months before the test. All subjects abstained from vasoactive substances, caffeine, alcohol and smoking from midnight prior to the test and did not train on the test day. For the oarsmen a standard 6-minute ergometer test, which simulated a 2,000 metre flat water-course was used. The boxers each took part in a bout of 6×5 min rounds. None were knocked out.

In the studies of the effects of D,L-α-aminoadipate on kainate excitotoxicity, $100 \mu g$ of the former compound was injected over a period of 2 min in a volume of $2 \mu l$ of phosphate-buffered saline, into the left striatum of sodium pentabarbitone anaesthetized rats using a stereotaxic apparatus. After a recovery time of 6 or 24 hours, the animals were sacrificed and striata were dissected. Histological examination was performed after transdermal infusion of the anaesthetized animals with 2% paraformaldehyde − 1% glutaraldehyde in 0.1 M phosphate buffer, pH 7.4. Coronal slices of 0.5 mm which included the striatum, corpus callosum and part of the cerebral cortex were prepared and incubated in perfusion chambers perfused at a rate of 0.5 − 1.0 ml/min with oxygenated Krebs' bicarbonate medium at 30 °C. After an initial 1 h incubation period, the drugs under study were added and incubation was continued for up to 60 min before the slices were prepared for histological or biochemical analysis. Citrate synthase activity was determined by the method of Schousboe (1982).

Antibodies were raised against purified ox liver glutamate dehydrogenase in New Zealand White rabbits. Purification, immunotitrations and Ouchterlony double-diffusion studies were carried out according to the procedures described by Mayer and Walker (1980). Subcellular fractionation of human leucocytes and rat brain were carried out using the procedures of Plaitakis et al. (1984) and Colon et al. (1986) with the following modifications. 50 mM phosphate buffer, pH 7.4, was used instead of 50 mM Tris-HCl buffer, since the mitochondrial enzyme is not stable in that buffer at low ionic strengths (see Couée and Tipton, 1988). Freeze-thaw cycles were avoided in the case of the rat brain preparation and only carried out once in the leucocyte preparation, since the variations of pH, which may accompany freezing and thawing, could affect the activity of the mitochondrial enzyme. In the case of the rat brain preparation the low-speed supernatant was centrifuged at 8500 g for 15 min prior to high-speed centrifugation in order to sediment any mitochondria left unbroken after the homogenisation procedure. The fractions from rat brain were treated with 0.1% (vol : vol) cetyltrimethylammonium bromide and kept at 0 − 4 °C for 60 min prior to assay. The high speed pellet from the leucocyte preparation was made 0.16% (vol : vol) with Triton X-100. Both pellets were suspended in 50 mM phosphate buffer, pH 7.4, prior to detergent treatment. Glutamate dehydrogenase activities in the brain preparations were assayed as described by McCarthy et al. (1980) in assay buffer containing 0.1% (vol : vol) Triton X-100 and 0.4 mM KCN. The activities of the leucocyte preparations were assayed in the presence of ADP, as described by Colon et al. (1986), because of the low activities present. Other assays used standard methods and distribution studies are presented as described by de Duve et al. (1955).

Results and discussion

Creatine kinase and head injury

Creatine kinase (EC 2.7.2.3; CK) is a dimeric enzyme which exists as three tissue-specific isoenzymes. CK-BB is found in brain, where it is specifically located in astrocytes, CK-MB is myocardial whereas CK-MM is present in skeletal muscle. Elevated levels of serum CK-MB have been used as a diagnostic marker for myocardial infarction (Roberts et al., 1976) and the amounts of this isoenzyme in serum have also been found to be elevated after mild exercise and sharply increased in marathon runners after competition (Robinson et al., 1982; Siegel et al., 1981). La Porta et al. (1981), who measured the total levels of CK in serum after strenuous exercise, suggested that leakage of the enzyme from cells might result from membrane damage due to hypoxia, the presence of metabolites or mechanical stress.

Due to its restricted distribution CK-BB has been used as a marker for subarachnoid haemorrhage and ischaemic brain damage (Phillips et al., 1982 a, 1983), head injury and recovery from concussion (Phillips et al., 1980; Thompson et al., 1980) and some neurological disorders (Bell et al., 1978). Brayne et al. (1982) reported significantly increased levels of CK-BB in serum in boxers after competition and smaller increases in cyclists after a 40 mile road race. Elevated serum levels of both CK-BB and CK-MB were reported in marathon runners immediately after a race (Phillips et al., 1982 b, 1983) and the levels of the former enzyme were comparable with those determined after minor head injuries.

The studies discussed above used polyclonal antibodies and it has been suggested that the apparently elevated CK-BB levels might have arisen from cross-reactivity with the MB isoenzyme released from heart or from the small pool of that enzyme in skeletal muscle (Kaste and Sherman, 1982). The possibility that the elevated CK-BB levels might result directly from myocardial damage is unlikely, since the levels of that isoenzyme in heart are extremely low, comprising only about $0-3\%$ of the total in normal adult Europeans and Americans (see Ogunro et al., 1977; Urdal et al., 1983). However more caution might be necessary in studies with Japanese subjects, since the myocardial CK-BB contribution appears to be higher (Kato and Shimizu, 1986).

In order to investigate this problem more precisely, we have examined the effects of strenuous exercise in the course of two very different types of sport, boxing and rowing, on the serum levels of CK-BB and CK-MB. The levels of neurone-specific enolase (NSE; $\gamma\gamma$-enclose) were also determined. The anti-CK-BB and anti-CK-MB monoclonal antibodies did not cross-react, to any detectable extent with CK-MB and CK-BB, respectively. Neither antibody reacted with the muscle isoenzyme, CK-MM. The results obtained showed that there was no significant change in the CK-BB, CK-MB or NSE levels in the rowers as a result of their exercise. In contrast the boxers showed significant increases in CK-BB (0.18 ± 0.04 before to 0.43 ± 0.1 after, $\mu g \, 1^{-1}$; mean $\pm$ s.e.m.) and NSE (7.89 ± 0.74 before to 19.0 ± 3.4 after, $\mu g \, 1^{-1}$, mean $\pm$ s.e.m.) but not in CK-MB levels. Thus, even in the short duration of the events studied here, boxing may lead to release of CK-BB due to brain damage without there being any significant release of the MB isoenzyme from the heart or skeletal muscle pools. Elevated levels of NSE might suggest that the damage resulting from boxing is not restricted to astrocytes. However the possibility that the latter enzyme might arise from damage to blood platelets, in which it is also located, cannot be excluded.

Clearly there is a distinction between the direct damage to the head that can result from boxing and the absence of any detectable astrocytic damage in rowers after strenuous exercise. The latter activity involves a smooth action with no obvious effects on the head. In contrast, marathon running might result in coup and contra-coup damage in the cranial cavity as descent is abruptly interrupted on a hard surface by each footfall, accounting for the elevated levels of CK-BB reported by Phillips et al. (1982 b, 1983). In studies using the monoclonal antibodies we were, however, unable to detect significant elevations of the concentrations of this enzyme in serum after the race. Although this may result from the greater selectivity of the monoclonal antibodies used in the present study, compared to the polyclonal antibodies used by Phillips et al. (1982 b, 1983), it should be remembered that in the time since the earlier studies were carried out there has been an increasing awareness of the importance of wearing correctly cushioned footwear during such events and this would protect the runner from injuries such as these described above.

Ischaemic brain damage

The results discussed above suggest that the release of CK-BB into the serum may be a useful marker for disease conditions involving astrocyte damage. The mechanisms of ischaemic brain damage have been reported to involve release of glutamate from nerves as a result of the reversal of the neuronal uptake system under hypoxic conditions (for review see Nichols, 1989) which in turn leads to excitotoxic damage. If this were a valid explanation for the phenomena, the use of glutamate-receptor agonists as excitotoxins should provide a model for the processes involved, just as the use of antagonists of the N-methyl-D-aspartate (NMDA) type of glutamate receptor can be used to afford some protection (see Huether and Bean, 1988). However a possible involvement of glial cells in excitotoxicity has also been reported (Lefebrve et al., 1987).

In order to investigate the mechanism involved, the effects of the glutamate analogue, D,L-α-aminoadipate (DL-AA) have been studied in vivo and in tissue slices (McBean, 1990). Although this compound is toxic to both neurones and glial cells at high concentrations, it gives a transient gliotoxicity at lower concentrations. This is illustrated in Fig. 1, where the direct intrastriatal injection of DL-AA caused a

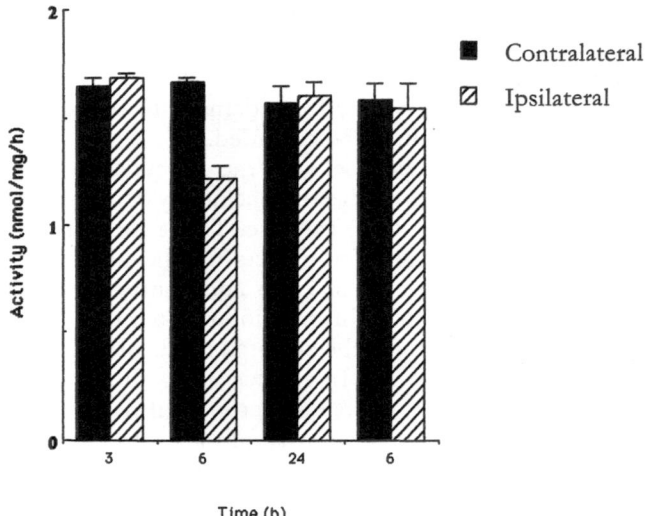

Time (h)

Fig. 1. Glutamine synthetase activity in rat striatal homogenates prepared after striatal injection of D,L-α-aminoadipate (DL-AA) or D,L-aminophosphonovalerate (APV). Rats were injected unilaterally into the left striatum in vivo with either 100 µg DL-AA or 61 µg APV and after the indicated times homogenates were prepared and assayed for glutamine synthetase activity. Results with DL-AA are shown in the first three data pairs, those for APV are shown in the last pair

significant decrease in the activity of the glial cell enzyme glutamine synthetase. This decrease was evident in striatal homogenates prepared 6 hours after in vivo injection of DL-AA, but the levels had returned to normal 24 hours after the injection. These results are consistent with those of Olney (1982) who showed that the effects of DL-AA on the morphology of retinal Muller cells were apparent after 6 but not 24 hours following intravitreal application. In tissue slices prepared 6 hours after in vivo injection of DL-AA there were no detectable changes in the morphology of the striatal neurones and there was a notable absence of glial cell proliferation around the site of injection. In contrast the NMDA receptor antagonist DL-aminophosphonovalerate (APV, 61 µg in 12 ml) caused no changes in glutamine synthetase activity six hours after intrastriatal injection (Fig. 1).

When slices were prepared from brain 6 hours after intrastriatal injection of DL-AA and incubated with 300 µM kainate for 40 min a widespread degeneration of striatal cells was evident, by microscopic examination, on the contralateral side but there was a marked survival of the neurones on the ipsilateral side, particularly those bordering the injection tract. In similar experiments it was found that prior injection of DL-AA did not afford protection of tissue slices against the neurotoxic effects of quinolinate (500 μM) or NMDA (500 μM). When the NMDA receptor antagonist APV was injected 6 hours before preparing slices there was no evident protection against kainate excitotoxicity.

The results suggest that glial cells are involved in the neurotoxic effects of kainate. This is consistent with the reports that glial cells contain a "neurotoxic activity" which can be released by potassium-ion depolarisation (Lefebrve et al., 1987). Whether this factor is glutamate itself is unclear, but it has been reported that the kainate-evoked glutamate release comes from a non-synaptosomal pool (Krespan et al., 1982).

The ability of DL-AA to protect against the toxicity of kainate but not that of quinolinate or NMDA indicates that different mechanisms are involved in these processes. Kainate is not an agonist of the NMDA receptor and the antagonist of that receptor APV does not protect against kainate toxicity. The apparent involvement of glial cells in the process suggests that more work on neurone-glial cell interactions may be necessary for a complete understanding of the mechanisms of this excitotoxicity. The possibility that such processes may be involved in neurodegenerative diseases suggest that glial cell activities may prove to be useful early markers for these conditions.

Olivopontocerebellar atrophy (OPCA)

In its adult-onset recessive form, but apparently not in the autosomal dominant form, OPCA has been shown to be associated with a partial deficiency of the enzyme glutamate dehydrogenase (GDH) (for review see Plaitakis and Berl, 1988). The brain areas undergoing degeneration in the course of the disease appear to depend on glutamatergic pathways. This led to the suggestion that the GDH deficiency leads to an impaired catabolism of glutamate and the accumulation of this neurotransmitter results in excitotoxic neuronal degeneration (Plaitakis and Berl, 1988).

Such subjects show elevated levels of plasma glutamate after a test dose of monosodium glutamate and the glutamate dehydrogenase levels in leucocytes and post-mortem brain samples have been found to be lower than normal (Plaitakis and Berl, 1988). It has been reported that there are two forms of glutamate dehydrogenase present in human leucocytes; the relatively well-characterized mitochondrial enzyme (see Couée and Tipton, 1988 for review) and an extra-mitochondrial, membrane-bound form (Plaitakis and Berl, 1988). Only the latter form of the enzyme appears to be missing in the leucocytes of patients suffering from OPCA. These two forms of GDH have been reported to be present in rat and human brain (Colon et al., 1986; Grossman et al., 1987) and it has recently been suggested that the membrane bound is associated with the synaptosomal membrane (Plaitakis and Berl, 1988). It is only this form of the enzyme that is lacking in post-mortem brain tissue from OPCA subjects.

Although the response of plasma glutamate levels after glutamate loading may provide a marker for OPCA susceptibility, direct measurement of the lymphocyte GDH levels might be more appropriate. The reported differences between the mitochondrial and membrane-bound GDH forms, which include different responses to activation by ADP in the presence of detergents and different thermal stabilities (see Plaitakis and Berl, 1988), are rather slight for straightforward discrimination. An immunological characterization of these two enzymes might thus be valuable for the development of a routine assay for this form of OPCA.

Antibodies were raised in rabbits to ox liver mitochondrial GDH purified by the procedure of McCarthy et al. (1980). Human leucocytes were fractionated into the high-speed supernatant and pellet fractions, reported to contain the mitochondrial and membrane bound GDH activities, respectively (Plaitakis et al., 1984). Figure 2 shows the distribution of GDH activities and those of the cytoplasmic enzyme, lactate dehydrogenase, and the mitochondrial inner-membrane and matrix enzymes, succinate dehydrogenase and citrate synthase in the two fractions.

Immunoprecipitation of the enzyme activities in the two fractions by the antibodies raised against the purified mitochondrial enzyme is shown in Fig. 3. When the high-speed pellet was incubated at a concentration of 3 mg/ml in 50 mM sodium phosphate buffer, pH 7.4, containing 2 mg/ml bovine serum albumin at 47.5 °C GDH activity was completely lost within 40 min. In contrast the activity in the high-speed supernatant, at a concentration of 3.5 mg/ml, was considerably more stable under these conditions with less than 50% being lost after 60 min.

When rat brain was fractionated by the procedure of Colon et al. (1986), modified as described above, similar results to those described for the activities in leucocytes were obtained. The activity in the high-speed supernatant, reported to correspond to the mitochondrial enzyme, was more stable to incubation at 45 °C, than the membrane-bound activity in the high-speed pellet. However the antibodies raised to the mitochondrial enzyme caused complete precipitation of both

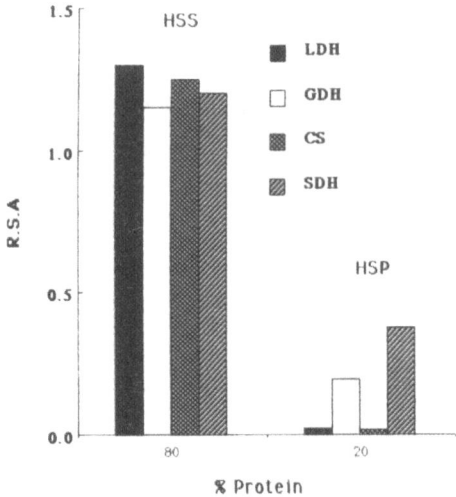

Fig. 2. Enzyme distributions in human leucocyte subcellular fractions. Leucocyte homogenates were separated into the high-speed supernatant (HSS) and pellet (HSP) fractions, as described in the text and these were assayed for protein content and for the activities of lactate and dehydrogenase (LDH), glutamate dehydrogenase (GDH), citrate synthase (CS) and succinate dehydrogenase (SDH). Results are shown in terms of relative specific activity (R.S.A.) values (de Duve et al., 1965)

activities. Furthermore Ouchterlony double-diffusion showed complete reaction of identity between the precipitin lines given by the activities in these two fractions.

Although the above results confirm the observations of Plaitakis et al. (1984) and Colon et al. (1986) that the solubilisable and membrane-bound GDH activities differ in their thermal stabilities, the use of antibodies raised against the mitochondrial enzyme failed to reveal any difference in the immunological recognition of these two enzyme forms in rat brain or human leucocytes. These results would suggest a close similarity, or even identity, between the polypeptides constituting the mitochondrial and membrane-bound forms of GDH. Since the mitochondrial enzyme is known to be capable of binding to outer-membrane phospholipids (Dodd, 1983; Couée and Tipton, 1989), it is possible that the particulate form of the enzyme may represent a proportion of the mitochondrial enzyme that is tightly-bound to membrane material. However the results of Colon et al. (1986) showing

G. J. McBean et al.

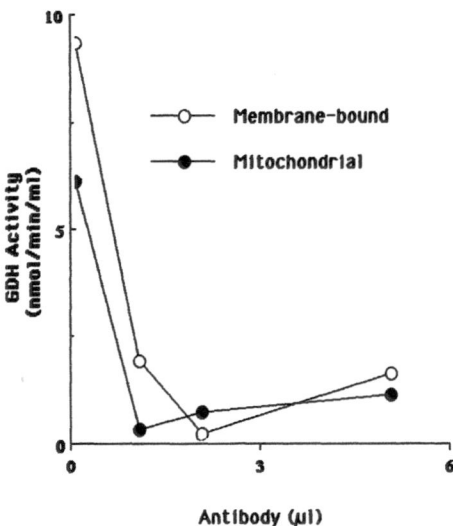

Antibody (µl)

Fig. 3. Immunoprecipitation of glutamate dehydrogenase from human leucocyte fractions. The solubilised membran-bound (HSP, 350 µg, ●) and mitochondrial (HSS, 70 µg, O) fractions of human leucocytes were incubated
with the indicated volumes of a 10-fold diluted preparation of antibodies
raised against purified ox liver glutamate dehydrogenase overnight. After
centrifugation the activity remaining in the supernatant was determined

different allosteric properties of purified preparations of the two forms
would suggest there to be some differences in their primary structures.
Clearly any such differences are not sufficient to be detected by this
immunological approach and studies at the nucleic acid level may be
required to resolve the problem and, perhaps, to provide a marker
for susceptibility towards this form of OPCA.

Acknowledgements

We are grateful to the Richmond Brain Research Foundation, to the
Health Research Board of Ireland and to Celltech Diagnostics Ltd.

References

Bell RD, Rosenberg RN, Ting R, Mukkerjee A, Stone MJ, Williams JT
(1978) Creatine kinase BB isoenzyme levels by radioimmunoassay in
patients with neurological disorders. Ann Neurol 3: 52–59
Brayne CEG, Dow L, Calloway SP, Thompson RJ (1982) Blood creatine
kinase isoenzyme BB in boxers. Lancet ii: 1308–1309

Colon A, Plaitakis A, Perakis A, Berl S, Clarke DD (1986) Purification and characterization of a soluble and a particulate glutamate dehydrogenase from rat brain. J Neurochem 46: 1811–1819

Couée I, Tipton KF (1988) Glutamate dehydrogenase. In: Kvamme E (ed) Glutamine and glutamate in mammals, vol 1. CRC Press, Boca Raton, pp 81–100

Couée I, Tipton KF (1989) The effects of phospholipids on the activation of glutamate dehydrogenase by L-leucine. Biochem J 261: 921–925

de Duve C, Pressman BC, Gianetto R, Wattiaux R, Applemans F (1965) Tissue fractionation studies 6. Intracellular distribution patterns of enzymes in rat liver tissue. Biochem J 60: 604–617

Dodd G (1973) The interaction of glutamate dehydrogenase and malate dehydrogenase with phospholipid membranes. Eur J Biochem 33: 418–427

Freund G (1985) Neuropathology of alcohol abuse. In: Tarter R, van Thiel DH (eds) Alcohol and the brain. Chronic effects. Plenum, New York, pp 3–17

Gusella J, Tanzi R, Bader P, Phelan M, Stevenson R, Hayden M, Hofman K, Faryniaz A, Gibbons K (1985) Deletion of Huntington's disease-linked G 8 (D 4510) locus in Wolf-Hirschlorn syndrome. Nature 318: 75–78

Grossman A, Rosenberg RN, Warmoth L (1987) Glutamate and malate dehydrogenase activities in Joseph disease and olivopontocerebellar atrophy. Neurology 37: 106–111

Huettner JE, Bean BP (1988) Block of N-methyl-D-aspartate current by the anticonvulsant MK-801: selective binding to open channels. Proc Natl Acad Sci USA 85: 1307–1311

Kaste M, Sherman DG (1982) Creatine kinase isoenzyme activities in marathon runners. Lancet ii: 327–328

Kato K, Shimizu A (1986) Highly sensitive immunoassay for human creatine kinase MM and MB isoenzymes. Clin Chim Acta 158: 99–108

Krespan B, Berl S, Nicklas WJ (1982) Alteration in neuronal-glial metabolism of glutamate by the neurotoxin kainic acid. J Neurochem 38: 509–518

La Porta MA, Linde HW, Bruce OL, Fitzsimons EJ (1978) Elevation of creatine phosphokinase in young men after recreational exercise. JAMA 239: 2685–2686

Lefebrve PP, Rogister B, Delree P, Leprince P, Selak I, Moonen G (1987) Potassium induced release of neurotoxic activity by astrocytes. Brain Res 413: 120–128

Mayer RJ, Walker JH (1980) Immunochemical methods in the biological sciences: enzymes and proteins. Academic Press, New York

McBean GJ (1990) Intrastriatal injection of DL-α-aminoadipate reduces kainate toxicity in vitro. Neuroscience 34: 225–234

McCarthy AD, Walker JM, Tipton KF (1980) Purification of glutamate dehydrogenase from ox brain and liver. Evidence that commercially-available preparations of the enzyme from ox liver have suffered proteolytic cleavage. Biochem J 191: 605–611

Nicholls DG (1989) Release of glutamate, aspartate and γ-aminobutyrate from isolated nerve terminals. J Neurochem 52: 331–341

Ogurno EA, Hearse DJ, Shillingford JP (1977) Creatine kinase isoenzymes. Their separation and quantitation. Cardiovasc Res 11: 94–102

Olney JW (1982) The toxic effects of glutamate and related compounds in the retina and brain. Retina 2: 341–359

Phillips JP, Jones M, Hitchcock R, Adams N, Thompson RJ (1980) Radioimmunoassay of serum creatine kinase BB as an index of brain damage after head injury. Br Med J 281: 777–779

Phillips JP, Horner EB, Horgan J, Teo K (1982 a) Brain type CPK-BB and cerebral ischaemia after subarachnoid haemorrhage. Lancet ii: 662–663

Phillips JP, Horner EB, Ohmann M, Horgan J (1982 b) Increased brain-type creatine phosphokinase in marathon runners. Lancet i: 1310

Phillips JP, Horner EB, Doorly T, Toland J (1983) Cerebrovascular accident in a 14 year old marathon runner. Br Med J 286: 351–352

Plaitakis A, Berl S (1988) Pathology of glutamate dehydrogenase. In: Kvamme E (ed) Glutamine and glutamate in mammals, vol 2. CRC Press, Boca Raton, pp 127–142

Plaitakis A, Berl S, Yahr MD (1984) Neurological disorders associated with a deficiency of glutamate dehydrogenase. Ann Neurol 15: 144–153

Roberts R, Sobel BE, Parker CW (1976) Radioimmunoassay for creatine kinase. Science 194: 855–857

Robinson B, Williams BT, Worthington DJ, Carter THN (1982) Raised creatine kinase activity and presence of creatine kinase MB isoenzyme after exercise. Br Med J 285: 1619–1620

Schousboe A (1982) Glial marker enzymes. Scand J Immunol 15: 339–356

Scott JM, Dinn J, Wilson P, Weir DG (1981) Pathogenesis of subacute combined degeneration: a result of methyl group deficiency. Lancet ii: 334–337

Siegel AJ, Silverman LM, Holman L (1981) Elevated creatine kinase MB isoenzyme levels in marathon runners. JAMA 246: 2049–2051

Strolin Benedetti M, Dostert P (1989) Monoamine oxidase: brain ageing and degenerative diseases. Biochem Pharmacol 38: 555–561

Thompson RJ, Graham JG, McQueen INF, Kynoch PAM, Brown KW (1980) Radioimmunoassay of brain-type creatine kinase-BB isoenzyme in human tissue and in patients with neurological disorders. J Neurol Sci 47: 241–254

Urdal P, Urdal K, Stromme JH (1983) Cytoplasmic creatine kinase isoenzeymes quantitated in tissue specimens obtained at surgery. Clin Chem 29: 310–313

Ward DC, Duvoisin RC, Ince SE, Nutt JD, Eldrige R, Calne DB (1984) Parksinon's disease in 65 pairs of twins and a set of quintuplets. Can J Neural Sci 11: 815–824

Senile dementia of Alzheimer's type and Parkinson's disease: neurochemical overlaps and specific differences

P. Riederer[1], E. Sofic[1], G. Moll[1], A. Freyberger[1], I. Wichart[2], W. Gsell[1], K. Jellinger[2], G. Hebenstreit[3] and M. B. H. Youdim[4]

[1] Clinical Neurochemistry, Department of Psychiatry, University of Würzburg, Federal Republic of Germany
[2] Ludwig-Boltzmann-Institute of Clinical Neurobiology, Lainz Hospital, Vienna,
[3] Department of Psychiatry, Landeskrankenhaus Mauer/Amstetten, Austria, and
[4] Department of Pharmacology, Technion, Haifa, Israel

Summary

Parkinson's disease (PD) is characterized by a clear rank order with respect to neuropathological findings. Both substantia nigra (SN) and locus coeruleus (LC) are severely damaged, whilst nucleus raphe dorsalis (NRD) and nucleus basalis of Meynert (NbM) are only moderately affected. In senile dementia of Alzheimer's type (SDAT) the rank order is less clearly defined, SN being slightly and LC, NRD, and NbM being moderately damaged. The loss of neurotransmitter concentrations found in PD and SDAT by neurochemical analysis only partly corresponds to the neuronal loss found in SN, LC, NDR, and NbM. This may indicate the ability of the neurotransmitter systems to compensate neuronal loss. Possible pathomechanisms underlying PD and SDAT are discussed.

Introduction

Parkinson's disease (PD) is a progressive neurodegenerative process the outstanding characteristic of which is a tremendous loss of dopaminergic neurons in the substantia nigra (SN) (Table 1) resulting in a considerable decrease of dopamine (DA) in nigro-striatal regions (Ehringer and Hornykiewicz, 1960). Other neurotransmitter systems like the noradrenergic, serotoninergic, and cholinergic systems also seem to be concerned in PD, but to a smaller extent (for reviews see Birkmayer and Riederer, 1985; Hornykiewicz and Kish, 1986). Senile dementia of Alzheimer's type (SDAT) is another neurodegenerative

Table 1. Neuropathological findings in Parkinson's disease (PD), dementia of Alzheimer's type (DAT) and Alzheimer's disease (AD). Results in % of controls

	Neuronal cell loss in different regions of brain stem			
	substantia nigra	locus coeruleus	nucleus raphe dors.	nucleus basalis M.
PD	79[a]	83[d]	45[g]	38[h]
DAT	20[b]	58[e]	44[g]	57[i]
AD	25[c]	68[f]	58[g]	72[j]

[a-j] Mean value of data from
[a] Mann and Yates (1983), Gibb and Lees (1987), Hirsch et al. (1988),
[b] Mann et al. (1983), Jellinger (1987 b), Gibb (1988),
[c] Mann et al. (1983), Jellinger (1987 b), Mann et al. (1987),
[d] Tomonaga (1983), Mann et al. (1983), Mann and Yates (1983),
[e] Mann et al. (1983), Tomlinson et al. (1981), Bondareff et al. (1982), Jellinger (1987 a),
[f] Jellinger (1987 a), Tomonaga (1983), Ingram et al. (1987),
[g] Jellinger (1989),
[h] Gaspar and Grey (1984), Whitehouse et al. (1983), Jellinger (1986 a),
[i] Jellinger (1987 c), Etienne et al. (1986), Mann et al. (1985), Ichimiya et al. (1986), Jellinger (1986 b),
[j] Etienne et al. (1986), Ichimiya et al. (1986), Jellinger (1986 b)

disease where the degenerative process is mainly localized in the hippocampus and neocortex, with major degeneration of the nucleus basalis of Meynert (NbM) (Table 1) and the septum, respectively, with additional loss of serotonin (5-HT), noradrenaline (NA) and – to some extent – DA. Therefore, we studied various biogenic neurotransmitters, choline acetyltransferase (CAT) activity and thermostability in different regions of brains from patients suffering from either SDAT, PD or PD plus SDAT in order to demonstrate overlaps and differences between SDAT, non-demented PD and PD plus SDAT by using brains from the same sources and which underwent the same neuropathological, histochemical and analytical procedures. For a more complete overview, results were also compared with previous data on PD.

Material and methods

If not indicated otherwise, data were obtained from analysis of brains and by the methods described below.

Brains from seven controls and from seven SDAT patients were obtained during autopsy and dissected by an experienced neuropathologist (K. J.). Post-mortem delay amounted to 6.3 ± 2.1 hours for controls and 3.7 ± 0.8

hours for SDAT. Controls were sex- (2 f, 5 m) and age-matched (controls: 76 ± 3 years; DAT 79 ± 3). Four brains from patients with PD plus SDAT (2 f, 2 m; 78 ± 3 years) and eight PD brains (6 f, 2 m; 73 ± 4 years) were obtained from the same source. Post-mortem delay ranged from 3 to 15 hours. In any case, diagnosis was verified by standard neuropathological and histochemical methods in order to differentiate between SDAT, PD and PD plus SDAT.

Neurotransmitters were determined by high pressure liquid chromatography coupled with either electrochemical (BAS-system; 5-HT, HVA, 5-HIAA) or coulometric (ESA-system; NA, DA) detection as described recently (Sofic, 1986).

Choline acetyltransferase activity was determined by a radiometric assay according to Fonnum (1975).

Thermostability and enthalpy of native brain tissues were analyzed by means of a Perkin Elmer DSC 7 differential scanning calorimeter.

Results

Post-mortem analysis of SDAT brains showed no significant changes in DA levels in basal ganglia, limbic system and cortical regions (Table 2). The data, however, possibly imply a slight deficit of DA in these regions. This may coincide with the moderate loss of dopaminergic cell bodies in the ventral tegmental area (50%, Mann et al., 1987). In contrast, PD is characterized by a tremendous, highly significant loss of DA levels in the basal ganglia (Table 2) and this parallels the extensive degeneration of the SN in PD (Table 1). Slightly reduced concentrations of this catecholamine were also found in cortical regions. The limbic system was found to be unaffected and this may coincide with an only moderate (55%) loss of dopaminergic neurons in the ventral tegmental area (Uhl et al., 1985). Considering the data for SDAT and PD in cortical regions, a decrease of DA should be expected. Astonishingly, an about 250% increase was found in PD plus SDAT.

With respect to homovanillic acid (HVA), slightly reduced concentrations may occur in the basal ganglia and limbic system of SDAT patients. In cortical areas decreased levels were found for SDAT, PD as well as for PD plus SDAT. Thus, HVA levels in PD plus SDAT resemble those found for either SDAT or PD alone.

The results obtained for NA levels in SDAT brains were similar to those found for DA concentrations. Though hardly any significant changes were found, the data — in general — imply a slight decrease of NA levels in basal ganglia, limbic system, and cortical regions in SDAT (Table 3) probably reflecting the moderate degeneration of the locus coeruleus (LC) in SDAT (Table 1). In contrast, in parkinsonian brains a significant loss of NA concentrations occurred in the basal ganglia,

Table 2. Concentration of dopamine [and its metabolite homovanillic acid (HVA)] in selected brain regions of patients with senile dementia of Alzheimer type (DAT) and/or Parkinson's disease (PD). Results in % of controls

Region	DAT		PD + DAT		PD	
	DA	HVA	DA	HVA	DA	HVA
Basal ganglia						
N. caudatus	87	107	—	—	10[a]+ +	—
Putamen	99	94	—	—	4[a]+ +	—
Limbic system						
N. amygdalae	49	39	—	—	112[a]	—
Hippocampus	47	45*	—	—	32[b]	—
Cortical regions						
Frontal C.	62	39*	210[c]	49	3+ +	46+
Parietal C.	46	37*	173[c]	53	115	46+
Temporal C.	45	40*	325[c]+	50	45	60
Occipital C.	75	45	242[c]	60	67	54

[a] Data from Birkmayer and Riederer (1975).
[b] Data from Scatton et al. (1984).
[c] Results may be affected by L-dopa treatment.
Statistical analysis: Wilcoxon rank sum test: * $p < 0.05$. Student's t-tests: + $p < 0.05$; + + $p < 0.001$

limbic system and cortical regions (Table 3). These results parallel the tremendous degeneration of LC in PD (Table 1). Though the loss of NA was significant in many brain regions, it did not reach the extent of dopaminergic degeneration found in PD. In PD plus SDAT, loss of NA in the cortex was found to be more similar to that in PD than to that in SDAT.

Little, if any, changes in 5-HT seem to occur in the basal ganglia and limbic system in SDAT brains (Table 4) reflecting the moderate loss of serotoninergic neurons in the nucleus raphe dorsalis (NRD) in SDAT (Table 1). However, a significant 70% loss of 5-HT was found in cortical regions (Table 4). On the other hand, a less severe, but significant loss of 5-HT was found in many regions of parkinsonian brains which is likely to mirror the moderate degeneration of the NRD in PD (Table 1). However, in cortical areas an increase in 5-HT concentrations was noted. Elevated 5-HT levels were also found in cortical

Table 3. Concentration of noradrenaline (NA) [and its metabolite 3-methoxy-4-hydroxyphenylglykol (MHPG)] in selected brain regions of patients with senile dementia of Alzheimer type (DAT) and/or Parkinson's disease (PD). Results in % of controls

Region	DAT	PD + DAT	PD	
	NA	NA	NA	MHPG
Basal ganglia				
N. caudatus	63	—	46[a]++	39[a]++
Putamen	59	—	56[a]+	20[a]++
Limbic system				
N. amygdalae	52*	—	30[a]++	63[a]
N. accumbens	104	—	62[a]++	35[a]++
Cortical regions				
Frontal C.	63	36	39++	—
Parietal C.	32	54	32+	—
Temporal C.	35*	23+	37+	—
Occipital C.	48	193	50+	—

[a] Data from Riederer et al. (1977).
Statistical analysis: Wilcoxon rank sum test: * $p < 0.05$. Student's t-tests: + $p < 0.05$; + + $p < 0.01$

regions from patients with PD plus SDAT, similar to the results obtained for PD.

5-HIAA levels in cortex were reduced in SDAT, PD, and PD plus SDAT to about 50% (Table 4). Thus, considering both 5-HT and 5-HIAA, the results for patients with PD plus SDAT are almost identical to those for PD alone.

The most prominent biochemical alteration in SDAT brains during our investigation was a tremendous decrease (75 − 90%) of CAT activity in frontal regions (Table 5) reflecting the degeneration of the nucleus basalis of Meynert (NbM) in SDAT (Table 1). Decreased CAT activity was also reported for the cortex of parkinsonian patients, but appears to be rather modest compared with the changes in SDAT. Loss of CAT activity in PD plus SDAT is more pronounced than in PD alone (Table 5).

The NbM, a locus of major cholinergic degeneration in SDAT, was analyzed by differential scanning calorimetry. Decreased ther-

Table 4. Concentration of 5-hydroxytryptamin (5-HT) [and its metabolite 5-hydroxyindolyl-acetic acid (5-HIAA)] in selected brain regions of patients with senile dementia of Alzheimer type (DAT) and/or Parkinson's disease (PD). Results in % of controls

Region	DAT		PD + DAT		PD	
	5-HT	5-HIAA	5-HT	5-HIAA	5-HT	5-HIAA
Basal ganglia						
N. caudatus	142	111	—	—	42[a]+++	—
Putamen	96	63	—	—	54[a]+++	—
Limbic system						
Gyrus cinguli	118	89	—	—	107[a]	—
N. amygdalae	85	54	—	—	70[a]+++	—
Cortical regions						
Frontal C.	30*	46	199[b]	48	182[b]	47
Parietal C.	23*	46*	173[b]	58	141[b]	37[++]
Temporal C.	33*	45*	263[b]	34[+]	246[b+]	54
Occipital C.	49	55	230[b]	36	165[b]	41

[a] Data from Riederer and Wuketich (1976).
[b] Results may be affected by L-dopa treatment.
Statistical analysis: Wilcoxon rank sum test: * $p < 0.05$. Student's t-tests: + $p < 0.05$; ++ $p < 0.01$; +++ $p < 0.001$

Table 5. Choline acetyltransferase activity (CAT) in cortical regions of patients with senile dementia of Alzheimer type (DAT) and/or Parkinson's disease (PD). Results in % of controls

	DAT	PD + DAT	PD
Frontal C.	16*	30[a+]	55[a+]
Parietal C.	26*	—	—
Temporal C.	9*	—	—
Occipital C.	13*	—	—

[a] Data from D'Amato et al. (1987).
* Wilcoxon rank sum test; $p < 0.01$.
+ Student's t-test; $p < 0.001$

Table 6. Thermostability and enthalpy of N. basalis M. (NbM) and substantia nigra (SN) in patients with senile dementia of Alzheimer's type or Parkinson's disease

	n	Sex	Age years	Δ H (J/g)	Thermostability (°C)
NbM					
Control	4	1 f, 3 m	68.0 ± 12.2	1554.6 ± 68.3	81.1 ± 10.8
DAT	4	1 f, 3 m	78.0 ± 4,7	1628.5 ± 85.4	68.4* ± 4.5

Wilcoxon rank sum test: * p = 0.0072.
Student's t-test: * p = 0.0093

	n	Sex	Age years	Δ H (J/g)	Thermostability (°C)
SN					
Control	6	4 f, 2 m	75.5 ± 5.7	− 43.7 (exotherm) ± 16.0	57.9 ± 6.5
PD	4	2 f, 2 m	76.5 ± 0,5	+ 82.3* (endotherm) ± 43.9	48.4** ± 4.0

Wilcoxon rank sum test: * p = 0.0095, ** p = 0.038.
Student's t-test: * p = 0.0001, ** p = 0.033

mostability was found in SDAT brains reflecting neuronal loss or/and changes in lipid-protein composition (Table 6). When the substantia nigra of parkinsonian patients was analyzed, both changes in enthalpy and reduction of thermostability were found compared to controls. Again, these alterations are indicative of neuronal loss, changes in lipid-protein composition or increased lipid peroxidation which was found to occur in SN in PD (Dexter et al., 1989).

Discussion

Summarizing all the data, there is a clear rank order in neuropathological terms in PD. Both SN and LC are heavily damaged to the same extent, whilst NRD and NbM are only moderately affected.

From a neurochemical point of view, a less impressive rank order can be deduced. Naturally, the degeneration of the SN is accompanied by a heavy loss of DA, however, the decrease of NA does not adequately reflect the adrenergic neuronal loss in LC compared to dopaminergic degeneration. Loss of 5-HT in some brain regions is comparable to NA, though cell loss is more pronounced in the LC than in the NRD. On the other hand, 5-HT seems to be increased in the cortex of parkinsonian brains. However, this result is not fully understood and could be a consequence of L-dopa treatment which is known to release 5-HT from vesicles. Both cholinergic cell bodies in the NbM and CAT activity are moderately reduced in PD.

In SDAT the rank order of neuropathological findings is less clearly defined than in PD. Only a slight loss of neurons in the SN is found, whilst moderate degeneration of LC, NRD, and NbM occurs, the loss of neurons in the LC and NbM being slightly more pronounced. In Alzheimer's disease, the degenerative process in the LC, NRD, and NbM is more severe, again LC and NbM being more concerned. Speaking in neurochemical terms, some correlation is found between neuropathological findings and post-mortem neurotransmitter analysis. No significant reduction in DA concentrations coincides with the neuropathology of the SN in SDAT. A moderate loss of noradrenergic neurons in the LC is mirrored by a slight to moderate reduction of NA levels. However, though the degenerative process in the NRD is comparable to the LC, a greater decrease of 5-HT than of NA in the cortex of SDAT brains is found. As reflected by the tremendous decrease of CAT activity, loss of acetylcholine is the outstanding biochemical characteristic of SDAT. The loss of CAT activity in SDAT is comparable to the reduction of DA concentrations in PD; however, the extent of neurodegeneration of the NbM in SDAT is only moderate and comparable to that of the LC and NRD, whilst degeneration of dopaminergic neurons in the SN in PD is extensive. On the whole, both in SDAT and PD, neurochemical findings only partially reflect the neurodegenerative processes in different transmitter systems. This lack of correlation may be indicative of the ability of the different systems to compensate the loss of neurons within a wide range.

Cases with PD plus SDAT should be expected to show additive changes in neurotransmitters compared to SDAT and PD alone. However, this was only found for NA in the cortex. Concentrations of 5-HT in cortical regions of SDAT are almost identical to those in PD, but not DAT. DA concentrations in PD plus SDAT were increased in cortex, though they were reduced in both SDAT and PD, whilst HVA levels were comparable in all three groups. This result is — so far – not fully understood; however, since only intravesicular DA is analyzed in post-mortem tissues, an increase of DA in PD plus SDAT may indicate diminished release from or elevated uptake into vesicles. It is possible that PD plus SDAT has features different from PD or SDAT alone. On the other hand, it cannot be excluded that L-dopa treatment is responsible for the rise of cortical DA levels. As evident from our tables, further investigations are necessary to evaluate the loss of neurotransmitters in PD plus SDAT. Probably it may also be helpful to create one more subgroup, namely DAT with extrapyramidal disturbances with an earlier onset of DAT which could be expected to be more similar to SDAT than to PD.

As evident from the changes in thermostability and enthalpy in the nucleus basalis of Meynert in SDAT and in the substantia nigra in PD, multiple biochemical deficits occur in these regions. The reasons for the degenerative process are so far not known, but especially the brain stem seems to be susceptible to toxic agents and radicals (Cadet, 1988). The discovery that 1-methyl-4-phenyl-1,2,3,6-tetrahydropyridine (MPTP) causes a parkinsonian syndrome in humans (Davies et al., 1979) and primates (Burns et al., 1983) has emphasized the possibility of a toxin being the causative agent in PD. Epidemiological data suggested a role of paraquat, a structural analogue of MPTP, in the etiology of PD, but the discussion is controversial. Oxygen-derived radicals may also be involved in the neurodegenerative process. In PD increased lipid peroxidation (Dexter et al., 1989), decreased activity of detoxifying enzymes such as catalase (Ambani et al., 1975) and glutathione (GSH) peroxidase (Kish et al., 1985) and reduced levels of GSH (Perry et al., 1982; Riederer et al., 1989) were found. Unfortunately, no such data are available for DAT at present. If indeed a toxin or oxygen-derived radicals are the trigger for the neurodegenerative process in SDAT and PD, the question arises why in SDAT mainly the cholinergic system is concerned, while in PD the dopaminergic system is the main target. So far we have no answer, but some modulating factors of environmental or genetic origin could be the reason for this difference. On the other hand it cannot be excluded that SDAT and PD have different underlying pathomechanisms. Though in SDAT the cholinergic system and in PD the dopaminergic system is the main target of neurodegeneration, other neurotransmitter systems are concerned as well. Whether this is a consequence of cholinergic or dopaminergic degeneration or indicative of an expanding generalized degenerative process is another unresolved question.

References

Ambani LM, Van Woert MH, Murphy S (1975) Brain peroxidase and catalase in Parkinson's disease. Arch Neurol 32: 114–118

Birkmayer W, Riederer P (1975) Responsibility of extrastriatal areas for the apperance of psychotic symptoms. J Neural Transm 37: 175–182

Birkmayer W, Riederer P (1985) Die Parkinson-Krankheit: Biochemie, Klinik, Therapie, 2. Aufl. Springer, Wien New York

Bondareff N, Mountjoy CQ, Roth M (1982) Loss of neurons of origin of the adrenergic projections to the cerebral cortex (nucleus locus coeruleus) in senile dementia. Neurology (NY) 32: 165–168

Burns BT, Chiueh CC, Markey SP, Ebert MH, Jacobowitz DM, Kopin I (1983) A primate model of parkinsonism: selective destruction of dopaminergic neurons in the pars compacta of the substantia nigra by N-methyl-4-phenyl-1,2,3,6-tetrahydropyridine. Proc Natl Acad Sci 80: 4546–4550

Cadet JL (1988) A unifying theory of movement and madness: involvement of free radicals in disorders of the isodentritic core of the brain stem. Med Hypotheses 27: 59–63

D'Amato RJ, Zweig RM, Whitehouse PJ, Wenk GL, Singer HS, Mayeux R, Price DL, Snyder S (1987) Aminergic systems in Alzheimer's disease and Parkinson's disease. Ann Neurol 22: 229–236

Davies GC, Williams AC, Markey SP, Ebert MH, Caine ED, Reichert CM, Kopin I (1979) Chronic parkinsonism secondary to intravenous injection of meperidine analogues. Psychiatry Res 1: 249–254

Dexter DT, Carter CJ, Wells FR, Javoy-Agid F, Agid Y, Lees A, Jenner P, Marsden CD (1989) Basal lipid peroxidation in substantia nigra is increased in Parkinson's disease. J Neurochem 52: 381–389

Ehringer H, Hornykiewicz O (1960) Verteilung von Noradrenalin und Dopamin (3-Hydroxytyramin) im Gehirn des Menschen und ihr Verhalten bei Erkrankungen des extrapyramidalen Systems. Klin Wschr 38: 1236–1239

Etienne P, Robitaille Y, Wood P, Gauthier S, Nair NPV, Quirion R (1986) Nucleus basalis neuronal loss, neuritic plaques and choline acetyltransferase activity in advanced Alzheimer's disease. Neuroscience 19: 1279–1291

Fonnum F (1975) A rapid radiochemical method for the determination of choline acetyltransferase. J Neurochem 24: 407–409

Gaspar P, Gray F (1984) Dementia in idiopathic Parkinson's disease. A neuropathological study of 32 cases. Acta Neuropathol (Berl) 64: 43–52

Gibb WRG (1988) The neuropathology of parkinsonian disorders. In: Jankovic J, Tolowa E (eds) Parkinson's disease and movement disorders. Urban & Schwarzenberg, Baltimore Munich, pp 205–223

Gibb WRG, Lees AJ (1987) The progression of idiopathic Parkinson's disease is not explained by age-related changes. Clinical and pathological comparisons with post-encephalitic parkinsonian syndrome. Acta Neuropathol (Berl) 73: 195–201

Hirsch E, Graybiel AM, Agid YA (1988) Melanized dopaminergic neurons are differentially susceptible to degeneration in Parkinson's disease. Nature 334: 345–348

Hornykiewicz O, Kish SJ (1986) Biochemical pathophysiology of Parkinson's disease. In: Yahr MD, Bergmann KJ (eds) Advances in neurology, vol 45. Raven Press, New York, pp 19–34

Ichimija Y, Arai H, Kosaka K, Iizuka R (1986) Morphological and biochemical changes in the cholinergic and monoaminergic system in Alzheimer-type dementia. Acta Neuropathol (Berl) 70: 112–116

Ingram VM, Koenig JH, Miller CH, Moore HE, Blanchard B, Perry DE (1987) The locus coeruleus: computer assisted 3-dimensional analysis of degeneration in Alzheimer's and Down's disease. In: Wurtman RJ, Corkin SH, Growden JH (eds) Alzheimer's disease: advances in basic research and therapy. Center for brain science and metabolism charitable trust, Cambridge, pp 435–440

Jellinger K (1986 a) Pathology of parkinsonism. In: Fahn S, Marsden CD, Jenner P, Teychenne R (eds) Recent developments in parkinsonism. Raven Press, New York, pp 33–66

Jellinger K (1986 b) Overview of morphological changes in Parkinson's disease. In: Yahr MD, Bergmann KJ (eds) Advances in neurology, vol 45. Raven Press, New York, pp 1–18

Jellinger K (1987 a) The pathology of parkinsonism. In: Marsden CD, Fahn S (eds) Movement disorders 2. Butterworth, London, pp 124–165

Jellinger K (1987 b) Neuropathological substrates of Alzheimer's and Parkinson's disease. J Neural Transm [Suppl] 24: 109–129

Jellinger K (1987 c) Quantitative changes in some subcortical nuclei in aging, Alzheimer's disease and Parkinson's disease. Neurobiol Aging 8: 556–561

Jellinger K (1989) Pathology of Parkinson's syndrome. In: Calne DB (ed) Handbook of experimental pharmacology, vol 88. Springer, Berlin Heidelberg New York, pp 47–112

Kish SJ, Mortio C, Hornykiewicz O (1985) Glutathione peroxidase activity in Parkinson's disease brain. Neurosci Lett 58: 343–346

Mann DMA, Yates PO (1983) Pathological basis for neurotransmitter changes in Parkinson's disease. Neuropathol Appl Neurobiol 9: 3–19

Mann DMA, Yates PO, Hawkes J (1983) The pathology of the human locus coeruleus. Clin Neuropathol 2: 1–7

Mann DMA, Yates PO, Marcyniuk B (1985) Correlation between senile plaque and neurofibrillary tangle counts in cerebral cortex and subcortical structures in Alzheimer's disease. Neurosci Lett 56: 51–55

Mann DMA, Tucken CM, Yates PO (1987) The topographic distribution of senile plaques and neurofibrillary tangles in the brain of non-demented persons of different ages. Neuropathol Appl Neurobiol 13: 123–139

Perry TL, Godin DV, Hansen S (1982) Parkinson's disease: a disorder due to nigral glutathione deficiency? Neurosci Lett 33: 305–310

Riederer P, Wuketich S (1976) Time course of nigrostriatal degeneration in Parkinson's disease. J Neural Transm 38: 271–301

Riederer P, Birkmayer W, Seemann D, Wuketich S (1977) Brain-noradrenaline and 3-methoxy-4-hydroxyphenylglycol in Parkinson's syndrome. J Neural Transm 41: 241–251

Riederer P, Sofic E, Rausch WD, Schmidt B, Reynolds GP, Jellinger K, Youdim MBH (1989) Transition metals, ferritin, glutathione, and ascorbic acid in parkinsonian brains. J Neurochem 52: 515–520

Scatton B, Javoy-Agid F, Rouquier L, Dubois B, Agid Y (1984) Neurochemistry of monoaminergic neurons in Parkinson's disease. In: Usdin E, Carlsson A, Dahlström A, Engel J (eds) Catecholamines: neuropharmacology and central nervous system — therapeutic aspects. A Liss, New York, pp 43–52

Sofic E (1986) Untersuchung von biogenen Aminen, Metaboliten, Ascorbinsäure und Glutathion mittels HPLC-ECD und deren Verhalten in ausgewählten Lebensmitteln und im Organismus von Mensch und Tier. Thesis, Technical University of Vienna

Tomlinson BE, Irving D, Blessed G (1981) Cell loss in the locus coeruleus in senile dementia of Alzheimer's type. J Neurol Sci 49: 419–428

Tomonaga M (1983) Neuropathology of the locus coeruleus: a semiquantitative study. J Neurol 230: 231–240

Uhl GR, Hedreen JC, Price DL (1985) Parkinson's disease: loss of neurons from the ventral tegmental area contralateral to therapeutic surgical lesions. Neurology 35: 1215–1218

Whitehouse PJ, Hendreen JC, White CL, Price DL (1983) Basal forebrain neurons in the dementia of Parkinson's disease. Ann Neurol 13: 243–248

Changes in brain energy metabolism and the early detection of Alzheimer's disease

S. Hoyer

Department of Pathochemistry and General Neurochemistry, University of
Heidelberg, Federal Republic of Germany

Summary

Among the early and most prominent disturbances of dementia of Alzheimer type are glycolytic glucose breakdown and pyruvate oxidation, associated with an excessive protein catabolism in the brain. It is suggested that the abnormality in intracellular glucose homeostasis is caused by a deficiency at the insulin/insulin receptor level in the neuron, indicating resemblance to non-insulin dependent diabetes mellitus in non-nervous tissues, and a genetic influence as well. Therefore, non-nervous markers may help to detect AD: reduced activity of phosphofructokinase in skin fibroblasts; abnormal glucose tolerance test; abnormal morphology in nasal epithelium.

Introduction

Normal brain function is strongly correlated with undisturbed cerebral circulation and the supply of the substrates oxygen and glucose, and with the undisturbed utilization of these two substrates. The oxidative brain metabolism provides the basis for energy formation necessary to maintain function and structure of this organ. Perturbation of the oxidative metabolism may thus give rise to the generation of abnormal clinical states.

Cell loss and the formation of paired helical filaments as well as neuritic plaques in the brain are considered to be hallmarks of dementia of Alzheimer type (DAT) (Terry et al., 1981; Tomlinson et al., 1970). However, they may be of value for morphological classifications only, but cannot be used for an early diagnostic procedure under in vivo conditions. On the other hand, unequivocal and reliable peripheral markers to rapidly assure the clinical diagnosis of DAT are lacking as yet. Since glucose and related brain metabolism play a pivotal role in maintaining normal neuronal function, it will be discussed whether or not early abnormalities in cerebral glucose and energy metabolism may serve as early markers of DAT.

Glucose and related metabolism in normal brain

With respect to oxidative brain metabolism, the mature, healthy, nonstarved mammalian brain uses glucose only to obtain energy as ATP to meet its functional and structural requirements (Gibbs et al., 1942; Gottstein et al., 1963; Hoyer, 1970; Siesjö, 1978). Cerebral glucose metabolism is controlled by means of various mechanisms: its uptake by an obviously insulin-dependent carrier-mediated transport mechanism at the blood-brain barrier (Bachelard, 1971; Hertz et al., 1981; Kahn, 1985); its glycolytic breakdown by the enzymes hexokinase and pyruvate kinase working in a concerted way under the control of phosphofructokinase (Newsholme and Start, 1973); its first step of oxidation by the multienzyme complex pyruvate dehydrogenase (Perry et al., 1980); its energy production by the respiratory chain together with feedback mechanisms influencing the activity of glycolysis.

The mitochondrial multienzyme complex pyruvate dehydrogenase (PDH) occupies a pre-eminent place in metabolism. It provides acetyl groups for subsequent oxidation and energy production on the one hand, and for acetylcholine synthesis on the other. The activity of the enzyme choline acetyltransferase (CAT) which catalyzes acetylcholine formation is closely linked functionally to the PDH complex (Gibson et al., 1975) and is stimulated by means of insulin (Kyriakis et al., 1987).

Furthermore, there is evidence that cytosolic Ca^{2+} increases the activity of the dephosphorylated (active) form of PDH after membrane depolarization, more extensively in brain mitochondria than in synaptosomes. Calcium accumulation by mitochondria was found to be closely linked to PDH activity when fuelled by pyruvate. Phosphorylation of the PDH α-subunit may interfere with the calcium-buffering capacity of mitochondria, indicating that the activity of PDH contributes to cellular Ca^{2+} homeostasis. On the other hand, PDH activity was markedly reduced when Ca^{2+} uptake by mitochondria was inhibited (Browning et al., 1981; Hansford and Castro, 1985). Independently of the intramitochondrial Ca^{2+} concentration, PDHC is activated by the stimulation of its phosphatase by means of insulin (Denton et al., 1986).

Under normal conditions, glucose carbon is rapidly transferred into amino acids via the tricarboxylic acid cycle and the γ-aminobutyric acid (GABA) shunt (Sacks, 1957, 1965). Glutamate, glutamine, aspartate, and GABA are formed most abundantly (Barkulis et al., 1960;

Geiger et al., 1960; Wong and Tyce, 1983). At least two compartments of these acidic, glucoplastic amino acids are assumed in brain, one of which is a storage compartment. These amino acids may partly serve as a fuel reserve under emergency conditions when glucose is lacking (see below). Moreover, glutamate/aspartate function as excitatory neurotransmitters active in nearly the whole of the brain but particularly in the cortical entorhinal afferents and the Schaffer collaterals both ending in the CA_1 subfield of the hippocampus, as well as in its mossy fibers (Monaghan et al., 1983; Strange, 1988). Cholinergic neurons within the septum, the nucleus of the diagonal band, the nucleus magnocellularis basalis, and the substantia innominata receive glutamatergic afferents, the first two from the hippocampus (Davies et al., 1984; Malthe-Sørensen et al., 1980; Walaas and Fonnum, 1980). Upon K^+-evoked depolarization, aspartate, glutamate, and GABA were preferentially released from neocortical tissue, the efflux of glutamate being calcium-dependent (Smith et al., 1983).

After its release from nerve endings, glutamate binds with high affinity to postsynaptic dendritic membranes (Rothman and Olney, 1986). The regional distribution of glutamate binding sites generally exhibits a marked heterogeneity, with the highest levels being found in the stratum moleculare and stratum radiatum of the hippocampus besides layers I and II in the frontal cortex (Greenamyre et al., 1984). More particularly, the distribution of different glutamate receptors was also found to be inhomogeneous in brain areas. The highest densities of N-methyl-D-aspartate (NMDA) receptors were found in stratum oriens and stratum radiatum of the hippocampal CA_1 subfield. In cerebral cortex, the highest levels were found in frontal, anterior cingulate and pyriform cortices with a dense band of sites in the cortical layers I, II and III, and Va. Glutamatergic quisqualate receptors are co-localized with NMDA binding sites indicating that these two glutamate receptors may work in concert. On the other hand, glutamatergic kainate binding sites displayed a distribution pattern complementary to that of NMDA sites with high densities in the termination zone of the mossy fibers in the stratum lucidum of the hippocampus, and in layers V and VI of the cerebral cortex (Cotman et al., 1987). Besides these three main glutamate receptor types (NMDA, quisqualate, kainate as agonists), a fourth one is known with L-2-amino-4-phosphonobutyrate (L-AP 4) as agonist (Cotman and Iversen, 1987).

Calcium and other divalent cations were found to regulate and to increase the maximum number of Na-independent glutamate binding sites without changing the affinity of the ligand for its receptor. Monovalent cations such as, for example, sodium and potassium inhibited this binding dose-dependently. The structural analogs of glutamate

(NMDA, quisqualate and kainate) increased membrane permeability for monovalent cations in hippocampal neurons. Fast excitatory transmission was mediated mainly by quisqualate and kainate receptors, whereas NMDA caused large openings for calcium influx (Jahr and Stevens, 1987). The response of NMDA was potentiated by means of glycine (Johnson and Ascher, 1987). Application of aspartate and glutamate to the neocortex and the CA_1 subfield of the hippocampus gave rise to a reduction of extracellular calcium which moved from the extracellular space into postsynaptic apical dendrites through channels activated by amino acid-dependent membrane depolarization (Zanotto and Heinemann, 1983). Excess glutamate increased oxygen consumption in hippocampal slices up to 200% of the resting level, indicating a hyperexcitability of the neurons and high energy utilization (Nishizaki et al., 1988).

During the past three years, it has been well documented that insulin and insulin receptors exist in the brain, the former in a concentration higher than in plasma. Brain insulin and its receptors appeared to be compartmented towards peripheral insulin/insulin receptors, thus indicating the independence of the latter compartment (Havrankova et al., 1978). Periventricular hypothalamic cells contain insulin mRNA (Young, 1986). Besides some areas of the olfactory system and the hypothalamus, insulin receptor density was found to be particularly high in the entorhinal cortex, ventral subiculum, nucleus amygdala, CA_1 subfield of the hippocampus, and lateral septum (Werther et al., 1987). Brain insulin receptors are composed of insulin-binding α-subunits, and of β-subunits, the phosphorylation of which is stimulated by insulin and the reactive component of which is a tyrosine kinase. The brain insulin receptor differed from insulin receptors in other tissues mainly in its α-subunit, which obviously represents a subtype with respect to insulin binding whereas the β-subunit appeared to be identical (Gammeltoft et al., 1984; Kahn, 1985; Roth et al., 1986).

Although the function of brain insulin and insulin receptors has not yet been elucidated, it may be supposed that insulin in the brain acts in the same way as in non-nervous tissue. Insulin has an effect on glucose transport across membranes, as suggested for the blood brain barrier (Hertz et al., 1981). As detailed above, insulin increased the activities of PDH (Denton et al., 1986), and CAT (Kyriakis et al., 1987). Furthermore, it inhibited the expression of phosphoenolpyruvate carboxykinase (Chu et al., 1987) which can be activated in emergency conditions such as ischemia (Hoyer and Krier, 1986). In the cell, insulin showed anabolic effects, and insulin deficiency induced catabolism with proteolysis and lipolysis (Kahn, 1985).

It thus becomes obvious that the cerebral glucose metabolism and its control, as well as the metabolism of releated compounds may play a central role in maintaining normal neuronal function. Disturbances in glucose metabolism and its control may, therefore, give rise to neuronal damage causing abnormal mental states, e.g. dementia of Alzheimer type (DAT). In the following sections, it will be discussed whether or not abnormalities in glucose and related brain metabolism contribute to changes indicative of early stages of DAT, and if so, whether they can be used as reliable markers of this brain disorder.

Glucose and related metabolism in DAT brain

In early-onset DAT, an early and predominant disturbance was characterized by a 44% reduction in the cerebral metabolic rate (CMR) of glucose whereas cerebral blood flow and CMR oxygen were un-altered (Hoyer et al., 1988). The diminished CMR glucose cannot be attributed to an insufficient glucose supply to the brain, since nor-moglycemia was observed. A diminution of CMR glucose of similar extent in Alzheimer patients with dominant inheritance had been re-ported (Polinsky et al., 1987).

CMR oxygen is not compromised in early-onset DAT and may be expected to be unchanged as can also be deduced from either the normal or elevated respiration rate and oxygen uptake rate of mito-chondria of freshly sampled homogenates of frontal DAT neocortex (Sims et al., 1985 b, 1987). Furthermore, the normal CO_2 production and the nearly normal ATP formation as measured in the same tissue samples, may indicate an undisturbed substrate oxidation in the tri-carboxylic acid cycle, the working enzymes of which are largely mi-tochondrial (Sims et al., 1983 b). Interestingly, the mitochondrial area in tangle-bearing pyramidal neurons of temporal cortex from DAT patients was found to be preserved, indicating a normally maintained oxidative metabolism (Sumpter et al., 1986). Therefore, the reduction of cerebral glucose utilization in DAT resulting in neuronal starvation may have other causes than a perturbation in substrate oxidation.

Several findings clearly demonstrated reduced activities of enzymes acting in glycolytic glucose breakdown. The flux-controlling enzyme phosphofructokinase has been found to be decreased to 10% of control values (Bowen et al., 1979; Iwangoff et al., 1980). Pyruvate formation may then diminish. This decrease and the reduced activity of the pyruvate dehydrogenase complex (PDHC) in DAT (Perry et al., 1980; Sorbi et al., 1983), may further decrease the formation of acetyl CoA for oxidation and acetylcholine production (Sims et al., 1980, 1983 a, 1985 b). Reduced PDHC activity may induce increased lactate for-mation, as was found in early-onset DAT (Hoyer et al., 1988). Since the abnormalities in glucose metabolism were intense even at the beginning of early-onset DAT, the primary metabolic disturbance may be suggested to consist in the glycolytic breakdown of glucose and its first step of oxidation at the PDHC level, whereas the subsequent oxidative breakdown in the tricarboxylic acid cycle may be assumed not to be altered (De Leon et al., 1988; Hoyer et al., 1988).

In early-onset DAT, the diminished CMR glucose was accompanied by a severe loss of amino acids and ammonia from the brain indicating catabolic proteolysis (Hoyer et al., 1988). In ante-mortem frontal and temporal cortices of DAT patients, an increased concentration of aspartate was found, whereas glutamate was reduced in temporal cortex but was unchanged in frontal cortex (Procter et al., 1988). These findings coincided with data from early-onset DAT patients in whom aspartate and glycine were released from the brain in significantly high concentrations whereas glutamate did not show to undergo any significant changes (Hoyer and Nitsch, 1989). From these results, it may be deduced that the deficiency in neuronal glucose may be partly substituted by endogenous glutamate which is used in the aspartate aminotransferase reaction. Aspartate may accumulate in and be released from the brain in high concentrations. When the neuron was starved of glucose under the different condition of arterial hypoglycemia causing slow waves-polyspikes in EEG, or depression of CNS functions in behaviour, respectively, CMR glucose fell to 54% of controls, whereas CMR oxygen and the concentration of energy-rich phosphates in brain cortex remained unchanged (Norberg and Siesjö, 1976). The brain cortical concentrations of glutamine and alanine decreased, and that of aspartate rose, whereas that of glutamate did not vary under the above conditions. In the hippocampus, glutamate and glutamine diminished, and aspartate increased (Butterworth et al., 1982). The concentration changes in the tissue were reflected in the amounts of amino acids released from the brain: the extracellular concentration of aspartate rose the most, followed by rather moderate increases in glutamate, taurine and γ-aminobutyric acid, whereas glutamine fell (Butcher et al., 1987). It may thus be concluded that, irrespective or the cause, the disturbed intracellular glucose homeostasis is followed by proteolytic changes with glutamate and aspartate being most severely involved.

From these early changes in DAT which may be assumed to be typical ones, as early changes in chronic diseases generally are, the question may be raised as to whether or not these abnormalities in cerebral glucose and related metabolism are reflected outside the brain, too, so as to be used as reliable peripheral markers in the diagnosis of early DAT.

Glucose metabolism related peripheral markers

Because morphological abnormalities are confined to the brain (Terry et al., 1981; Tomlinson et al., 1970), and as supported by the clinical syndrome of dementia, DAT appears as a uniquely neuropsychiatric brain disorder. However, different metabolic variations attributable to DAT have also been found in peripheral tissues, thus supporting the view that DAT may be a systemic disorder with its most predominant abnormalities in the brain. With respect to glucose metabolism, fibroblast phosphofructokinase activity was reduced in Alzheimer's disease and Down's syndrome (Sorbi and Blass, 1983). However, in a repeated study, this finding could only be partially confirmed (Sims and Blass, 1986). On the other hand, a reduced glycolytic flux and thus a decreased activity of phosphofructokinase may be deduced from the significant increase in fibroblast lactate production in patients suffering from morphologically confirmed DAT (Sims et al., 1985 a), the lactate production pointing to a reduced activity of the PDH complex.

As pointed out above, there is a close relationship between both glucose and Ca^{2+}-metabolism mediated by glutamate/aspartate. Because of the increased extracellular aspartate concentration in DAT, the neuronal Ca^{2+} homeostasis may be severely impaired. In cultured skin fibroblasts, total bound calcium nearly doubled in DAT as compared to controls (Peterson and Goldman, 1986), which may explain the reduced calcium uptake in the same kind of cells (Peterson et al., 1985). On the other hand, cytosolic free calcium was found to be decreased. The molecular basis of this finding remained unexplained as yet (Peterson et al., 1986).

Other data pointing to an abnormal regulation of glucose metabolism were reported by Bucht et al. (1983). In a retrospective study, no diabetes mellitus coexisted with DAT. On the other hand, fasting blood sugar was lower, and insulin levels were higher in the oral glucose tolerance test in DAT patients than in elderly healthy controls. The oral glucose tolerance test area was smaller in DAT as compared to vascular dementia patients and controls. These data may indicate an insulin resistance at the insulin receptor or postreceptor level.

At first sight, two more findings may be considered as being curious when attributed to glucose metabolism: olfactory tests deficits and abnormal morphology in nasal epithelium. Early-stage DAT patients who underwent the smell identification test and the olfactory match-to-sample discrimination task revealed impaired abilities to identify common odors and made significantly more errors as compared to controls (Kesslak et al., 1988).

Nasal epithelium neurons which are of central origin and which derived from autopsied material of DAT patients showed increased reactivity to neurofilament antibodies and abnormal neuronal structures. They also were strongly immunoreactive for phosphorylated forms of neurofilaments and exhibited extensive neurite accumulation (Talamo et al., 1989).

These two findings may point to a severe functional abnormality in the olfactory cortex obviously due to severe and maximal morphological involvement in DAT (Pearson et al., 1985). Interestingly, highest densities of insulin receptors were found in olfactory structures (Werther et al., 1987). If insulin in the brain functions on glucose metabolism as in non-nervous tissues, an early perturbation at the insulin/insulin receptor level in the brain may be assumed to occur in the pathogenesis of DAT (Hoyer, 1988). The insulin/insulin receptor mediated control of glucose metabolism may then become deficient causing the abnormalities in cerebral glucose utilization as detailed above. These pathobiochemical changes in glucose metabolism and related pathways are found to be most prominent in the brain, but may also become detectable outside the brain as reliable peripheral markers for DAT.

References

Bachelard HS (1971) Specific and kinetic properties of monosaccharide uptake into guinea pig cerebral cortex in vitro. J Neurochem 13: 213–222

Barkulis SS, Geiger A, Kawikata Y, Aguilar V (1960) A study of the incorporation of ^{14}C-derived from glucose into free amino acids of the brain cortex. J Neurochem 5: 339–348

Bowen DM, White P, Spillane JA, Goodhardt MJ, Curzon G, Iwangoff P, Meier-Ruge W, Davison AN (1979) Accelerated ageing or selective neuronal loss as an important cause of dementia? Lancet i: 11–14

Browning M, Baudry M, Bennett WF, Lynch G (1981) Phosphorylation — mediated changes in pyruvate dehydrogenase activity influence pyruvate-supported calcium accumulation by brain mitochondria. J Neurochem 36: 1932–1940

Bucht G, Adolfsson R, Lithner F, Winblad B (1983) Changes in blood glucose and insulin secretion in patients with senile dementia of Alzheimer type. Acta Med Scand 213: 387–392

Butcher SP, Sandberg M, Hagberg H, Hamberger A (1987) Cellular origins of endogeneous amino acids released into the extracellular fluid of the rat striatum during severe insulin-induced hypoglycemia. J Neurochem 48: 722–728

Butterworth RF, Merkel AD, Landreville F (1982) Regional amino acid distribution in relation to function in insulin hypoglycaemia. J Neurochem 38: 1483–1489

Chu DTW, Stumpo DJ, Blackshear PJ, Granner DL (1987) The inhibition of phosphoenolpyruvate carboxykinase (guanosine triphosphate) gene expression by insulin is not mediated by protein kinase C. Mol Endocrinol 1: 53–59

Cotman CW, Iversen LL (1987) Excitatory amino acids in the brain-focus on NMDA receptors. Trends Neurosci 10: 263–265

Cotman CW, Monaghan DT, Ottersen OP, Storm-Mathisen J (1987) Anatomical organization of excitatory amino acid receptors and their pathways. Trends Neurosci 10: 273–280

Davies SW, McBean GJ, Roberts PJ (1984) A glutamatergic innervation of the nucleus basalis/substantia innominata. Neurosci Lett 45: 105–110

DeLeon MJ, George AE, Marcus DL, Miller JD (1988) Positron emission tomography with the deoxyglucose technique and the diagnosis of Alzheimer's disease. Neurobiol Aging 9: 90–92

Denton RM, McCormack JG, Thomas AP (1986) Hormonal regulation of intramitochondrial metabolism. Biol Chem Hoppe-Seyler 367 [Suppl]: 64

Gammeltoft S, Kowalski A, Fehlmann M, van Obberghen E (1984) Insulin receptor in rat brain: insulin stimulates phosphorylation of its receptor β-subunit. FEBS Lett 172: 87–90

Geiger A, Kawikata Y, Barkulis SS (1960) Major pathways of glucose utilization in the brain in brain-perfusion experiment in vivo and in situ. J Neurochem 5: 323–338

Gibbs EL, Lennox WG, Nims LF, Gibbs FA (1942) Arterial and cerebral venous blood. Arterial-venous differences in man. J Biol Chem 144: 325–332

Gibson GE, Jope R, Blass JP (1975) Decreased synthesis of acetylcholine accompanying impaired oxidation of pyruvic acid in rat brain minces. Biochem J 148: 17–23

Gottstein U, Bernsmeier A, Sedlmeyer I (1963) Der Kohlenhydratstoffwechsel des menschlichen Gehirns. I. Untersuchungen mit substratspezifischen enzymatischen Methoden bei normaler Hirndurchblutung. Klin Wschr 41: 943–948

Greenamyre JT, Young AB, Penny JB (1984) Quantitative autoradiographic distribution of L-(^{3}H) glutamate-binding sites in rat central nervous system. J Neurosci 4: 2133–2144

Hansford RG, Castro F (1985) Role of Ca^{2+} in pyruvate dehydrogenase interconversion in brain mitochondria and synaptosomes. Biochem J 227: 129–136

Havrankova J, Schmelchel D, Roth J, Browstein M (1978) Identification of insulin in rat brain. Proc Natl Acad Sci USA 75: 5737–5741

Hertz MM, Paulson OB, Barry DI, Christiansen JS, Svendsen PA (1981) Insulin increases glucose transfer across the blood-brain barrier. J Clin Invest 67: 597–604

Hoyer S (1970) Der Aminosäurenstoffwechsel des normalen menschlichen Gehirns. Klin Wschr 48: 1239–1243

Hoyer S (1988) Glucose and related brain metabolism in dementia of Alzheimer type and its morphological significance. Age 11: 158–166

Hoyer S, Krier C (1986) Ischemia and the aging brain. Studies on glucose and energy metabolism in rat cerebral cortex. Neurobiol Aging 7: 23–29

Hoyer S, Nitsch R (1989) Cerebral excess release of neurotransmitter amino acids subsequent to reduced cerebral glucose metabolism in early-onset dementia of Alzheimer type. J Neural Transm 75: 227–232

Hoyer S, Oesterreich K, Wagner O (1988) Glucose metabolism as the site of the primary abnormality in early-onset dementia of Alzheimer type? J Neurol 235: 143–148

Iwangoff P, Armbruster R, Enz A, Meier-Ruge W, Sandoz P (1980) Glycolytic enzymes from human autoptic brain cortex: normally aged and demented cases. In: Roberts PJ (ed) Biochemistry of dementia. Wiley, Chichester, pp 258–262

Jahr CE, Stevens CF (1987) Glutamate activates multiple single channel conductances in hippocampal neurons. Nature 325: 522–525

Johnson JW, Ascher P (1987) Glycine potentiates the NMDA response in cultured mouse brain neurons. Nature 325: 529–531

Kahn CR (1985) The molecular mechanism of insulin action. Ann Rev Med 36: 429–451

Kesslak JP, Cotman CW, Chui HC, van den Noort S, Fang H, Pfeffer R, Lynch G (1988) Olfactory tests as possible probes for detecting and monitoring Alzheimer's disease. Neurobiol Aging 9: 399–403

Kyriakis JM, Hausman RE, Peterson SW (1987) Insulin stimulates choline acetyltransferase activity in cultured embryonic chicken retina neurons. Proc Natl Acad Sci USA 84: 7463–7467

Malthe-Sørensen D, Skrede K, Fonnum F (1980) Calcium dependent release of D-^{3}H-aspartate from the dorsal septum after electrical stimulation of the fimbria in vitro. Neuroscience 5: 127–133

Monaghan DT, Holets VR, Toy DW, Cotman CW (1983) Anatomical distributions of four pharmacologically distinct ^{3}H-L-glutamate binding sites. Nature 306: 176–179

Newsholme EA, Start C (1973) Regulation in metabolism. Wiley, Chichester, pp 100–104

Nishizaki T, Yamauchi R, Okada Y (1988) Enhancement of the oxygen consumption in the hippocampal slices of the guinea pig induced by glutamate and its related substances. Neurosci Lett 85: 61–64

Norberg K, Siesjö BK (1976) Oxidative metabolism of the cerebral cortex of the rat in severe insulin-induced hypoglycaemia. J Neurochem 26: 345–352

Pearson RCA, Esiri MM, Hiorns RW, Wilcock GK, Powell TPS (1985) Anatomical correlates of the distribution of the pathological changes in the neocortex in Alzheimer disease. Proc Natl Acad Sci 82: 4531–4534

Perry EK, Perry RH, Tomlinson BE, Blessed G, Gibson PH (1980) Coenzyme A acetylating enzymes in Alzheimer's disease: possible cholinergic "compartment" of pyruvate dehydrogenase. Neurosci Lett 18: 105–110

Peterson C, Goldman JE (1986) Alterations in calcium content and biochemical processes in cultured skin fibroblasts from aged and Alzheimer donors. Proc Natl Acad Sci USA 83: 2758–2762

Peterson C, Gibson GE, Blass JP (1985) Altered calcium uptake in cultured skin fibroblasts from patients with Alzheimer's disease. N Engl J Med 312: 1063–1065

Peterson C, Ratan RR, Shelanski ML, Goldman JE (1986) Cytosolic free calcium and cell spreading decrease in fibroblasts from aged and Alzheimer donors. Proc Natl Acad Sci USA 83: 7999–8001

Polinsky RJ, Noble H, DiChiro G, Nee LE, Feldman RG, Brown RT (1987) Dominantly inherited Alzheimer's disease: cerebral glucose metabolism. J Neurol Neurosurg Psychiatry 50: 752–757

Procter AW, Palmer AM, Francis PT, Lowe SL, Neary D, Murphy E, Doshi R, Bowen DM (1988) Evidence of glutamatergic denervation and possible abnormal metabolism in Alzheimer's disease. J Neurochem 50: 790–802

Roth RA, Morgan DO, Beaudoin J, Sara V (1986) Purification and characterization of the human brain insulin receptor. J Biol Chem 261: 3753–3757

Rothman SM, Olney JW (1986) Glutamate and the pathophysiology of hypoxic-ischaemic brain damage. Ann Neurol 19: 105–111

Sacks W (1957) Cerebral metabolism of isotopic glucose in normal human subjects. J Appl Physiol 10: 37–44

Sacks W (1965) Cerebral metabolism of doubly labeled glucose in humans in vivo. J Appl Physiol 20: 117–130

Siesjö BK (1978) Brain energy metabolism. Wiley, Chichester, pp 1–55

Sims NR, Blass JP (1986) Phosphofructokinase activity in fibroblasts from patients with Alzheimer's disease and age- and sex-matched controls. Met Brain Dis 1: 83–90

Sims NR, Bowen DM, Smith CCT, Flack RHA, Davison AN, Snowdon JS, Neary D (1980) Glucose metabolism and acetylcholine synthesis in relation to neural activity in Alzheimer's disease. Lancet i: 333–336

Sims NR, Bowen DM, Allen SJ, Smith CCT, Neary D, Thomas DJ, Davison AN (1983a) Presynaptic cholinergic dysfunction in patients with dementia. J Neurochem 40: 503–509

Sims NR, Bowen DM, Neary D, Davison AN (1983b) Metabolic processes in Alzheimer's disease: adenine nucleotide content and production of $^{14}CO_2$ from (U-^{14}C) glucose in vitro in human neocortex. J Neurochem 41: 1329–1334

Sims NR, Finegan JM, Blass JP (1985a) Altered glucose metabolism in fibroblasts from patients with Alzheimer's disease. N Engl J Med 313: 638–639

Sims NR, Finegan JM, Bowen DM, Blass JP (1985b) Mitochondrial function in Alzheimer's disease measured in vitro using neocortical tissue homogenates. J Neurochem 44 [Suppl]: S 192

Sims NR, Finegan JM, Blass JP, Bowen DM, Neary D (1987) Mitochondrial function in brain tissue in primary degenerative dementia. Brain Res 436: 30–38

Smith CCT, Bowen DM, Davison AN (1983) The evoked release of endogenous amino acids from tissue prisms of human neocortex. Brain Res 269: 103–109

Sorbi S, Blass JP (1983) Fibroblast phosphofructokinase in Alzheimer's disease and Down's syndrome. Banbury Report 15: Biological aspects of Alzheimer's disease. Cold Spring Laboratory, pp 297–307

Sorbi S, Bird ED, Blass JP (1983) Decreased pyruvate dehydrogenase complex activity in Huntington and Alzheimer brain. Ann Neurol 13: 72–78

Strange PG (1988) The structure and mechanisms of neurotransmitter receptors. Implications for the structure and function of the central nervous system. Biochem J 249: 309–318

Sumpter PQ, Mann DMA, Davies CA, Yates PO, Snowdon JS, Neary D (1986) An ultrastructural analysis of the effects of accumulation of neurofibrillary tangles in pyramidal neurons of the cerebral cortex in Alzheimer's disease. Neuropathol Appl Neurobiol 12: 305–319

Talamo BR, Rudel RA, Kosik KS, Lee VMY, Neff S, Adelman L, Kauer JS (1989) Pathological changes in olfactory neurons in patients with Alzheimer's disease. Nature 337: 736–739

Terry RD, Peck A, DeTeresa R, Schechter R, Horoupian DS (1981) Some morphometric aspects of the brain in senile dementia of Alzheimer type. Ann Neurol 10: 184–192

Tomlinson BE, Blessed G, Roth M (1970) Observations on the brains of demented old people. J Neurol Sci 11: 205–242

Walaas I, Fonnum F (1980) Biochemical evidence for glutamate as a transmitter in hippocampal efferents to the basal forebrain and hypothalamus in the rat brain. Neuroscience 5: 1691–1698

Werther GA, Hogg A, Oldfield BJ, McKinley MJ, Figdor R, Allen AM, Mendelsohn FAO (1987) Localization and characterization of insulin receptors in rat brain and pituitary gland using in vitro autoradiography and computerized densitometry. Endocrinology 121: 1562–1570

Wong KL, Tyce GM (1983) Glucose and amino acid metabolism in rat brain during sustained hypoglycemia. Neurochem Res 8: 401–415

Young WS (1986) Periventricular hypothalamic cells in the rat brain contain insulin mRNA. Neuropeptides 8: 93–97

Zanotto L, Heinemann U (1983) Aspartate and glutamate induced reactions in extracellular free calcium and sodium concentration in area CA_1 of "in vitro" hippocampal slices of rats. Neurosci Lett 35: 79–84

Choline metabolism in Alzheimer's disease: hints as to possible markers

A. Nordberg

Department of Pharmacology, Uppsala University, Sweden

Summary

For diagnostic and therapeutic purposes, it would be very important to have reliable early markers for Alzheimer's disease. Since there's still a lack of agreement between clinical diagnosis and histopathological examinations, we must try to find new diagnostic markers which allow us to characterize the course of the disease and determine the specificity of the dementia. Cholinergic deficits characterize Alzheimer brains and, recently, attention has been focused on the presence of cholinergic receptors on peripheral elements and on their role as peripheral markers for Alzheimer's disease. A decrease in the number of muscarinic and nicotinic receptors has been measured in lymphocytes from Alzheimer patients. Another interesting finding is the presence of cholinergic neuron antibodies in serum and CSF (see McRae, this symposium).

In vivo imaging techniques, such as positron emission tomography (PET), are promising techniques for the study of cholinergic receptor activity in human brain. In the attempt to visualize nicotinic receptors in vivo, the (+) and the (−) isomers of ^{11}C-nicotine were given intravenously to Alzheimer patients and a lower uptake of both isomers (especially the (+) form) has been observed in the cortical areas of the brain compared to age-matched healthy volunteers. This observation prompts further PET studies in patients with a different progression of Alzheimer's disease, as well as in other types of dementia and following different therapeutic strategies. An in vitro acetylcholine release model using autopsy human brain tissue allows studies on functional cholinergic activity and its interaction with potential drugs in pathological tissue. This model will be a useful tool for the development of new drugs of potential use in the treatment of Alzheimer's disease.

Introduction

Alzheimer's disease and senile dementia of Alzheimer type (AD/SDAT) are the most common types of primary dementias with an insidious onset and gradual regression of intellectual functions. For a definitive diagnosis both a clinical and a histopathological diagnosis is required, though a lack of agreement still exists between these two forms of examination (Todorov et al., 1975), which emphasizes the need for improved diagnostic resolution. Novel therapeutic strategies in AD/SDAT also highlight the need for early markers of the disease.

Neurochemical studies in AD/SDAT brains indicate deficiencies in several transmitter systems (for review see Hardy et al., 1985; Gottfries, 1985). The most consistent and pronounced changes have been found to occur in the cholinergic nervous system (for review see Perry, 1986). The acetylcholine synthesing enzyme choline acetyltransferase (CAT) was the first cholinergic marker reported to be changed in AD/SDAT (Davies and Maloney, 1976; Perry et al., 1977; White et al., 1977). Later, deficits in several presynaptic markers, such as high affinity choline uptake (Rylett et al., 1983), acetylcholine (ACh) synthesis (Sim et al., 1980), release (Nilsson et al., 1986), and number of nicotinic receptors (Nordberg and Winblad, 1986 b) have been observed in AD/SDAT brains. The muscarinic cholinergic receptors which are mainly supposed to be postsynaptically located are preserved in AD/SDAT (for review see Nordberg and Winblad, 1986 a) but a loss of muscarinic M_2 receptors cannot be excluded (Mash et al., 1985). Figure 1 illustrates a cholinergic synapse and its pre- and postsynaptic

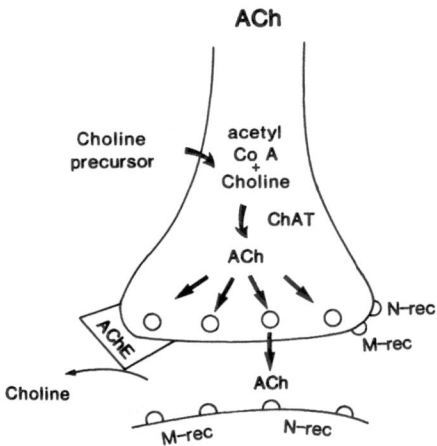

Fig. 1. Schematical drawing of a cholinergic synapse with its presynaptic and postsynaptic elements. *ChAT* choline acetyltransferase; *ACh* acetylcholine; *AChE* acetylcholinesterase; *M-rec* muscarinic receptor; *N-rec* nicotinic receptor

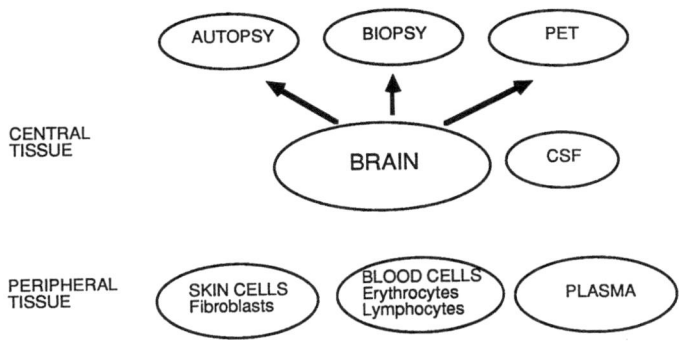

Fig. 2. Neural and non-neural tissues and techniques used in neurotransmitter studies in AD/SDAT

components. Neurochemical data for AD/SDAT brains have mostly been obtained from in vitro studies in autopsy or biopsy brain tissues and restricted to transmitter levels, enzyme activities, and receptor densities. New neurochemical techniques are being developed which will enable us to measure the functional physiological mechanisms in diseased brain tissues. Brain imaging techniques, such as positron emission tomography (PET), are promising techniques for in vivo studies of brain transmitter and receptor activity in AD/SDAT patients (Burns et al., 1989). For the cholinergic system some attempts have already been made (see below).

There is growing evidence that AD/SDAT might not be exclusively confined to the brain. Major efforts have been made to investigate whether human peripheral tissue elements such as blood cells (Adem et al., 1986; Zubenko et al., 1987) and skin cells (Peterson et al., 1986) can be used as models. The advantage of peripheral tissues is that they can easily be obtained from patients. It is still an open question whether changes in peripheral markers can correlate with the progress of the disease.

Figure 2 illustrates different tissues and techniques used in cholinergic transmission studies in AD/SDAT. This presentation will focus on the measurement of cholinergic markers in lymphocytes and brain of AD/SDAT patients using in vitro and in vivo techniques.

Choline markers in non-neural tissues of AD/SDAT patients

Choline

Endogeneous choline (Ch) content in plasma of AD/SDAT patients has been measured in several studies. Although findings are

Table 1. Changes of cholinergic markers in neural and non-neural tissues of patients with AD/SDAT

	Non-neural tissue			Neural tissue	
	Lymphocytes	Erythrocytes	Plasma	Brain	CSF
Ch content			↑		↓→↑
Ch uptake	↑			↓	
ACh synthesis				↓	
ACh release				↓	
AChE activity		↓		↓	↓→
ChE activity			↑		
ChAT activity				↓	
Musc. rec.	↓			↑↓→	
Nic. rec.	↓			↓	

↑ increase, ↓ decrease, → no change

somewhat conflicting, observations indicate an elevated level of Ch in the plasma of elderly compared to young individuals (Sherman et al., 1986) (Table 1). No distinction, however, could be made between AD/SDAT patients and healthy age-matched controls. The temperature dependent accumulation of choline in red blood cells was also higher in elderly and AD/SDAT patients compared to young volunteers (Glen et al., 1981; Sherman et al., 1986). It may therefore be assumed that observations probably indicated an aging process rather than a pathophysiological process in AD/SDAT.

Cholinesterase activity

Plasma cholinesterase activity has been reported to increase (Smith et al., 1982) and erythrocyte acetylcholinesterase activity to decrease (Chippenfield et al., 1981) in Alzheimer patients. Subsequent studies showed no changes either in plasma cholinesterase or erythrocyte acetylcholinesterase activities (Marquis et al., 1984; Adem et al., 1986 a), whereas a significant lower plasma cholinesterase activity was measured in patients with Parkinson's disease (Adem et al., 1986 a). These observations should be compared with decreased (Soininen et al., 1984) or unchanged (Davies, 1979; Wood et al., 1982) acetylcholinesterase activity in the CSF and loss of acetylcholinesterase activity (Pope et al., 1965; Davies, 1979; Perry, 1986; Attack et al., 1983) and increase in butyrilcholinesterase activity (Perry, 1980) in autopsy AD/SDAT brain tissue. A decreased ratio of CSF acetylcholinesterase to butyrilcholinesterase activity has been reported (Arendt et al., 1984). Although CSF acetylcholinesterase activity correlates with certain cognitive dysfunctions (Soininen et al., 1984) it probably is a poor marker for the progression of the AD/SDAT disease (Riekkinen et al., 1987).

Cholinergic receptors

The presence of muscarinic binding sites (Strom et al., 1981; Adem et al., 1986 b) and nicotinic binding sites (Adem et al., 1986 a) has been demonstrated on intact lymphocytes and lysed lymphocyte membranes. The muscarinic binding sites on lymphocytes may have functional relevance since they are modulated by GTP proteins (Bering et al., 1987). It is also possible that they represent a subset of muscarinic receptors (Adem et al., 1986 a; Bering et al., 1987). Evidence for functionally distinct nicotinic receptors on human lymphocytes had been provided already in 1979 (Richman and Arnason, 1979). The number of both muscarinic and nicotinic binding sites was found to be decreased (Adem et al., 1986 a) in lymphocytes from AD/SDAT patients compared to age-matched controls, whereas no change was observed in patients with multi-infarct dementia (MID). In lymphocytes from patients with Parkinson's disease the muscarinic but not the nicotinic binding sites were reduced (Adem et al., 1986 a). Our observations concerning the muscarinic receptors in lymphocytes from patients with Alzheimer's and Parkinson's diseases have recently been confirmed by Ravizza et al. (1988). In a comparative study of institutionalized and non-institutionalized patients, a significant reduction in muscarinic and nicotinic binding sites was only found in the lymphocytes from institutionalized AD/SDAT patients while no changes in receptor binding sites were found in the lymphocytes from non-institutionalized patients, who were not so heavily affected by the disease (Nordberg et al., 1987). In an ongoing longitudinal study we are currently investigating a larger patient population. Receptor changes are monitored and correlated with severity and progression of AD/SDAT. The possibility that lymphocyte cholinergic receptors can act as an antemortem marker in AD/SDAT deserves further investigation.

Positron emission tomography studies in AD/SDAT brains

Positron emission tomography (PET) is a non-invasive in vivo imaging techniques which gives cross-sectional images of tissue ra-

dioactivity in the brain, following administration of short-lived positron emitting tracers. PET studies in AD/SDAT patients with labelled deoxyglucose have shown abnormalities in the temporal and
parietal cortex (Friedland et al., 1985; Jagust et al., 1988). Very few
PET studies have been performed regarding AD/SDAT and functional
neurotransmitter activity. In order to study the cholinergic system in
brain, PET studies with ^{11}C-Ch were performed in humans (Gauthier
et al., 1985). ^{11}C-Ch has the disadvantage that a very small proportion
of the ^{11}C-Ch compound is taken up by the brain after intravenous
injection. Ch is also partly metabolised to acetylcholine but also phos-
phorylcholine. Holman et al. (1985) using SPECT technique and ^{123}I-
3-quinuclidinyl-4-iodobenzilate observed signs of preservation of the
muscarinic receptors in the brain of one Alzheimer patient. This is in
agreement with earlier in vitro binding studies (Nordberg and Winblad, 1986 b). Attempts to measure nicotinic receptors by PET technique
and ^{11}C-nicotine have been made in monkeys and healthy volunteers

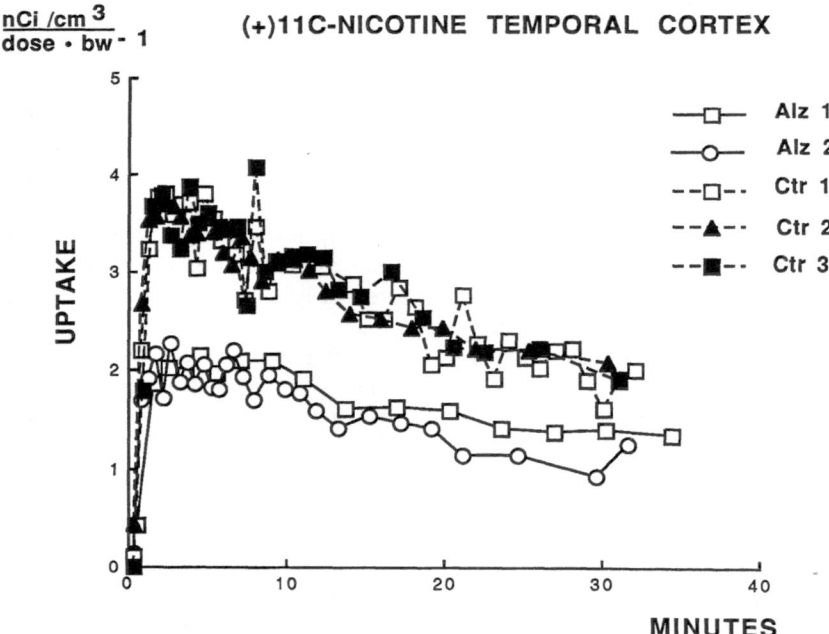

Fig. 3. Uptake and time course of ^{11}C-radioactivity in temporal cortex following intravenous injection of (+) ^{11}C-nicotine to Alzheimer patients (mean
age 66 ± 2 ys, duration of disease 4 ± 1 ys) and age-macthed healthy volunteers

(Nordberg et al., 1989 a; Nybäck et al., 1989). Following intravenous injection of a tracer dose, ^{11}C-nicotine is rapidly taken up by the brain and distributed to various parts of the brain with a distribution pattern similar to the one observed by in vitro nicotinic receptor binding techniques. When the two isomers of ^{11}C-nicotine, the (−) (L) and (+) (R) forms, are given intravenously to Alzheimer patients, a lower uptake of both isomers is observed in areas such as frontal and temporal cortex, compared to age-matched healthy volunteers (Nordberg et al., 1990). Especially, a much lower uptake of (+) ^{11}C-nicotine to cortical regions is observed in Alzheimer patients compared to controls (Fig. 3). Preliminary parallel studies using ^{11}C-butanol as a flow marker points to the conclusion that the difference observed in (+) ^{11}C-nicotine uptake between Alzheimer patients and controls is due to a change in nicotinic receptor binding rather than simply to a change in blood flow. The observation of a lower uptake of (−) and especially (+) ^{11}C-nicotine in Alzheimer brains compared to age-matched controls might indicate a change in the subtypes of nicotinic receptors. In studies of postmortem AD/SDAT brain tissue a shift in the ratio of high to low affinity sites has been observed (Nordberg et al., 1988 a). Further studies are needed to clarify the underlying mechanism for the difference in uptake of (+) and (−) ^{11}C-nicotine in Alzheimer brains revealed by PET and its significance as a marker for nicotinic receptor changes.

Functional transmitter studies in autopsy AD/SDAT brains

Metabolically and functionally preparations can now be obtained from human brain after considerable postmortem delay and used for functional transmitter studies (for review see Dodd et al., 1988). The agonal state of the tissue is of the outmost importance. Normally, viable preparations can be obtained from control human brain tissue up to 24 h postmortem if the subject suddenly died. In AD/SDAT subjects, whose death is often slow, tissue respiration declines faster postmortem.

A model has been developed which allows measurement of synthesis and release of acetylcholine (ACh) from thin human autopsy brain slices (Nilsson et al., 1986). The release of ACh is enhanced with high potassium concentrations (35 mM) and lowered in the absence of calcium. A significantly lower potassium-induced release of ACh is observed in AD/SDAT cortical tissues compared to control tissues (Nilsson et al., 1986). The amount of ACh released in different human cortices positively correlates with the number of nicotinic and muscarinic receptors measured in the same preparation (Nordberg et al., 1987; Nordberg and Nilsson-Håkansson, 1989). A parallel reduction in potassium-evoked ACh release and in the number of nicotinic bind-ing sites is measured in the frontal cortex of AD/SDAT brains, while no such parallel change can be found between ACh release and number of muscarinic receptors. The potassium-evoked release of ACh is mark-edly diminished in cortical autopsy AD/SDAT tissue and comparable with earlier observations for the ACh synthesis in brain biopsies (Sim et al., 1980). An advantage of this release model is that the mechanisms of potential cholinergic drugs can be evaluated in AD/SDAT brain tissue. This is important since no reliable animal model is yet available.

Among the different therapeutic agents with various cholinergic effects that have been clinically tested in Alzheimer's disease, only the acetylcholinesterase inhibitors physostigmine and tetrahydroaminoac-ridine (THA) have shown some positive effects. Therefore, further investigations of their underlying mechanism of action are justified. The in vitro ACh release model is quite suitable. In cortical tissues from control subjects both physostigmine and THA decrease ACh release. This effect is expected since it can be explained by a negative feedback mechanism via presynaptic muscarinic autoreceptors. In AD/SDAT cortical tissue however, the acetylcholinesterase inhibitors en-hance the release of ACh and restore it to control levels (Nilsson et al., 1987). Thus both physostigmine and THA facilitate the release of ACh in AD/SDAT tissue while it has an opposite effect in control tissue. Further studies indicate that both muscarinic and nicotinic receptors must be directly/indirectly involved in the mechanism of action of the acetylcholinesterase inhibitors in AD/SDAT brain tissue (Nordberg et al., 1988 b, 1989 b, c; Nordberg and Nilsson-Håkansson, 1990). We believe that the ACh release method in human autopsy brain tissue is a valuable method for studying the effects and underlying mechanisms of some drugs of potential use in neurodegenerative disorders such as Alzheimer's disease.

Acknowledgements

This study was supported by grants from the Medical Research Council, The Swedish Tobacco Company, Loo and Osterman's Foundation, and Stohne's Foundation.

References

Adem A, Nordberg A, Bucht G, Winblad B (1986 a) Extraneuronal cholin-ergic markers in Alzheimer's and Parkinson's diseases. Prog Neuro-psychopharmacol Biol Psychiatry 10: 247–257

Adem A, Nordberg A, Slanina P (1986 b) A muscarinic receptor type in human lymphocyte in comparison of ^{3}H-QNB binding to intact lym-phocytes and lysed lymphocyte membranes. Life Sci 38: 1359–1368

Attack JR, Perry EK, Bonham JR, Perry RH, Tomlinson BE, Blessed G, Fairbain A (1983) Molecular forms of AChE in senile dementia of Alzheimer type: selective loss of the intermediate (10 S) form. Neurosci Lett 40: 199–204

Bering B, Moises HW, Muller WE (1987) Muscarinic cholinergic receptors on intact human lymphocytes — properties and subclass characterization. Biol Psychiatry 22: 1451–1458

Burns A, Tune L, Steele C, Folstein M (1989) Positron emission tomography in dementia: a clinical review. Int J Ger Psych 4: 67–72

Chipperfield B, Newman PM, Moyes ICA (1981) Decreased erythrocyte cholinesterase activity in dementia. Lancet ii: 199

Davies P (1979) Neurotransmitter-related enzymes in senile dementia of Alzheimer's type. Brain Res 171: 319–327

Davies P, Maloney AJF (1976) Selective loss of central cholinergic neurons in Alzheimer's disease. Lancet ii: 1403

Dodd PR, Hambley JW, Cowburn RF, Hardy J (1988) A comparison of methodologies for the study of functional transmitter neurochemistry in human brain. J Neurochem 50: 1333–1345

Friedland RP, Budinger TF, Koss E, Ober BA (1985) Alzheimer's disease: anterior-posterior and lateral hemispheric alterations in cortical glucose utilization. Neurosci Lett 53: 235–240

Gauthier S, Diksic M, Yamamoto L, Tyler F, Feindel W (1985) Positron emission tomography with ^{11}C-choline in human subjects. Can J Neurol Sci 12: 214

Glen AIM, Yates CM, Simpson J, Christie JE, Shering A, Whalley LJ, Jellineh EH (1981) Choline uptake in patients with Alzheimer pre-senile dementia. Psychol Med 11: 469–476

Gottfries CG (1985) Alzheimer's disease and senile dementia: biochemical characteristics and aspects of treatment. Psychopharmacology 86: 245–252

Hardy JA, Adolfsson R, Alafuzoff I, Bucht G, Marcusson J, Nyberg P, Perdahl E, Wester P, Winblad B (1985) Transmitter deficits in Alzheimer's disease. Neurochem Int 7: 545–563

Holman BL, Gibson RE, Hill TC, Eckelman WC, Albert M, Rebz RC (1985) Muscarinic acetylcholine receptors in Alzheimer's disease. In vivo imaging with iodine [123]-labeled I-quinuclidinyl-4-iodobenzilate and emission tomography. JAMA 254: 3063–3066

Jacust WJ, Friedland RR, Budinger TF, Koss E, Ober B (1988) Longitudinal studies of regional cerebral metabolism in Alzheimer's disease. Neurology 38: 909–912

Marquis JK, Vollicer L, Direnfeld LK, Freeman M (1984) Assay of cholinesterase in plasma, erythrocytes and cerebrospinal fluid (CSF) of SDAT patients and normal controls. In: Wurtman RJ, Corkin SJ, Growdon JH (eds) Alzheimer's disease: advances in basic research and therapies. Springer, Wien New York, pp 161–182

Mash DC, Flynn DD, Potter LT (1985) Loss of M 2 muscarinic receptors in the cerebral cortex in Alzheimer's disease and experimental cholinergic denervation. Science 228: 1115–1117

Nilsson L, Nordberg A, Hardy J, Wester P, Winblad B (1986) Physostigmine restores ^{3}H-acetylcholine efflux from Alzheimer brain slices to normal level. J Neural Transm 67: 275–285

Nilsson L, Adem A, Hardy J, Winblad B (1987) Do tetrahydroaminoacridine (THA) and physostigmine restore acetylcholine in AD/SDAT brains via nicotinic receptors? J Neural Transm 70: 357–368

Nordberg A, Winblad B (1986 a) Brain nicotinic and muscarinic receptors in normal aging and dementia. In: Fisher A, Hanin I, Lachman C (eds) Alzheimer's and Parkinson's diseases: strategies for research and development. Plenum Press, New York, pp 95–108 (Advances in behavioral biology, vol 29)

Nordberg A, Winblad B (1986 b) Reduced number of ^{3}H-nicotine and ^{3}H-acetylcholine binding sites in the frontal cortex of Alzheimer brains. Neurosci Lett 72: 115–119

Nordberg A, Nilsson-Håkansson L (1990) Modulation of cholinergic activity in Alzheimer brains by potential drugs. In: Bunney WE, Hippius H, Laakman G (eds) Neuropsychopharmacology proceedings of the XVIth CINP Congress (in press)

Nordberg A, Adem A, Nilsson L, Winblad B (1987) Cholinergic deficits in CNS and peripheral non-neuronal tissue in Alzheimer dementia. In: Dowdall M, Hawthorne J (eds) Cellular and molecular basis of cholinergic function. Ellis Horwood, Chichester, pp 858–868

Nordberg A, Adem A, Hardy J, Winblad B (1988 a) Changes in nicotinic receptor subtypes in temporal cortex of Alzheimer brains. Neurosci Lett 86: 317–321

Nordberg A, Nilsson L, Adem A, Hardy J, Winblad B (1988 b) Effect of THA on acetylcholine release and cholinergic receptors in Alzheimer brains. In: Giacobini E, Becker R (eds) Current research in Alzheimer therapy — cholinesterase inhibitors. Taylor & Francis, New York, pp 247–257

Nordberg A, Hartvig P, Lundqvist H, Antoni G, Ulin J, Långström B (1989 a) Uptake and distribution of $(+) - (R)$ and $(-) - (S)$-N-(methyl-^{11}C)-nicotine in the brain of Rhesus monkeys — an attempt to study nicotinic receptors in vivo. J Neural Transm [PD-Sect] 1: 195–205

Nordberg A, Nilsson-Håkansson L, Adem A, Hardy J, Alafuzoff I, Lai Z, Herrera-Marschitz, Winblad B (1989 b) In: Nordberg A, Fuxe K, Holmstedt B, Sundvall A (eds) Nicotinic receptors in the CNS — their role in synaptic transmission. Prog Brain Res 79: 353–362

Nordberg A, Nilsson-Håkansson L, Adem A, Lai Z, Winblad B (1989 c) Multiple actions of THA on cholinergic neurotransmission in Alzheimer brains. In: Iqbal H, Wiesnewski H, Winblad B (eds) Alzheimer's disease and related disorders. Alan Liss, New York, pp 1169–1178

Nordberg A, Hartvig P, Lilja A, Viitanen M, Amberla K, Lundqvist H, Andersson Y, Ulin J, Winblad B, Långström B (1990) Decreased uptake and binding of ^{11}C-nicotine in brain of Alzheimer patients as visualized by positron emission tomography. J Neural Transm [P-D Sect] 2: 215–224

Nybäck H, Nordberg A, Långström B, Halldin C, Åhlin A, Schwan C-G, Sedvall G (1989) Attempts to visualize nicotinic receptors in the brain of monkey and man by positron emission tomography. In: Nordberg A, Fuxe K, Holmstedt B, Sundwall A (eds) Nicotinic receptors in the CNS — their role in synaptic transmission. Prog Brain Res 79: 313–319

Perry E (1980) The cholinergic system in old age and Alzheimer's disease. Age Aging 9: 1–8

Perry E (1986) The cholinergic hypothesis: ten years on. Br Med Bull 42: 63–69

Perry EK, Perry RH, Gibson RH, Blessed G, Tomlinson BE (1977) A cholinergic connection between normal aging and senile dementia in the human hippocampus. Neurosci Lett 6: 85–89

Peterson C, Ratan RR, Shelanski ML, Goldman JE (1986) Cytosolic free calcium and cell spreading decrease in fibroblasts from aged and Alzheimer donors. Proc Natl Acad Sci USA 83: 7999–8001

Pope A, Hess HH, Levin E (1965) Neurochemical pathology of the cerebral cortex in presenile dementias. Trans Am Neuro Assoc 89: 15–16

Ravizza L, Ferrero P, Eva C, Rocca P, Tarenzi L, Benna P (1988) Peripheral cholinergic changes and pharmacological aspects in Alzheimer's disease. In: Giacobini E, Becker R (eds) Current research in Alzheimer therapy. Taylor & Francis, New York, pp 355–363

Richman DP, Arnason BGW (1979) Nicotinic acetylcholine receptor: evidence for a functionally-distinct receptor on human lymphocytes. Proc Natl Acad Sci 76: 4632–4635

Riekkinen PJ, Laulumaa V, Sirviö J, Soininen H, Helkala EL (1987) Recent progress in research in Alzheimer's disease. Med Biol 65: 83–88

Rylett RJ, Bull MJ, Colhoun EH (1983) Evidence for high affinity choline transport in synaptosomes prepared from hippocampus and neocortex of patients with Alzheimer's disease. Brain Res 289: 169–175

Sherman K, Gibson GE, Blass JP (1986) Human red blood cells choline uptake with age and Alzheimer's disease. Neurobiol Aging 7: 205–209

Sim NR, Smith CCT, Davison AN, Bowen DM, Flack RHA, Snowden JS (1980) Glucose metabolism and acetylcholine synthesis in relation to neuronal activity in Alzheimer's disease. Lancet i: 333–336

Smith RC, Ho BT, Hsu L, Vroulis G, Claghorn J, Schoolar J (1982) Cholinesterase enzymes in the blood of patients with Alzheimer's disease. Life Sci 30: 543–546

Soininen HS, Jolkkonen JT, Reini-Kainen KJ, Haolnen TO, Riekkinen PJ (1984) Reduced cholinesterase activity and somatostatin-like immuno-reactivity in cerebrospinal fluid of patients with dementia of the Alzheimer type. J Neurol Sci 63: 167–172

Strom TB, Lane MA, George R (1981) The parallel, time dependent bio model change in lymphocyte-mediated cytotoxicity after lymphocyte activation. J Immunol 127: 705–710

Todorov AB, Go RCP, Constantinidis J, Elston RC (1975) Specificity of the clinical diagnosis of dementia. J Neurol Sci 26: 81–98

White P, Goodhardt MJ, Keet J, Hiley CR, Carrasco LH, Williams IEI (1977) Neocortical cholinergic neurons in elderly people. Lancet i: 668–670

Wood PL, Etienne P, Lal S, Gauthier S, Cajal S, Nair NPV (1982) Reduced lumbar CSF somatostatin levels in Alzheimer's disease. Life Sci 31: 2073–2079

Zubenko GS, Cohen BM, Boller F, Malinakova I, Keefe N, Chojnacki B (1987) Platelet membrane abnormality in Alzheimer's disease. Ann Neurol 22: 237–244

The presence of antibrain antibodies in the CSF of some Alzheimer disease patients: correlation with CSF parameters

A. McRae[1, 2], **K. Blennow**[3], **A. Wallin**[3], **P. Fredman**[3], **C. G. Gottfries**[3], and **A. Dahlström**[1]

[1] Institute of Neurobiology, University of Göteborg, Göteborg, Sweden
[2] U-259, Bordeaux, France
[3] St. Jörgens Hospital, Göteborg, Sweden

Summary

Our recent investigations demonstrated that antibodies in the cerebrospinal fluid (CSF) of some AD patients recognize cholinergic neuronal populations in the rat central nervous system (CNS). The present study correlated immunocytochemical markings produced by AD and control CSF samples in a cholinergic brain region with clinical and CSF protein parameters. There was a higher incidence of positive markings produced by AD CSF compared to control CSF. No correlation was noted with basic clinical data. However, the blood-brain barrier function was significantly correlated with immunocytochemical markings produced by AD CSF.

Introduction

Alzheimer's disease (AD) is a neurodegenerative disorder characterized by loss of memory, and impairment of other higher cortical functions. Although a number of factors have been considered to play a role in the pathogenesis of AD, to date the etiology of the disease is obscure. Recent evidence suggests that the pathogenesis of AD may be related to aberrations in the immune system. This concept is supported by the presence of amyloid, immunoglobulins and complement in senile plaques (Ishii and Haga, 1976) and findings of antibrain antibodies directed against cholinergic neurons in the sera (Chapman et al., 1986; Fillit et al., 1985) and CSF (McRae-Degueurce et al., 1987) in AD patients. To further investigate the involvement of immunological factors in the pathogenesis of AD the present study was carried out to evaluate antibrain antibodies in the CSF of AD patients and correlate their presence with clinical and CSF parameters.

Material and methods

Patients and controls

The Alzheimer's disease (AD) group consisted of 68 patients consecutively admitted to the Department of Psychiatry and Neurochemistry St. Jörgen's Hospital. The clinical investigation included medical history, physical, neurological and psychiatric examination, screening laboratory tests, EEG and CT-scan of the brain. Patients with major psychiatric or neurological disorders, alcohol or drug abuse or major head trauma were excluded from the study. The diagnostic criteria were those of "probable Alzheimer's disease", outlined in the NINCDS-ADRDA Work Group (McKahn et al., 1984). The severity of dementia was evaluated according to DSM-III-R (Am Psychiatry Ass, 1987). In the AD group clinical vascular factors (mild hypertension, mild non-insulin dependent diabetes, mild ischaemic heart disease) were recorded. This course of action was taken since vascular factors in AD patients have been suggested to influence the blood-brain barrier (BBB) function (Blennow et al., submitted). The control group (n = 13) was obtained from persons admitted for minor urologic surgery under spinal anaesthesia. Individuals with histories or signs of psychiatric or neurological diseases and individuals with malignant diseases, systemic disorders (e.g. rheumatoid arthritis), or infectious diseases were excluded. The cognitive status was examined using the "Mini-Mental State" (Folstein et al., 1975). Scores of 28 − 30 were inclusion criteria for controls. Table 1 depicts the clinical data and CSF and serum parameters for the 2 groups of patients.

Analytical methods of serum and CSF

Lumbar punctures were performed in the morning under standard conditions in the L 3/L 4 or L 4/L 5 interspace. A volume of 12 ml was collected and gently mixed to avoid gradient effects. A blood sample was taken at the same time. CSF samples from both AD patients and controls with more than 500 erythrocytes/µl were excluded.

Quantitative determinations of albumin and IgG in serum and CSF was performed by rocket immunoelectrophoresis (Laurell, 1972). The albumin ratio, calculated as the ratio of CSF albumin to serum albumin, was used as an indicator for BBB function. The IgG index was calculated according to the formula: CSF/serum ratio for IgG divided by CSF/serum ratio for albumin and was employed as an indicator for intrathecal IgG synthesis (Link and Tibbling, 1977).

Immunocytochemical approach

Male adult rats (Sprague Dawley) were perfused with a mixture of 5% glutaraldehyde and allyl alcohol (1 M) in a cacodylate buffer (0.5 M, pH 12)

Table 1. Clinical data and CSF parameters in AD and controls

	Alzheimer's disease (n = 68)	Controls (n = 13)	Significance*
Clinical data			
Age (yrs)	70.8 ± 8.1	64.6 ± 6.7	p < 0.01
Severity			
mild	25 (37%)		
moderate	28 (41%)		
severe	15 (22%)		
Sex			
males	29 (43%)	13 (100%)	p < 0.001
females	39 (57%)	0 (0%)	
CSF parameters			
S-IgG (g/l)	9.9 ± 1.9	10.7 ± 1.8	NS
CSF-IgG (mg/l)	37.0 ± 15.5	28.7 ± 10.7	NS
Albumin ratio	7.2 ± 2.7	5.5 ± 1.8	p < 0.05
IgG index	0.52 ± 0.09	0.49 ± 0.05	NS

(Values in mean ± SD, or number and percentage of patients.)
* Wilcoxon 2-sample test

as previously described by McRae-Degueurce et al. (1987). Our former findings indicated that the CSF from some AD patients contained an antibody-like factor which recognized cholinergic-like neurons located in the medial septum (MS) region (McRae-Degueurce et al., 1987). For this reason the MS region was selected to further evaluate the presence of this antibody in the CSF samples from these patients. Serial frozen sections (10 μm) of the MS region were mounted on gelatine coated glass slides and pre-incubated with 2% normal goat serum for 30 minutes. The serial sections of the MS were incubated overnight with the CSF from patients (1 : 1) diluted in TRIS metabisulfite buffer (pH 7.6) containing 1% normal goat serum and 0.5% Triton-X-100. Control sections were incubated with only biotinylated anti-human IgG serum, excluding the CSF. The sections were processed for immunocytochemistry using the avidin biotin peroxidase complex method (Vector Lab, Burlingame, California).

The immunocytochemical stainings were rated as positive, weak, negative, or unspecific in a blinded fashion: the observer did not have knowledge of the clinical diagnosis.

Correlations

The ratings of the immunocytochemical reaction produced by the CSF samples were then correlated with some basic clinical and CSF parameters. The following parameters were selected: (*A*) *basic clinical data* 1) age and sex, 2) duration of disease, 3) severity, 4) history of disease with involvement of the immune system (e.g. major infectious and autoimmune disease), 5) vascular factors consisting of hypertension, minor/major coronary stroke, heart disease, diabetes mellitus and peripheral arteriosclerosis; (*B*) *protein analysis* of the CSF and serum (IgG levels, CSF/serum albumin ratio, the CSF IgG index, and signs of IgG production within the CNS).

Statistics

For comparison between quantitative parameters the Wilcoxon 2-sample test was used for comparisons of two groups and the Kruskal-Wallis test (Lehman, 1975) was used for comparisons of three or more groups.

The investigation was approved by the Ethical Committee in Göteborg and all patients had given their consent to the study.

Results

The results from the immunocytochemical investigation revealed that incubations of the MS with CSF samples from these patients produced 4 different types of staining patterns in this region: positive, weak, negative and unspecific. A *positive reaction* was considered as one that distinctly labeled neurons limited to the MS region. This type of staining labeled more neuronal elements than small sized cells. The neuronal cell body appeared to be labeled and it was rare to observe labeled processes. No labeling was observed in the striatum. Most samples labeled cortical cells and occasionally cortical neuronal processes. A *weak reaction* was judged as one which produced a faint and hazy labeling of neurons limited to the MS region. A *negative staining* was devoid of an immunocytochemical reaction in the MS. An *unspecific*

Table 2. Number and percentage of immunocytochemical reactions produced by CSF from Alzheimer's disease patients and controls

	Negative	Weak	Positive	Unspecific	Significance*
Alzheimer's disease	23 (34%)	17 (25%)	23 (34%)	5 (7%)	p < 0.05
Controls	10 (77%)	1 (8%)	2 (15%)	0	p < 0.05

* Kruskal-Wallis test

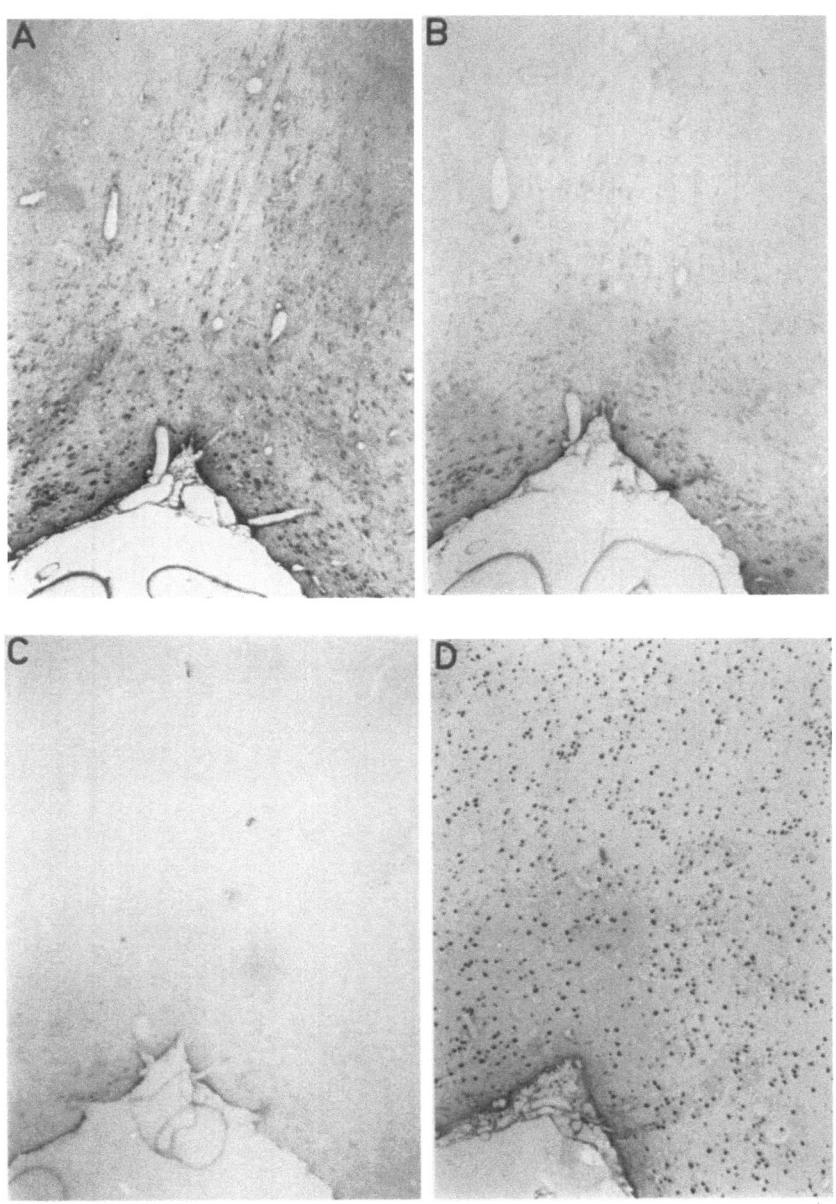

Fig. 1. Photomicrographs of cryostat sections (10 μm) of the medial septum region following incubations with AD patient CSF samples. This Fig. depicts examples of the different immunocytochemical stainings produced by AD CSF samples. **A** Positive, **B** Weak, **C** Negative, or **D** Unspecific. The immunocytochemical procedure is described in the text. Magnification × 64

Table 3. Relation between immunocytochemical results and basic clinical data or CSF parameters in Alzheimer patients

	Negative (n = 23)	Weak (n = 17)	Positive (n = 23)	Non-specific (n = 5)	Significance*
Clinical data					
Age (years)	69.6 ± 8.7	71.1 ± 7.3	70.0 ± 8.2	78.8 ± 4.8	NS
Sex					NS
males	11 (48%)	6 (35%)	11 (48%)	1 (20%)	
females	12 (52%)	11 (65%)	12 (52%)	4 (80%)	
Severity					NS
mild	10 (43%)	7 (41%)	8 (35%)	0	
moderate	10 (43%)	3 (18%)	11 (48%)	4 (80%)	
severe	3 (13%)	7 (41%)	4 (17%)	1 (20%)	
CSF parameters					
S-IgG (g/l)	9.4 ± 1.7	9.1 ± 1.6	10.4 ± 1.9	12.2 ± 2.4	p < 0.05
CSF-IgG (mg/l)	25.3 ± 9.7	37.5 ± 16.6	44.2 ± 11.8	55.4 ± 11.4	p < 0.0001
Albumin ratio	5.3 ± 1.5	7.3 ± 2.4	8.6 ± 2.8	9.3 ± 3.4	p < 0.0001
IgG index	0.51 ± 0.05	0.56 ± 0.10	0.52 ± 0.10	0.51 ± 0.03	NS

Values are given in mean ± SD or number and percentage of patients.
* Kruskal-Wallis test

reaction produced a dotted type of labeling which virtually stained every cell on the entire section including the striatum and cortex. Examining the immunocytochemical staining produced by the CSF from the different patient groups revealed distinct markings in the MS region. Examples of these immunocytochemical markings are depicted in Fig. 1. Table 2 compares the immuncytochemical reactions produced by AD and control patient CSF and demonstrates that the immunocytochemical reaction produced by the patient CSF varies. There was a statistically significant difference between immunocytochemical reactions produced by the AD group compared to the control group.

As indicated in Table 3, no significant differences were found when the immunocytochemical markings were compared to basic clinical data. However (Table 3) when the CSF markings were compared to protein analysis in the CSF and serum 3 parameters were highly significantly correlated with the immunocytochemical ratings, namely: serum IgG, CSF-IgG and CSF/S albumin-ratio. The CSF IgG index as a parameter for IgG synthesis within the CNS was not correlated to the markings produced by CSF.

Discussion

Our previous findings have suggested that the antibody in the CSF from some AD patients recognizes cholinergic neuronal populations in the rat brain (McRae-Degueurce et al., 1987). Recently we have extended these findings by demonstrating that the CSF antibody recognizes cultured cholinergic neurons (Dahlström et al., 1990). These immunocytochemical investigations add support to the ongoing concept that immunological factors may be playing a role in the pathogenesis of this disease. Even so in these previous studies no correlation was made between the presence of an antibody and the disease process. The results from this investigation further suggest that the presence of an IgG in the CSF of some AD patients is correlated with BBB function. This is of interest since some investigators have suggested that the BBB may be damaged in patients with AD (Wisniewski and Kozlowski, 1982). Furthermore recent studies by Filit et al. (1985) and Filit et al. (1987) have demonstrated the existence of anti-cholinergic and anti-vascular antibodies in the serum of some AD patients. The immunocytochemical results from this investigation suggest similar findings in the CSF from some AD patients. Firstly, there was a higher incidence of positive immunocytochemical reactivity in a cholinergic CNS region produced by AD CSF compared to controls. Secondly, this investigation revealed that serum IgG, the CSF/serum albumin ratio as a parameter for BBB function, and the CSF IgG level were highly significantly correlated with the immunocytochemical ratings (Table 3). Our findings could be interpreted to mean that antibodies produced outside the CNS, pass through the BBB and once inside the CNS interact with components of the neuronal network. With an increased serum level of IgG or an increased BBB permeability an increased amount of antibody (represented by an increased CSF IgG level) is available, thus producing a more pronounced immunocyto-chemical reaction. The higher frequency of this IgG in some AD patients compared to controls suggests that the IgG could play a role in the pathogenesis of this neurodegenerative disease (Table 1). This is further supported by the fact that the clinical vascular factors recorded in AD patients did not influence the frequency of appearance of the different CSF immunocytochemical staining patterns.

Conclusions

Questions remain concerning the relevance of these antibodies in the CSF to AD. Are they early markers for the disease, or indicators of a degeneration process, or do they delineate a sub-population of AD patients that could benefit from novel treatment strategies? Further experimentation in progress should provide answers to these questions and help to establish how antibodies existing in the CSF of AD patients are related to the disease process.

Acknowledgements

Supported by the Swedish MRC (2207), by Riksbankens Jubileumsfond, by Handl. Hj. Svensson's Foundation, Hans & Loo Osterman's Foundation, Axel & Margareta Ax:son Johnsson's Stiftelse för Allmännyttiga Ändamål, Stiftelsen Gamla Tjänarinnor, The Swedish Medical Association, and Göteborg's Medical Society.

References

American Psychiatry Association Task Force on Nomenclature and Statistics (1980) Diagnostic and Statistical Manual of Mental Disorders (DSM-III), 3rd edn. American Psychiatry Association, Washington DC
Blennow K, Wallin A, Fredman P, Karlsson I, Gottfries CG, Svennerholm L Blood-brain barrier disturbances in Alzheimer's disease patients are related to vascular factors. (Submitted)
Chapman J, Korczyn AD, Hareuveni M, Michaelson DM (1986) Antibodies to cholinergic cell bodies in Alzheimer's disease. In: Fisher A, Hanin I, Lachman C (eds) Alzheimer's and Parkinson's diseases: strategies for research and development. Plenum Press, New York, pp 329–336

Dahlstrom A, Wigander A, Lundmark K, Gottfries CG, McRae A (1990) Investigations on autoantibodies in Alzheimer's and Parkinson's diseases, using defined neuronal cultures. J Neural Transm [Suppl] 29: 195–206

Fillit HM, Luine VN, Reisberg B, Amador R, McEwen B, Zabriskie JB (1985) Studies of the specificity of antibrain antibodies in Alzheimer's disease. In: Hutton JT, Kenny AD (eds) Senile dementia. Alan R Liss, New York, pp 307–318

Filit HM, Kemeny E, Luine V, Weksler ME, Zabriskie JB (1987) Antivascular antibodies in the sera of patients with senile dementia of the Alzheimer's type. J Gerontol 42: 180–184

Folstein M, Folstein S, McHugh P (1975) "Mini-Mental State": a practical method for grading the cognitive state of patients for the clinician. J Psychiatr Res 12: 189–198

Ishii T, Haga S (1976) Immunoelectron microscopic localization of immunoglobulins in amyloid fibrils of senile plaques. Acta Neuropathol 36: 243–249

Laurell CB (1972) Electroimmunoassay. Scand J Clin Lab Invest 29: 21–37

Lehman EL (1975) Nonparametrics. Holden Day, San Francisco

Link H, Tibbling G (1977) Principles of albumin and IgG analyses in neurological disorders. II. Relation of the concentration of the proteins in serum and cerebrospinal fluid. Scand J Clin Lab Invest 37: 391–396

McKahn G, Drachman D, Folstein M, Katzman R, Price D, Stadlan E (1984) Clinical diagnosis of Alzheimer's disease: report of the NINCDS-ADRA Work Group under the auspices of Department of Health and Human Services Task Force on Alzheimer's disease. Neurology 34: 939–944

McRae-Degueurce A, Bööj S, Haglid K, Rosengren L, Karlsson JE, Karlsson I, Wallin A, Svennerholm L, Gottfries CG, Dahlström A (1987) Antibodies in the cerebrospinal fluid of some Alzheimer patients recognize cholinergic neurons in the rat central nervous system. PNAS 84: 9214–9218

Wisniewski HM, Kozlowski PB (1982) Evidence for blood-brain barrier changes in senile dementia of the Alzheimer's type. Ann NY Acad Sci 396: 119–127

Increased monoamine oxidase activity and vitamin B-12 deficiency in dementia disorders

L. Oreland[1], Y. Hiraga[1,2], S. S. Jossan[1], B. Regland[3], and C. G. Gottfries[3]

[1] Department of Medical Pharmacology, University of Uppsala, Sweden
[2] Zeria Pharmaceutical Co. Ltd., Japan
[3] Department of Psychiatry and Neurochemistry, University of Gothenburg, Sweden

Summary

MAO-B activity increases with age in the human brain and is further increased in AD/SDAT, Huntington's chorea, and Parkinson's disease. The cause is likely to be a reactive gliosis, presumably astrocytosis.

Demented patients with low serum B-12 levels have high platelet MAO-B activity. Evidence is presented favouring the hypothesis that low B-12, causing an increased level of homocysteic acid, an excitotoxic compound, is involved in the etio-pathogenesis of neurodegenerative states.

Introduction about monoamine oxidase

Monoamine oxidase (MAO; E.C. 1.4.3.4.) is the most well-studied enzyme associated with central monoamine transmitter systems. Based upon its sensitivity to clorgyline it is divided into two forms; a highly sensitive A-form and a less sensitive B-form (Johnston, 1968). The oxidative deamination of serotonin and noradrenaline in the brain is mainly brought about by MAO-A. Since dopamine is an equally well preferred substrate for MAO-A and -B, the oxidation of this transmitter substance is catalyzed by the two forms in proportion to their concentrations and to the concentration of dopamine exposed to the respective form (Oreland et al., 1983 a). MAO is present in almost all tissues in the body and, at the subcellular level, is localized mainyl to the outer mitochondrial membrane and to the nuclear membrane. Some MAO activity can also be found in the endoplasmic reticulum, where it is synthesized. In the mammalian brain both MAO-A and -B are present, whereas only the B-form is found in human platelets (Fowler and Oreland, 1982). The biochemical characteristics of MAO-B in human brain and platelets are very similar, although probably not identical (Fowler and Oreland, 1982).

When considering the physiological function of the MAO localized in the nervous tissue, it is of importance to distinguish between MAO activities in different compartments, such as extra- and intra-neuronally. In the monoamine system the intra-neuronal portion of the enzyme is small, less than 1% of the total amount (Oreland et al., 1983 a). Therefore it is not measurable with the conventional high substrate concentration ($> 10 \mu M$) techniques used with brain tissue homogenates, which mainly estimate extra-monoamine-synaptosomal MAO. However, in vivo, the monoamine membrane pumps increase the intra-neuronal monoamine substrate concentration hundredfolds, which increases the relative importance of this intra-axonal oxidative deamination. By using such a system in vitro with physiologically low amine concentrations ($0.1 \mu M$), allowing the membrane pumps to increase the intra-monoamine-synaptosomal concentration and by using various specific uptake blockers, an estimate of the contribution of the intra-neuronal (intrasynaptosomal) deamination in the single monoamine system can be obtained (Demarest et al., 1980; Fagervall and Ross, 1986; Oreland et al., 1983 a; Stenström et al., 1987). E.g. in Oreland et al. (1983 a) are given the proportions of extra- and intrasynaptosomal oxidation of dopamine by MAO-A and -B in rat and human nucleus caudatus.

Brain MAO and aging

It has long been known that there is a selective increase in the activity of MAO-B with age in the human brain (Fowler et al., 1980; Mann and Stanley, 1984). Several investigators (Leung et al., 1981; Mantle et al., 1976; Strolin-Benedetti and Keane, 1980) also found a selective increase in MAO-B activity in the rat brain, while Petkov et al. (1987) and Navarro et al. (1987) found an increase in MAO-A activity. Arai and Kinemuchi (1988), however, in a recent paper showed that MAO-A in the rat forebrain decreased, while MAO-A activity in the striatum increased with age. The rate of increase in human brain MAO-B activity varies in the different regions of the brain (Table 1) and the increase was found to be paralleled by an increasing concentration of otherwise unchanged MAO-B molecules (Fowler et al., 1980).

In contrast to MAO-B activity, MAO-A activity is remarkably steady with age in the human brain (Fowler et al., 1980), while there are reports both of an increase and a decrease in the rat brain (see above).

Table 1. Increase in MAO-B activity in various regions of the human brain (controls without neurological or psychiatric disorders)

Brain region	Increase in MAO-B activity (%/decade)
Centr. Semiov. (white matter)	30%
Hypothalamus	30***
N. Accumbens	28–30**
Thalamus	25–30**
N. Amygd.	25–30**
S. Nigra	23–29**
Putamen	18–29***
N. Caud.	18–24***
Hippocampus	19**
Cort. G. Cinguli	17–18*
Pons	15*
Cort. Precentralis	8–11*
C. Frontalis	8*
Medulla Oblong.	4

Data from Fowler et al. (1980) and Gottfries et al. (in preparation).
* p < 0.05, ** p < 0.01, *** p < 0.001 as compared with control

In another study (Oreland and Gottfries, 1986), MAO activities in white matter from centrum semiovale were analyzed in control brains. Mean MAO-A activity was in the order of $^1/_3$ that in nucleus caudatus. In contrast to the absence of any relation between MAO-A and age when the various regions consisting mainly of gray matter were investigated, a significant increase in MAO-A with age of about 10% per decade (p = 0.02) was found in the white matter. White matter MAO-B activity was about half of that in the nucleus caudatus and increased by about 30% per decade (p = 0.001) (Table 1).

When the "low substrate technique" (see above) for estimation of extra- and intra-synaptosomal MAO activities was applied on young (7 weeks) and old (107 weeks) rats, it was found that, in the striatal tissue, the total oxidation rate of dopamine was decreased by approximately 13%. This decrease was mainly due to a decreased oxidation rate within the DA synaptosomes by MAO-A (from 76% to 64%). There was, however, a significant increase in extra-synaptosomal MAO-B acitity of about 30% (from 3% to 4%) (Arai et al., 1985). The increase in extra-synaptosomal MAO-B activity explains the increase in MAO-B activity found when a conventional method is used (e.g. Strolin-Benedetti and Keane, 1980), since intra-synaptosomal contribution to the activity measured in this way is negligible.

A reasonable explanation of the results with brain tissue MAO-A and -B in relation to aging is that the degeneration of nerve terminals reflected in many other biochemical variables (see Gottfries, 1985), at least in the rat brain, can also be measured as a decrease in intra-neuronal MAO activity. This activity, which in the dopaminergic and noradrenergic cells mainly derives from the A-form of the enzyme (Demarest et al., 1980; Fagervall and Ross, 1986; Oreland and Engberg, 1986), is relatively high in the rat brain and relatively low in the human brain, which may explain why the decrease with age is usually recognized in the rat brain, but rarely so in human brain tissue. The increase in MAO-B activity is extra-neuronal in origin and is supposed to occur as a result of a reactive gliosis (Demarest et al., 1980; Oreland et al., 1980, 1983; Riederer and Jellinger, 1983). Gliosis has been reported to develop with aging (Brizzee et al., 1976; Geinisman et al., 1977). There is, however, no definite answer as to which type of glial cells might be connected with the increased MAO-B activity. Astrocytes have been shown to be rich in MAO-B activity (Levitt et al., 1982) and astrocytosis has been demonstrated at least in senile brains (Ravens and Calvo, 1966; Schechter et al., 1981).

Brain MAO after experimental lesions

The hypothesis that the increase in MAO-B activity with age indeed reflects an increase in the proportion of a cell population, probably glial, rich in MAO-B activity, finds some further support in experiments involving lesions of various kinds. Thus, after hemitransection of the rat brain, with the resulting loss of the transected neurons, it was found, in a series of reports, that there was a selective increase in MAO-B activity, as well as in MAO-B concentration, on the operated side (Fowler et al., 1979; Oreland et al., 1980; Stenström et al., 1985). An increase in fibrous astrocytes is usually seen in cerebral scar formations (Duffy et al., 1980).

Also biochemical lesions have resulted in a selective increase in cerebral MAO-B activity (Melamed et al., 1985; Schoepp and Azzaro, 1983). Francis et al. (1985) injected kainic acid locally into rat striata and found a persistent 15 – 20% loss of MAO-A activity, whereas MAO-B activity initially decreased, but then increased to more than twice the control value after about 2 months. Although in their paper there was no direct evidence for an astroglial proliferation, kainic acid had previously been shown to induce growth of astrocytes (Coyle et al., 1978). Recently it has been shown that also the selective cholinergic neurotoxin AF64A could induce a selective increase in the activity of the B-form of MAO. In this case, however, the increase

Table 2. Effect of long-term nitrous oxide (N_2O) exposure (4 weeks, about 20%) on brain monoamine oxidase (MAO) activity in rats (n = 7)

	Specific MAO activity (nmol/mg protein/min)			
	MAO-A (Mean ± S.E.M.)		MAO-B (Mean ± S.E.M.)	
	Control	N_2O	Control	N_2O
Striatum	1.29 ± 0.06	1.40 ± 0.06	0.61 ± 0.02	0.67 ± 0.01*
Hippocampus	2.01 ± 0.09	1.84 ± 0.04	0.89 ± 0.02	0.85 ± 0.02
Cortex	1.80 ± 0.09	1.61 ± 0.08	0.77 ± 0.02	0.81 ± 0.01 [+]
n	7	7	7	7

Data from Hiraga et al. (in manuscript).
[+] $0.05 < p < 0.10$, * $p < 0.05$ as compared with control

was obvious already after 2.5 weeks (Jossan et al., 1989 a). Of particular interest for the possible relation between vitamin B-12 deficiency and neurodegnerative changes (see below), are recent studies on the effect of chronic administration of nitrous oxide (N_2O) on brain MAO activities (Hiraga et al., in manuscript). Nitrous oxide induces a change in the valency state of cobalt in vitamin B-12, thereby inactivating the compound. The activity of the brain enzyme methionine synthetase, for which vitamin B-12 is a cofactor, was dramatically reduced by 4 weeks of continuous exposure (about 80%) and there was also a slight increase in striatal MAO-B activity (Table 2).

Thrombocyte MAO activity and aging

There are some reports about an increase in thrombocyte MAO activity with age, however, of a modest order of magnitude (Robinson et al., 1972). There are also a number of reports on the absence of such a correlation (e.g. Danielczyk et al., 1988; Murphy et al., 1976). Thus, if there is an increase with normal aging, it occurs at a very low rate.

Brain MAO activity in primary degenerative disorders

Biochemical studies

In our first study on MAO activities in the brains from patients with Alzheimer's disease or senile dementia of Alzheimer's type (AD/

SDAT), we found a significant selective increase in MAO-B activity in two out of the four regions investigated when compared with age-matched controls (Adolfsson et al., 1980). In a later study, so far only preliminarily reported (Oreland and Gottfries, 1986), this findings was confirmed in the only region studied, the nucleus caudatus. When MAO acitivity in white matter from centrum semiovale was investigated, there was no difference with regard to MAO-A activity, while there was a dramatic increase of about 70% with regard to MAO-B activity in comparison with the controls. A kinetic analysis revealed that there was no significant difference between white matter MAO-B in the AD/SDAT brains and the control brains with regard to K_m values towards β-phenylethylamine. The V_{max} values, however, differed fivefold with higher activities for the AD/SDAT brains (Oreland and Gottfries, 1986). There are a number of studies showing gliosis and in particular astrocytosis in Alzheimer's disease (Ravens and Calvo, 1966; Schechter et al., 1981). Microglia and/or astrocytes have recently been suggested to have a regulatory function for amyloid deposition in the brain (Hauw et al., 1988).

Autoradiographical studies

Histochemical and immunocytochemical techniques have been applied to investigate the cellular localization of MAO in the brain (Kishimoto et al., 1983; Levitt et al., 1982; Westlund et al., 1985). Fur-

Table 3. Percent increase in MAO-B binding sites in various parts of the brain from Alzheimer's patients compared with controls

Brain region	MAO-B increase (%)	Level of significance
Caudate nucleus (head)	33	0.001
Caudate nucleus (tail)	49	0.001
Globus pallidus (lateral)	23	0.001
Globus pallidus (medial)	25	0.001
Putamen	41	0.001
Frontal cortex	28	0.001
Occipital cortex	40	0.02
Temporal cortex	26	0.02
Corpus callosun	71	0.01
Frontal white matter	54	0.01
Occipital white matter	110	0.001

The increase in human brain MAO-B binding sites was measured in three controls, aged 71, 62 and 70 years and three patients with Alzheimer's disease, aged 80, 77 and 78 years, respectively

thermore, positron emission tomography using [^{11}C]-labelled suicide inactivators of both forms of MAO have been utilized to study the regional distribution of the enzyme in the human brain (Fowler et al., 1987). [^{3}H]- or [^{11}C]-labelled, selective irreversible inhibitors of MAO can also be used to investigate, in detail, the regional distribution of both forms of MAO in brain and other tissues by means of autoradiography. We have developed an autoradiographical method to visualize the topographical localization of MAO-B in large cryosections from postmortal human brain using [^{3}H]-labelled l-deprenyl with the main purpose of investigating the changes in MAO-B activity associated with aging and Alzheimer's disease (Jossan et al., 1989 b).

In Table 3, the increase in the specific binding of [^{3}H]l-deprenyl in various regions of the brain in the Alzheimer's patients is shown as percentage of the binding in the control brains. The rate of increase in MAO-B binding sites varied in different regions of the human brain. White matter had the highest increase while cerebral cortex and globus pallidus had the lowest increase.

Thrombocyte MAO activity in primary dementia disorders

Patients suffering from AD/SDAT have, in several studies, been shown to have higher mean thrombocyte MAO activity than age-matched controls (Adolfsson et al., 1980; Alexopoulos et al., 1984, 1987; Danielczyk et al., 1988; Smith et al., 1982). This increase was due to an increased V_{max} value, rather than to a change in the properties of the enzymes active site (as estimated by K_m value) both in male and in female patients (Oreland and Gottfries, 1986). Furthermore, Danielczyk et al. (1988) found a positive correlation between the degree of dementia and MAO activity. Oreland and Shaskan (1983) and Alexopoulos et al. (1987) have speculated whether high platelet MAO activity is caused by the disease or rather is a constitutional trait reflecting a predisposition to the development of a dementia syndrome. For the subgroup of patients with AD/SDAT, in which high thrombocyte MAO activity is associated with a functional deficit of vitamin B-12 (see below), a causal connection, however, seems obvious. Recently, another "radical" hypothesis has also been presented (Konradi et al., 1986). This hypothesis is based on the ability of hydrogen peroxide to enhance MAO-B activity together with its alleged neurodegenerative ability.

MAO activities in other neuro-degenerative disorders

A selective increase in MAO-B activity in the brains from patients with Huntington's chorea (Mann et al., 1980, 1986; Riederer and Jellinger, 1983) and possibly also Parkinson's disease (Riederer and Jellinger, 1983; Schneider et al., 1981) has been reported. This would further support the idea that an increase in brain MAO-B activity is a rather non-specific marker for degenerative processes, which might induce a reactive gliosis. In their 1986 paper, Mann et al. argue that the finding of Lange et al. (1976) of no change in the absolute number of glial cells in the striatum in Huntington's disease would support the idea that the increase in MAO-B in the caudate is intraneuronal. There are, however, reports of gliosis in the striatum in Huntington's disease (see Riederer and Jellinger, 1983).

With regard to thrombocyte MAO activity in Huntington's disease, Mann and Chiu (1978) reported a higher activity as compared to controls. Also Norman et al. (1987) found higher activity in patients with Huntington's disease than in controls, though with some reservations for the possible influence of environmental factors. Of particular interest in this context is the finding of Belendiuk et al. (1980) of higher thrombocyte MAO activity in the at-risk offspring of patients with Huntington's disease. This finding seems to indicate that, at least in this variety of degenerative disorder, high thrombocyte MAO activity serves as a marker for increased disposition.

Also in Parkinson's disease an increased thrombocyte MAO activity has been reported by Danielczyk et al. (1988), though with reservations for the possibility that drugs might have affected the result. In their material, increased thrombocyte MAO activity was found regardless of whether dementia was present or not. The latter finding is not surprising, since thrombocyte MAO activity ought rather to be correlated to the degree of predisposition or to the factors inducing the neurodegenerative changes than to a specific clinical symptom. A likely explanation for high thrombocyte MAO activities in patients with Parkinson's disease would be the over-representation of individuals with a premorbid personality profile (Poewe et al., this volume) typical of "high thrombocyte MAO activity" probands (Schalling et al., 1987) among such patients.

Another disorder, which is of considerable interest in connection with AD/SDAT, is Down's syndrome (Schapiro et al., 1988; Yatham et al., 1988; Delabar et al., this volume). The two disorders share some of the characteristic histopathological changes in the brain. With regard to thrombocyte MAO, however, the activity was found to be considerably lower than in controls (Fowler et al., 1981). An interesting opportunity for further studies on brain MAO in various neurodegenerative disorders would be to investigate brains from patients with Down's syndrome.

Vitamin B-12 deficiency in neurodegenerative disorders

Vitamin B-12 is an essential nutrient, and is considered to be necessary for maintaining normal function of the nervous system. The brain contains considerably high concentrations of vitamin B-12, which seems to be almost uniformly distributed throughout the brain (Inada et al., 1982 a).

Neuropathological findings

Schrumpf and Bjelke (1970) found that the levels of vitamin B-12 in CSF of patients with severe atrophy were slightly lower than those in a group of patients with moderate atrophy. This is well in accordance with the results of Inada et al. (1982 b), who found a decrease of vitamin B-12 levels in demented brains with severe neuronal loss, myelin degeneration, brain atrophy, and ventricular dilatation in comparison with controls. Furthermore, Kosaka et al. (1989) recently reported on significantly lower levels of methyl-cobalamin in some brain areas of patients with senile dementia of the Alzheimer type.

Structural *white-matter* changes in senile dementia of the Alzheimer type (SDAT) are rarely discussed. White-matter changes, however, seem to be frequent in SDAT. Microscopically they involve a partial loss of axons, myelin sheaths, and oligodendro-glial cells, accompanied by a mild reactive gliosis and a fibrohyaline arteriolosclerosis (Englund et al., 1989). Low centrum semiovale concentrations of white-matter lipids in SDAT have been reported by Gottfries et al. (1985). In a discussion about an association between the white-matter changes and brain nutritional deficiency of vitamin B-12, it is interesting that vitamin B-12 deficiency is known to cause a "subacute combined degeneration" (SACD), most often seen in cases of myelopathy, but also in cases of encephalopathy in patients with pernicious anaemia (Adams and Kubic, 1944; Ferraro et al., 1945). Notably, very similar lesions are often seen also in patients with AIDS dementia (Petito et al., 1985).

Agamanolis et al. (1978) performed *experimental B-12 deficiency* experiments in rhesus monkeys. Prolonged deprivation of vitamin B-12 made it quite clear that experimental SACD is primarily a demyelinating condition. At a later stage, there was degeneration and loss of axons and marked gliosis. These findings were confirmed by Scott et al. (1981), who administered nitrous oxide, known to cause a functional B-12 deficiency (above and Banks et al., 1986), to rhesus monkeys. Scott et al. (1981) also showed that dietary methionine had a preventive effect on the degenerative changes in the spinal cord. From this experiment the attention was focused on methionine as an important factor for myelin production in the white matter.

Siddons et al. (1975) found no pathological brain lesions in baboons, who had been deprived of dietary vitamin B-12 for 2.5 to 4 years. After administration of cobalamin analogues to the vitamin B-12 depleted baboons, however, neuropathological examination disclosed several striking abnormalities on microscopy of the cerebral hemispheres and spinal cords. There were also extensive scattered areas of demyelination and fibrillary gliosis.

Clinical findings

Vitamin B-12 deficiency leads to two major clinical syndromes. One is the classical syndrome of megaloblastic pernicious anaemia with a defective proliferation of all rapidly dividing cells, with such sequelae as glossitis and hypospermia. The other is the neurological syndrome, often with serum B-12 levels only slightly lowered and without megaloblastic anaemia. In fact, several recent studies show that borderline levels of vitamin B-12 require attention and should not be dismissed by the physician (Carmel and Karnaze, 1986; Carmel et al., 1987; Lindenbaum et al., 1988; Magnus, 1986). Anyhow, low B-12 levels in serum are known to be common in patients with senile dementia and confusion (Cole and Prchal, 1984; Droller and Dossett, 1959; Regland et al., 1988). Also in the cerebrospinal fluid low B-12 levels have been reported in cases with SDAT (Tiggelen et al., 1983). It seems likely that the main cause for a vitamin B-12 deficiency is an atrophic gastritis (Regland et al., in manuscript).

Deficiency of vitamin B-12 affects the metabolism of methylmalonic acid (MMA), as well as the methylation of homocysteine to methionine, and leads to increased serum levels of MMA and homocysteine and decreased levels of methionine. Lindenbaum et al. (1988) recently reported on vitamin B-12 deficiency and increased levels of MMA and homocysteine in patients with various neuropsychiatric disorders, including dementia; MMA and homocysteine in serum were normalized after vitamin B-12 therapy. With regard to methionine, Reynolds et al. (1989) have reported on significantly lowered levels of S-adenosylmethionine in CSF of patients with Alzheimer's disease.

Relationship between platelet MAO activity and vitamin B-12

In 1980 Glover et al. presented data on highly increased thrombocyte MAO activities in patients with megaloblastic anaemia. They also observed a significant correlation between thrombocyte MAO activity and the severity of bone marrow megaloblastic change, assessed by the deoxyuridine suppression test and bone marrow morphology.

In a retrospective study (Regland et al., 1988), we found a significantly negative correlation between serum B-12 levels and platelet MAO activity (r = 0.33; p < 0.002) in patients with dementia disorders.

To further test the causal relationship between platelet MAO activity and vitamin B-12 status, we investigated 14 patients with SDAT and relatively low serum B-12 levels, and 4 patients with pernicious anaemia (Regland et al., 1989 a). The patients with SDAT were treated intensively with intramuscular injections of hydroxocobalamin. In all SDAT patients with high thrombocyte MAO activity a considerable decrease in this activity was obtained within a short time after treatment. The patients with pernicious anaemia had a very high thrombocyte MAO activity, which was normalized after adequate treatment.

The "excitatory amino acid" hypothesis

Recently the role of excitatory amino acids, especially glutamate, in neurodegenerative processes has been extensively studied. With respect to neuronal toxicity of glutamate, N-methyl-D-aspartate (NMDA) receptors are the most important among the 3 subtypes (NMDA, kainic acid, quisqualic acid) of glutamate receptors (Robinson and Coyle, 1987). Excessive release of glutamate causes influx of cations such as Ca^{++}, K^+ and Na^+ into nerve cells followed by abnormal cell firing. After these processes, gradual dying of neurons occurs. It has been suggested that such a series of neurodegenerative processes is expected to be involved in cerebral ischaemia and neurodegenerative disorders such as Alzheimer's, Huntington's and Parkinson's disease and amyotrophic lateral sclerosis (ALS). In brains from patients with Alzheimer's disease, L-[³H]glutamate or [³H]TCP (which is also a selective ligand of NMDA receptors) binding, was significantly reduced (Greenamyre et al., 1987; Maragos et al., 1987; Simpson et al., 1988). Furthermore, the number of presynaptic uptake sites of glutamate, which was estimated with D-[³H]aspartate, was also decreased in patients with Alzheimer's disease (Cowburn et al., 1988; Cross et al., 1987; Simpson et al., 1988). A similar decrease in NMDA receptors and presynaptic uptake sites was observed in brains from Huntington's disease patients (Cross et al., 1986; Young et al., 1988). These findings indicate that abnormal functioning of glutamatergic systems may play an important role in neurodegenerative processes.

Another line of evidence for the excitatory amino acid hypothesis was obtained from the experiments with selective NMDA antagonists. To date several selective NMDA receptor antagonists (e.g. CGS 19755 and MK-801), which are effective even after systemic administration, have been synthesized. These substances have been shown to have a protective effect on cell loss in experimental cerebral ischaemia, induced by carotid or middle cerebral artery occlusion in rats, gerbils and cats (Boast et al., 1988; Ozyurt et al., 1988; Park et al., 1988) and in NMDA-induced neurotoxicity (Foster et al., 1987, 1988; Olney et al., 1987 a).

In addition to the evidence described above, the possibility has been suggested that some endogeneous agonists of NMDA receptors, such as quinolinic acid and L-homocysteic acid (HCA) may be produced in the brain and cause nerve cell degeneration (Do et al., 1988; Robinson and Coyle, 1987). HCA even seems to be a neurotransmitter, since it can be released from brain slices in a transmitter-like fashion (Do et al., 1986 a, b). An endogeneous production of HCA in the brain is especially interesting in relation to our working hypothesis about the etiopathological mechanisms behind Alzheimer's disease. Thus, HCA shows a more potent exciting action on neurons than does glutamate (Do et al., 1988); furthermore, the neurotoxic potency of HCA has been demonstrated in cultured cortical neurons (Kim et al., 1987) and in the retina (Olney et al., 1987 b).

Our working hypothesis

We have shown, as mentioned above, that increased thrombocyte MAO activity in patients with primary dementia disorders is connected with some deficiency in the utilization of vitamin B-12 (Regland et al., 1988). Furthermore, in the experimental vitamin B-12 deficiency model with long-term N_2O treatment, we found an increase in brain MAO-B activity (Table 2; Hiraga et al. in manuscript). These findings, in combination with the methionine transsulfuration metabolic pathways shown by Do et al. (1988), form the basis of our current working hypothesis about the etiopathological mechanisms of primary dementia disorders. Homocysteine is metabolized to 3 different metabolites; (1) to methionine by methionine synthetase, (2) to cystathionine by cystathionine β-synthetase, (3) to HCA through homocysteine sulfinic acid (Do et al., 1988). If long-term N_2O exposure blocks the metabolic pathway to methionine as a result of methionine synthetase inhibition, the synthesis rate of HCA may be supposed to be accelerated. Then, as a consequence of such accelerated HCA synthesis, excessive cell firing and progressive neuronal death may occur. This assumption is supported by the findings of increased levels of homocysteine in plasma and brain tissue in patients with cystathionine β-synthetase deficiency (Sprince et al., 1969), in patients with serum folate or cobalamine (vitamin B-12) deficiency (Kang et al., 1987; Stabler et al., 1988) and in rats with folate deficiency (Lin et al., 1989). There are, indeed, some studies confirming low levels of vitamin B-12 and methylcobalamin in the brains of patients with Alzheimer's disease (Inada et al., 1982; Kosaka et al., 1989). Some clinical support for the hypothesis was provided by Lindenbaum et al. (1988), who showed increased levels of serum homocysteine in patients with neuropsychiatric disorders. Furthermore, the levels of homocysteine were decreased after vitamin B-12 supplementation.

If our hypothesis of a key role of a vitamin B-12 functional deficiency in a subgroup of patients with primary dementia disorders should be correct, the decrease in methionine synthetase activity should also result in low levels of methionine. A state of methionine deficiency could theoretically result in a wide variety of symptoms, but of immediate interest with regard to primary dementia disorders is a recent report about low levels of S-adenosylmethionine in CSF from patients with Alzheimer's disease and an improvement in a variety of symptoms in a pilot double-blind cross-over study on the effect of oral administration of S-adenosylmethionine (Reynolds et al., 1989).

References

Adams RD, Kubic CS (1944) Subacute degeneration of the brain in pernicious anaemia. N Engl J Med 231: 1–9

Adolfsson R, Gottfries CG, Oreland L, Wiberg Å, Winblad B (1980) Increased activity of brain and platelet monoamine oxidase in dementia of Alzheimer type. Life Sci 27: 1029–1034

Agamanolis DP, Victor M, Harris JW, Hines JD, Chester EM, Kark JA (1978) An ultrastructural study of subacute combined degeneration of the spinal cord in vitamin B-12 deficient rhesus monkeys. J Neuropathol Exp Neurol 37: 273–299

Alexopoulos GS, Lieberman KW, Young RC, et al (1984) Platelet MAO activity and age at onset of depression in elderly depressed women. Am J Psychiatry 141: 1276–1278

Alexopoulos GS, Young RC, Leiberman KW, Shamoian CA (1987) Platelet MAO activity in geriatric patients with depression and dementia. Am J Psychiatry 144: 1480–1483

Arai Y, Stenström A, Oreland L (1985) The effect of age on extra-and intraneuronal MAO-A and -B activities in the rat brain. Biogen Amin 2: 65–71

Arai Y, Kinemuchi H (1988) Differences between monoamine oxidase concentrations in striatum and forebrain of aged and young rats. J Neural Transm 72: 99–105

Banks RGS, Henderson RJ, Pratt JM (1968) Reactions of gases in solution. Part III. Some reactions of nitrous oxide with transition-metal complexes. J Chem Soc (A): 2886–2889

Belendiuk K, Belendiuk GW, Freedman DX (1980) Blood monoamine metabolism in Huntingtons's disease. Arch Gen Psychiatry 37: 325–332

Boast CA, Gerhardt SC, Pastor G, Lehmann J, Etienne PE, Liebman JM (1988) The N-methyl-D-aspartate antagonists CGS 19755 and CPP reduce ischemic brain damage in gerbils. Brain Res 442: 345–348

Brizzee KR, Ordy JM, Hansche J, Kaack B (1976) Quantitative assessment of change in neuron and glial cell packing density and lipofuscin accumulation with age in the cerebral cortex of a non-human primate (*Macaca mulatta*). In: Terry, Gershon (eds) Neurobiology of aging, vol 3. Raven Press, New York, pp 229–244

Carmel R, Karnaze DS (1986) Physician response to low serum cobalamin levels. Arch Int Med 146: 1161–1165

Carmel R, Sinow RM, Karnaze DS (1987) Atypical cobalamin deficiency. Subtle biochemical evidence of deficiency is commonly demonstrable in patients without megaloblastic anemia and is often associated with protein-bound cobalamin malabsorption. J Lab Clin Med 109: 454–463

Cole MG, Prchal JF (1984) Low serum vitamin B-12 in Alzheimer type dementia. Age Ageing 13: 101–105

Cowburn R, Hardy J, Roberts P, Briggs R (1988) Presynaptic and postsynaptic glutamatergic function in Alzheimer's disease. Neurosci Lett 86: 109–113

Coyle J, Molliver M, Kuhar M (1978) In situ injections of kainic acid: a new method for selectively lesioning cell bodies while sparing axons of passage. J Comp Neurol 180: 301–323

Cross AJ, Slater P, Reynolds GP (1986) Reduced high-affinity glutamate uptake sites in the brains of patients with Huntington's disease. Neurosci Lett 67: 198–202

Cross AJ, Slater P, Simpson M, Royston C, Deakin JFW, Perry RH, Perry EK (1987) Sodium dependent D-[³H]aspartate binding in cerebral cortex in patients with Alzheimer's and Parkinson's diseases. Neurosci Lett 79: 213–217

Danielczyk W, Streifler M, Konradi C, Riederer P, Moll G (1988) Platelet MAO-B activity and the psychopathology of Parkinson's disease, senile dementia and multi-infarct dementia. Acta Psychiatr Scand 78: 730–736

Demarest KT, Smith DJ, Azzaro AJ (1980) The presence of the type A form of monoamine oxidase within nigrostriatal dopamine-containing neurons. J Pharmacol Exp Ther 215: 461–468

Do KQ, Herrling PL, Streit P, Turski WA, Cuénod M (1986a) In vitro release and electrophysiological effects in situ of homocysteic acid, an endogenous N-methyl-(D)-aspartic acid agonist, in the mammalian striatum. J Neurosci 6: 2226–2234

Do KQ, Mattenberger M, Streit P, Cuénod M (1986b) In vitro release of endogenous excitatory sulfur-containing amino acids from various rat brain regions. J Neurochem 46: 779–786

Do KQ, Herrling PL, Streit P, Cuénod M (1988) Release of neuroactive substances: homocysteic acid as an endogenous agonist of the NMDA receptor. J Neural Transm 72: 185–190

Droller H, Dossett JA (1959) Vitamin B-12 levels in senile dementia and confusional states. Geriatrics 14: 367–373

Duffy PE, Rapport M, Graf L (1980) Glial fibrillary acidic protein and Alzheimer-type senile dementia. Neurology 30: 778–782

Englund E, Brun A, Gustafsson L (1989) A white-matter disease in dementia of Alzheimer's type – clinical and neuropathological correlates. Int J Geriatr Psychiatry 4: 87–102

Fagervall I, Ross SB (1986) A and B forms of monoamine oxidase within the monoaminergic neurons of the rat brain. J Neurochem 47: 569–576

Ferraro A, Arieti S, English WH (1945) Cerebral changes in the course of pernicious anaemia and their relationship to psychiatric symptoms. J Neuropathol Exp Neurol 217–239

Foster AC, Gill R, Kemp JA, Woodruff GN (1987) Systemic administration of MK-801 prevents N-methyl-D-aspartate-induced normal neuronal degeneration in rat brain. Neurosci Lett 76: 307–311

Foster AC, Gill R, Woodruff GN (1988) Neuroprotective effects of MK-801 in vivo: selectivity and evidence for delayed degeneration mediated by NMDA receptor activation. J Neurosci 8: 4745–4754

Fowler CJ, Callingham BA (1979) The inhibition of rat heart type A monoamine oxidase by clorgyline as a method for the estimation of enzyme active centres. Mol Pharmacol 16: 546–555

Fowler CJ, Oreland L (1982) Human platelet MAO – some biochemical findings. In: Kamijo K, Usdin E, Nagatsu T (eds) Monoamine oxidase: basic and clinical frontiers. Excerpta Medica, Amsterdam, pp 28–39

Fowler CJ, Wiberg Å, Oreland L, Marcusson J, Winblad B (1980) The effect of age on the activity and molecular properties of human brain monoamine oxidase. J Neural Transm 49: 1–20

Fowler CJ, Wiberg Å, Gustavson KH, Winblad B (1981) Platelet monoamine oxidase in Down's syndrome. Clin Genet 19: 307–311

Fowler JS, MacGregor RR, Wolf AP, Arnett CD, Dewey SL, Schlyer D, Christman D, Logan J, Smith M, Sachs H, Aquilonius SM, Bjurling P, Halldin C, Hartvig P, Leenders KL, Lundqkvist H, Oreland L, Ståhlnacke GG, Långström B (1987) Mapping human brain monoamine oxidase A and B with ^{11}C-labelled sucide inactivators and PET. Science 235: 481–485

Francis A, Pearce LB, Roth JA (1985) Cellular localization of MAO A and B in brain: evidence from kainic acid lesions in striatum. Brain Res 334: 59–64

Geinisman Y, Bondareff W, Dodge JT (1977) Age-related loss of dendritic branches and spines in the molecular layer of the rat dentate gyrus. Anat Rec 187: 186

Glover V, Sandler M, Hughes A, Hoffbrand AV (1980) Platelet monoamine oxidase activity in megaloblastic anaemia. J Clin Pathol 33: 963–965

Gottfries CG (1985) Alzheimer's disease and senile dementia: biochemical characteristics and aspects of treatment. Psychopharmacology (Berlin) 86: 245–252

Gottfries CG, Karlsson I, Svennerholm L (1985) Senile dementia – a "white matter" disease? In: Gottfries CG (ed) Normal aging, Alzheimer's disease and senile dementia. Editions de l'Université de Bruxelles, Brussels, pp 111–118

Greenamyre JT, Penney JB, D'Amato CJ, Young AB (1987) Dementia of the Alzheimer's type: changes in hippocampal L-[³H]glutamate binding. J Neurochem 48: 543–551

Hauw JJ, Duyckaerts C, Delaere P, Chaunu MP (1988) Alzheimer's disease, amyloidosis, microglia and astrocytes. Rev Neurol 144: 155–157

Hiraga Y, Jossan SS, Gottfries CG, Regland B, Oreland L (1989) Increase in rodent brain MAO-B activity after exposure to nitrous oxide. Manuscript in preparation

Inada M, Toyoshima M, Kameyama M (1982 a) Brain content of cobalamin and its binders in elderly subjects. J Nutr Sci Vitaminol 28: 351–357

Inada M, Toyoshima M, Kameyama M (1982 b) Cobalamain contents of the brains in some clinical and pathologic states. Int J Vit Nutr Res 52: 423–429

Johnston JP (1968) Some observations upon a new inhibitor of monoamine oxidase in brain tissue. Biochem Pharmacol 17: 1285–1297

Jossan SS, Hiraga Y, Oreland L (1989 a) The cholinergic neurotoxin ethylcholine mustard aziridinium (AF64A) induces an increase in MAO-B activity in the rat brain. Brain Res 476: 291–297

Jossan SS, d'Argy R, Gillberg PG, Aquilonius SM, Långström B, Halldin C, Bjurling P, Stålnacke CG, Fowler J, MacGregor R, Oreland L (1989 b) Localisation of monoamine oxidase B in human brain by autoradiographic use of ¹¹C-labelled l-deprenyl. J Neural Transm 77: 55–64

Kang SS, Wong PW, Norusis M (1987) Homocysteinemia due to folate deficiency. Metabolism 36: 458–462

Kishimoto S, Kimura H, Maeda T (1983) Histochemical demonstration for monoamine oxidase (MAO) by coupled peroxidation method. Cell Mol Biol 29: 61–69

Kim JP, Koh J, Choi DW (1987) L-homocysteate is a potent neurotoxin on cultured cortical neurons. Brain Res 437: 103–110

Konradi C, Riederer P, Youdim MBH (1986) Hydrogen peroxide enhances the activity of monoamine oxidase type-B but not of type-A: a pilot study. J Neural Transm [Suppl] 22: 61–73

Kosaka K, Arai H, Kobayashi K, Inada M (1989) Changes of methylcobalamin in Alzheimer-type dementia brains. Dementia 3: 219–222

Lange H, Thorner G, Hoff A, Schroder KF (1976) Morphometric studies of the neuropathological changes in choreic diseases. J Neurol Sci 28: 401–425

Leung TKC, Lai JCK, Lim L (1981) The regional distribution of monoamine oxidase activities towards different substrates: effects in rat brain of chronic administration of manganese chloride and of ageing. J Neurochem 36: 2037–2043

Levitt P, Pintar JE, Breakfield XO (1982) Immunocytochemical demonstration of monoamine oxidase B in brain astrocytes and serotonergic neurons. Proc Natl Acad Sci USA 79: 6385–6389

Lin JY, Kang SS, Zhou J, Wong PWK (1989) Homocysteinemia in rats induced by folic acid deficiency. Life Sci 44: 319–325

Lindenbaum J, Healton EB, Savage DG, Brust JCM, Garrett TJ, Podell ER, Marcell PD, Stabler SP, Allen RH (1988) Neuropsychiatric disorders caused by cobalamin deficiency in the absence of anemia or macrocytosis. N Engl J Med 318: 1720–1728

Magnus EM (1986) Cobalamin and unsaturated transcobalamin values in
•   pernicious anaemia. Relation to treatment. Scand J Haematol 36: 457–465

Mann JJ, Chiu E (1978) Platelet monoamine oxidase activity in Huntington's chorea. J Neurol Neurosurg Psychiatry 41: 809–812

Mann JJ, Stanley M (1984) Postmortem monoamine oxidase enzyme kinetics in the frontal cortex victims and controls. Acta Psychiatr Scand 69: 135–139

Mann JJ, Stanley M, Rossor M, Gershon S (1980) Mental symptoms in Huntington's disease and a possible primary aminergic neuron lesion. Science 210: 1396–1371

Mann JJ, Kaplan RD, Bird ED (1986) Elevated postmortem monoamine oxidase B activity in the caudate nucleus in Huntington's disease compared to schizophrenics and controls. J Neural Transm 65: 277–283

Mantle TJ, Garrett NJ, Tipton KF (1976) The development of monoamine oxidase in rat liver and brain. FEBS Lett 64: 227–230

Maragos WF, Chu DCM, Young AB, D'Amato CJ, Penney JB (1987) Loss of hippocampal [^{3}H]TCP binding in Alzheimer's disease. Neurosci Lett 74: 371–376

Melamed E, Youdim MBH, Rosenthal J, Wester P, Winblad B (1985) In vivo effect of MPTP on monoamine oxidase activity in mouse striatum. Brain Res 359: 360–363

Murphy DL, Wright C, Buchsbaum N, Nichols A, Costa JL, Wyatt RJ (1976) Platelet and plasma amine oxidase activity in 680 normals: sex and age differences and stability over time. Biochem Med 16: 254–265

Navarro HA, Aloyo VJ, Rush ME, Walker RF (1987) Serotonin pharmacodynamics in hypothalamic tissues from young and old female rats. Brain Res 421: 291–296

Norman TR, Chiu E, French MA (1987) Platelet monoamine oxidase activity in patients with Huntington's disease. Clin Exp Pharmacol Physiol 14: 547–550

Olney J, Price M, Salles KH, Labruyere J, Frierdich G (1987a) MK-801 powerfully protects against N-methyl aspartate neurotoxicity. Eur J Pharmacol 141: 357–361

Olney JW, Price MT, Salles KS, Labruyere J, Ryerson R, Mahan K, Frierdich G, Samson L (1987b) L-Homocysteic acid: an endogeneous excitotoxic ligand of the NMDA receptor. Brain Res Bull 19: 597–560

Oreland L, Shaskan EG (1983) Some rationale behind the use of monoamine oxidase activity as a biological marker. Trends Pharmacol Sci 4: 339–341

Oreland L, Engberg G (1986) Relation between brain MAO activity and the firing rate of locus coeruleus neurons. Naunyn-Schmiedebergs Arch Pharmacol 333: 235–239

Oreland L, Gottfries CG (1986) Platelet and brain monoamine oxidase in aging and in dementia of Alzheimer's type. Prog Neuro Psychopharmacol Biol Psychiatry 10: 533–540

Oreland L, Fowler CJ, Carlsson A, Magnusson T (1980) The effect of hemitransection of rats upon the brain monoamine oxidase MAO-A and MAO-B activity. Life Sci 26: 139–146

Oreland L, Arai Y, Stenström A (1983 a) The effect of deprenyl on intra- and extraneuronal dopamine oxidation. Acta Neurol Scand [Suppl] 95: 81–85

Oreland L, Arai Y, Stenström A, Fowler CJ (1983 b) Monoamine oxidase activity and localisation in the brain and activity in relation to psychiatric disorders. In: MAO in psychiatric research. Mod Probl Pharmacopsychiatry 19: 246–254

Ozyurt E, Graham DI, Woodruff GN, McCulloch J (1988) Protective effect of the glutamate antagonist, MK-801 in focal cerebral ischaemia in the cat. J Cereb Blood Flow Metab 8: 138–143

Park CK, Nehls DG, Graham DI, Teasdale GM, McCulloch J (1988) The glutamate antagonist MK-801 reduces focal ischemic brain damage in the rat. Ann Neurol 24: 543–551

Petito CK, Navia BA, Cho ES, Jordan BD, George DC, Price RW (1985) Vacuolar myelopathy pathologically resembling subacute combined degeneration in patients with the acquired immunodeficiency syndrome. N Engl J Med 312: 874–879

Petkov VD, Stancheva SL, Petkov VV, Alova LG (1987) Age-related changes in brain biogenic monoamines and monoamine oxidase. Gen Pharmacol 18: 397–401

Ravens JR, Calvo W (1966) Neurological changes in the senile brain. In: Lüthy, Bishoff (eds) Proceedings of the 5th International Congress of Neuropathology. Excerpta Medica, Amsterdam, pp 506–512

Regland B, Gottfries CG, Oreland L, Svennerholm L (1988) Low B-12 levels related to high activity of MAO in platelets in patients with dementia disorders. Acta Psychiatr Scand 78: 451–457

Regland B, Gottfries CG, Lindstedt G (1989 a) Dementia patients with low serum cobalamins: relationship to atrophic gastritis (submitted)

Regland B, Gottfries CG, Oreland L (1989 b) Platelet MAO-B activity in senile dementia of Alzheimer type: a direct relationship to vitamin B-12 deficiency. [Abstract at the congress on "Aging of the brain and dementia: ten years later" in Florence (Italy) May 31–June 3 held by the World Federation of Neurology]

Reynolds EH, Godfrey P, Toone BK, Carney MWP (1989) S-Adenosylmethionine and Alzheimer's disease (abstract). Neurology 39 [Suppl 1]: 397

Riederer P, Jellinger K (1983) Neurochemical insights into monoamine oxidase inhibitors, with special reference to deprenyl (selegiline). Acta Neurol Scand [Suppl] 95: 43–55

Robinson DS, Nies A, Davis JN (1972) Ageing, monoamines and mono-amine-oxidase levels. Lancet i: 290–291

Robinson MB, Coyle JT (1987) Glutamate and related acidic excitatory neurotransmitters: from basic science to clinical application. FASEB J 1: 446–455

Schalling D, Åsberg M, Edman G, Oreland L (1987) Markers for vulnerability to psychopathology: temperament traits associated with platelet MAO activity. Acta Psychiatr Scand 76: 172–182

Schapiro MB, Ball MJ, Grady CL, Haxby JV, Kaye JA, Rapoport SI (1988) Dementia in Down's syndrome: cerebral glucose utilization, neuropsychological assessment and neuropathology. Neurology 38: 938–942

Schechter R, Yen SC, Terry RD (1981) Fibrous astrocytes in senile dementia of the Alzheimer type. J Neuropathol Exp Neurol 40: 95–101

Schneider G, Oepen H, von Wedel HR (1981) Aktivat in verschiedenen Hirngebieten und Körperorganen von Patienten mit Mb Huntington und Mb Parkinson. Arch Psychiatr Nervenk 230: 5–15

Schoepp DD, Azzaro AJ (1983) Effects of intrastriatal kainic acid injection of [^{3}H]dopamine metabolism in rat striatal slices: evidence for postsynaptic glial cell metabolism by both the type A and B forms of monoamine oxidase. J Neurochem 40: 1340–1348

Schrumpf E, Bjelke E (1970) Vitamin B-12 in the serum and the cerebrospinal fluid. Acta Neurol Scand 46: 243–248

Scott JM, Dinn JJ, Wilson P, Weir DG (1981) Pathogenesis of subacute combined degeneration: a result of methyl group deficiency. Lancet ii: 334–337

Siddons RC, Spence JA, Dayan AD (1975) Experimental vitamin B-12 deficiency in the baboon. In: Meldrum BS, Marsden CD (eds) Advances in neurology, vol 10. Raven Press, New York, pp 239–252

Simpson MDC, Royston MC, Deakin JFW, Cross AJ, Mann DMA, Slater P (1988) Regional changes in [^{3}H]D-aspartate and [^{3}H]TCP binding sites in Alzheimer's disease brains. Brain Res 462: 76–82

Smith RC, Ho BT, Kralik P, Vroulis G, Gordon J, Wolff J (1982) Platelet monoamine oxidase in Alzheimer's disease. J Gerontol 37: 572–574

Sprince H, Parker CM, Josephs JA, Magazino J (1969) Convulsant activity of homocysteine and other short-chain mercaptoacids: protection therefrom. Ann NY Acad Sci 166: 323–325

Stabler SP, Marcell PD, Podell ER, Allen RH, Savage DG, Lindenbaum J (1988) Elevation of total homocysteine in the serum with cobalamine or folate deficiency detected by capillary gas chromatography-mass spectrometry. J Clin Invest 81: 466–474

Stenström A, Arai Y, Oreland L (1985) Intra- and extraneuronal MAO-A and -B activities after central axotomy (hemitransection) on rats. J Neural Transm 61: 105–113

Stenström A, Hardy J, Oreland L (1987) Intra- and extra-dopamine-synaptosomal localization of monoamine oxidase in striatal homogenates from four species. Biochem Pharmacol 36: 2931–2935

Strolin-Benedetti M, Keane PE (1980) Differential changes in monoamine oxidase A and B activity in the aging rat brain. J Neurochem 35: 1026–1032

Tiggelen CJM, Peperkamp JPC, Tertoolen JFW (1983) Vitamin B-12 levels of cerebrospinal fluid in patients with organic mental disorder. J Orthomol Psychiatry 12: 305–311

Westlund KL, Denney RM, Kochersperger LM, Rose RM, Abell CW (1985) Distinct monoamine oxidase A and B populations in primate brain. Science 230: 181–183

Yatham LN, McHale PA, Kinsella A (1988) Down's syndrome and its association with Alzheimer's disease. Acta Psychiatr Scand 77: 38–41

Young AB, Greenamyre JT, Hollingsworth Z, Albin R, D'Amato C, Shoulson I, Penney JB (1988) NMDA receptor losses in putamen from patients with Huntington's disease. Science 241: 981–983

Round table on Alzheimer's disease

Participants: C.-G. Gottfries (moderator), L. Amaducci, A. Brun, M. Da Prada, D. W. K. Kay, D. L. Murphy, S. Sorbi.

Gottfries: Summarizing the day very roughly, I think we can say that we don't have yet the perfect marker for Alzheimer's disease, and that for several reasons. But what I would like to comment first is the rationale for making epidemiological studies when we can hardly define the disorder, as is the case in Alzheimer's dementia. In fact, Alzheimer's disease is a very heterogeneous disease. Therefore, what can be the rationale for making epidemiological investigations for such a disorder and how to go further from here? I wonder if Dr. Kay could start the discussion.

Kay: I really wanted to talk about something slightly different which has a bearing on this. I wonder whether I could approach this problem by making other remarks and then perhaps we can come around to discuss epidemiology. Talking about markers, I'd just like to start by a remark made by Sir M. Roth. He remarked that the most powerfully discriminating ante-mortem marker of all is the clinical diagnosis provided, of course, this is based on adequate observations and information. This, I think, certainly doesn't apply to early dementia assuming that's mild dementia. And it also seems to me that the clinical approach has been already put to the test in assessing dementia of advanced age.

I would like to mention the results obtained by Larsson and colleagues in Sweden and published in their massive work in 1963. I regard that as the start of the study of senile dementia. They did a detailed family study of quite a large number of probands, and came to the very interesting conclusion that Alzheimer's disease and senile dementia were not so connected in any way. And that, in fact, was the current idea. And then quite soon in the following years, it came to be generally agreed that it would be best to look upon these conditions as a single disorder; this was partly based on the similarities in the neuropathology and partly on the finding that the clinical features that were thought to be typical of Alzheimer's disease were also to be found in patients of senile dementia if you looked carefully enough.

And these features were the instrumental type of disorders sometimes called disorders of higher function, aphasias, agnosias, and apraxias. Subsequently, when people reported studies of senile dementia or Alzheimer's disease, they usually reported senile dementia of Alzheimer's type in which a pathology was often presumed to be of Alzheimer's type. But the other aspect, the clinical findings, often seems not to be fully described. And then the next stage, that seemed to me important, was a paper by Breitner and Folstein in which they returned to the importance of the Alzheimer's type features and proposed that agraphia would be a good thing to look at, really proposed it as a marker for a sort of genuine Alzheimer's disease, agraphia being an easy way to measure the presence or absence of this instrumental disorder.

Coming back now to the present times, what interests me very much is Prof. Gottfries' observations and also some observations made yesterday, in which we were led to consider the possibility that neither idiopathic Parkinson's disease nor idiopathic Alzheimer's disease might possibly exist; that would be an extreme position; but, at any rate, there are many difficulties before you get down to these core conditions, idiopathic Parkinson's and Alzheimer's diseases. And the suggestion that Prof. Gottfries made, was that, at any rate, if senile dementia of Alzheimer's type does exist, at least it is often misdiagnosed. Personally, I haven't really the belief that Alzheimer's disease is overdiagnosed; I tend to feel it's underdiagnosed. This raises the question of what has happened to the condition which one can call "senile dementia" not of Alzheimer's tpye. Is there really a senile dementia which is not of Alzheimer's type? I think this is of great importance for prevalence and incidence studies, because it is now quite clear that the prevalence or incidence of Alzheimer's disease increases markedly with age and reaches a high point in about the 9th decade. So it seems to me that the question we need to answer is: what proportion of late-onset cases really do have typical Alzheimer's features? What is their cause? Do they differ from those without Alzheimer's type features, and what is the neuropathology? I don't think that we have a clear answer to this question and people who have looked at this have come up with the notion that there are two types of Alzheimer's disease, type A and type B, in which there are differences depending on the age of onset or even the age of death of the patients. But they haven't made any attempt to relate this to the clinical features, as far as I'm aware.

And then, of course, if there is a case of non-Alzheimer's type senile dementia, what are the features of that condition? In what way is it similar to, or different from, age-related memory impairment?

Only a little while ago, it seemed possible that molecular genetic studies were going to show that there was a gene on chromosome 21, but now it seems that only a proportion of cases of families, and mostly those with early onset, have a gene of chromosome 21 and the rest do not. So, we seem to be right back to the position of Larsson and his colleagues. What we urgently need is further clinical, pathological, and chemical observations particularly in late-onset cases. Of course the research into the possible markers is made more difficult by the fact that Alzheimer's disease appears to be heterogeneous. But at the same time, further research into possible markers may help to provide the solution, always provided that we're getting adequate clinical description of the patients. I don't think that age of onset itself is an adequate way to separate Alzheimer's type and non-Alzheimer's type senile dementia patients. Age of onset is unreliable and not always easy to assess. What I suggest we need is to try and link up studies into markers with a careful clinical description which would include a full description of whether the patients satisfy the criteria for the classical Alzheimer's type disease.

Gottfries: Thank you very much, you really answered my question, too. Prof. Amaducci, what do you say about the epidemiologic studies?

Amaducci: This morning I showed in one slide that an etiological hypothesis stems from different sources, which are descriptive epidemiology, clinical observation, and basic research. With all those, you identify some putative etiological hypotheses that you have to test in the two methodological approaches of analytical epidemiology: retrospective case control studies and prospective longitudinal studies. However, after the case control study is completed, you may find out that your working hypothesis is wrong and you might have to start a new case control study or prospective study. Sometimes there are very simple questions that you will never be able to answer when you're doing a case control study to identify the risk factors. For instance there are traditional parameters, such as smoking or drinking; but, although the family knows how many bottles of wine they drink in the week, nobody can say who drinks the wine. In a longitudinal study it is easier to identify the right questions. I think that, if there are major prevalence data in a certain population, then you may identify the possible risk factor. But, if not, then you have to carry out very expensive, time-consuming, longitudinal studies. Sometimes it is a matter of resources, not a matter of ideas. In conclusion, as mentioned by Dr. Kay, I would say that the first sign or marker in any disease is clinical observation. There are very simple tests on psychomotor activity, that may give you very important hints that something is really going on and may be one of the indications of the beginning of the disease. The problem with those tests is that they are not yet completely standardized and the interrate reliability between clinicians is extremely poor. I think the problem of the clinicians is that they're not able to agree on the standardization of simple neuropsychological tests.

Sorbi: Just to add some of our data to the problem of heterogeneity of Alzheimer's disease, I would just like to comment some results on lactic production and glycolytic enzymes activities in leukocytes and skin fibroblasts in two families with Alzheimer's disease. In both families a genetic defect on chromosome 21 was observed. As you know, in the New England Journal of Medicine it has recently been published that lactic production may be increased in fibroblasts from Alzheimer patients. In one family, there was no change in lactic production between patients and controls, whereas in the other family there was an increase in lactic production in the patients. This shows that there is some difference between families with the same genetic lesion of chromosome 21. We also studied all the glycolytic enzymes, and found a decrease in hexokinase activity in one family, but not in the other. This shows that metabolic changes, which have been clearly demonstrated in the brain of patients with Alzheimer's disease, are not confined to the brain and also that the phenotypical expression of the genetic defect may be different within the same subpopulation of patients.

Concerning the differentiation between presenile dementia and senile dementia of the Alzheimer's type, Prof. Gottfries suggested in his lecture that there is a difference based on clinical features. I must confess that, to me, this does not appear to be a very strong argument, because many diseases manifest in a different way at different ages. One can take as examples rubella or hypothyroidism, whose manifestation is age-dependent. So, it doesn't surprise me at all if the same dementing process will manifest differently, perhaps at different rates, perhaps with different expressions, in younger and in older people. I don't think that this should be an argument for separating presenile dementia and senile dementia of Alzheimer's type. The best argument against separation of presenile and senile dementia lies in the epidemiological data that we have. Prof. Amaducci showed that if you look at the incidence rates for presenile and senile dementia, they are, in fact, continuous. There is no reason to believe there are two different processes. Of course, you can say that there is heterogeneity in each, but if there is heterogeneity, it is not age-specific according to the epidemiological data.

Gottfries: I agree with you that the onset of age is only one variable. I think, indeed, that the findings of Larsson and colleagues in Sweden really confirm that these were two age-related, independent disorders. If you study the familial aggregation of the disorders, they really behave like two disorders separated from each other. Dr. Kay have you got any comments on this?

Kay: The conclusions that Larsson and his coworkers drew are certainly that they were separate disorders. But other people have reworked the data, which they published in great detail, and have shown that you could reach an opposite conclusion. So, careful reworking of the data is compatible with the opposite, and I don't know which is the correct view; I think it's still open.

Gottfries: Prof. Amaducci, do you want to comment, too?

Amaducci: We certainly could agree that there are differences because age is only one parameter. But the explanation could be simply related to the severity of the disease. It might be that the genetic component is stronger in the early form, and that the environmental component is much stronger in the late form in the sense that phenotypic expression appears at a later time. As you know, homozygous twins have only a 60% concordance. So, I think that the environmental factors may play an important role. As we were saying this morning, the amyloid process can induce overproduction of amyloid either because of an activated gene or because environment can stimulate a promotor and eventually lead to the same result but at a later time. The severity of the disease may better explain the qualitative difference between the two forms of the disorder.

Brun: I entirely agree with Prof. Amaducci that the difference between presenile and senile dementia is not qualitative from a pathoanatomic point of view. We observe exactly the same changes in the two cases, although in the senile variety focalization is much less pronounced and it is more severe in the early form. There may be another difference, and that pertains to the white matter. The changes are more frequent in the senile variety of the disease and that may be connected with vascular factors in general.

Gottfries: May I suggest that we switch over to brain imaging. I would just like to raise this question: why has the white matter not been pointed out before? Neuropathologists have been discussing senile plaque and fibrillary tangles for decades, but the problem of white matter changes has not been tackled until the last few years. Why is it so, and how can we use these changes in brain imaging? Prof. Brun, how should we interpret the white matter changes? Are they epiphenomena or are they of pathogenetic importance?

Brun: It's a difficult question. We assume they are due to hypoperfusion in the concerned areas. They are very often combined with Alzheimer's disease. The amyloid angiopathy does not concern the white matter as a whole, but only the cortex and the meninges; but still it may influence the regulation of the perfusion of the brain. And there are also the various transmitter deficits in Alzheimer's disease, with possible disturbances of vascular innervation. So, there may be some factors connecting Alzheimer's disease and the white matter lesions. This is important because, if the white matter changes are not connected with Alzheimer's disease, there will be a common cerebrovascular lesion making most cases of Alzheimer's disease mixed cases.

Gottfries: Dr. Murphy, can we use brain imaging to separate a group which Hachinsky calls leucoaraiosis and would it be rational to do that?

Murphy: I've been talking with some people who are doing imaging studies but I'm not convinced at this point that there is enough data to do that.

Lieberman: Only the widespread use of NMR has made everybody very conscious of white matter changes. Using NMR, you see these unidentified bright objects (UBO's) in non-demented and in demented people, as well. There is no clear understanding of what these UBO's are. So, any attempt to relate these white matter changes to the ones seen in Alzheimer's disease is premature. For instance, the diagnosis of Binswanger's encephalopathy ten years ago was a rarity. No one knew it was a pathological entity, and now with NMR every fifth day you're diagnosing the Binswanger's. So, I think the first question really is: what do these UBO's represent? Until we know that, I think it's premature to make speculations about their possible involvement in Alzheimer's.

Gottfries: My question to Prof. Brun was that the neuropathologists have not found the white matter changes until now and they are not using NMR; Alzheimer discovered the senile plaque and fibrillary tangles in 1906, but Prof. Brun reported on the white matter changes in 1976. Why haven't they been discovered earlier? Don't you think that it is of value to try to delimit this group and study it, as we have studied entities according to the senile plaque and fibrillary tangles?

Lieberman: In one of the very few post-mortem studies in which a brain with a known UBO has undergone pathological examination, these UBO's were examined and the result was not very clear. Some of these UBO's were not even seen by the naked eye.

Brun: You are quite right; we don't see them with the naked eye and that's what makes our study blind and reliable. We saw that there was something wrong with the white matter. The new brains were sampled and studied biochemically and microscopically and we correlated the biochemical findings with the microscopical findings. So, you are certainly right in the sense that you don't see them grossly, and that is one reason why they have not been observed too frequently in the past. But they can be all intensities. So, I wonder to what extent NMR sensitivity allows that method to discover these changes.

Lieberman: I think it's even more complex, because it also depends upon how NMR is performed. In other words, depending upon the field strength of the magnet used, the cut used, and the various lengths of the T 2 used, these UBO's can be seen or not. So, even in the MR literature there is a lot of confusion because there is no uniformity as to how we're going to look for them. I think the first step should be to set some uniform NMR standard.

Gottfries: Anyone else who wants to comment upon white matter or brain imaging used as markers for early-onset Alzheimer dementia?

Fariello: I just wondered if anyone has any experience with ^{31}P MR imaging. I think that this is something that should give us some information in terms of the energy consumption in the brain, probably more refined than the one that we have obtained so far.

Lieberman: The problem of images with phosphorous is that there are just few institutions that do them. We haven't started to do them in a comprehensive way and there are a couple of problems. First, the resolution with phosphorous is not good. The pictures seen are very fuzzy. The results with this technique are premature.

Brun: Just a short comment. That touches upon my thoughts about the threshold levels. Can we still not definitely decide the metabolic levels in the cortex that acts normally and, on the other hand, the one area that has started to fail? I'm thinking about the Down series that we are studying. We plan to correlate the flow values determined with the xenon method with the pathology, in order to grade the severity of the encephalopathy. But, so far, we haven't reached that stage.

Gottfries: Leaving brain imaging and going to laboratory findings, I would like to ask Dr. Nordberg if lymphocytes and cholinergic receptors on the lymphocytes can be good markers to be used in clinical practice.

Nordberg: What I presented here was just the start of a study. It will be interesting to try to follow the progress of the disease in certain patients and see if there is a change in the number of cholinergic receptors. In fact, we are doing this in a small sample of patients and there is some evidence of changes in those patients who have clinical progression of the disease. It will be very important to carry out these studies on a very large number of individuals and try to divide them into different subclasses. This is just an early attempt.

Gottfries: Prof. Oreland, what about the use of platelet MAO as biological marker for some forms of dementia?

Oreland: I would like to say that times are premature to come up with a definite statement. Platelet MAO as such can hardly ever be used as a marker because there is such a great overlap between controls and patients. But, it might be interesting to investigate whether an effect of B_{12} on the slightly elevated MAO activity might be a marker for some kind of B_{12} deficiency. If so, it might be a valuable tool for diagnosis of this particular subgroup, if it exists.

Gottfries: Prof. Riederer, do you have any comments on this?

Riederer: According to my experience, I would say that platelet MAO-B activity is not a good measure. In earlier times a number of investigators have shown that there is a change in MAO-B activity in psychiatric diseases, such as depression and schizophrenia. I think, MAO-B platelet activity would be an unspecific marker. It would be nice to know why MAO-B activity is increased in such conditions, but as long as we don't know that, I wouldn't use it as a marker.

Strolin-Benedetti: I remember some papers concerning MAO-B activity in women compared to men. And as today I was impressed by Prof. Amaducci's remark concerning the prevalence of Alzheimer's in white females and even more in black females, in order to answer the question of whether MAO activity in platelet is a reliable parameter, I wonder if MAO-B activity has now been better studied in the different male and female, white and black population.

Oreland: Unfortunately, I don't know anything about black and white, females and males. But to reply to Prof. Riederer, I would say that, in my mind, a correlation between B_{12} levels and platelet MAO activity is quite different from the previous results which established a correlation between personality traits and platelet MAO activity. In the latter circumstances platelet MAO activity obviously expressed its native levels. We have for example genetic heritability factor of 0.98 in a twin study. But the relation between MAO-B activity and B_{12} is interesting because, if we supplement this deficiency of B_{12}, we could use the change in activity as a measure.

Murphy: Just in response to Dr. Strolin-Benedetti's question, I would like to say that in our study of 680 normals, which we undertook about $10-12$ years ago, we did find very definitely a 13% or so higher values of platelet MAO activity in females, and we didn't find differences between black and white females. I think one way to think about this on a continuum hypothesis idea, is that perhaps there is an aggregation of risk factors. If MAO should in fact be considered a potential risk factor, as it is something that does increase with age and is higher in Alzheimer's patients, it could be an additive factor with age. But, concerning blood cell markers in general, one of the reasons why we gave up trying to use them is that many parameters can influence the production and manifestation of activities in blood cells; except for the cases, that we've just studied in the last year, of individuals who lack the gene for MAO and have totally zero MAO-B activity, in whom I think it's a definite marker, anything else is going to be in a hazy field for some time.

Youdim: As Dr. Murphy said, trying to measure MAO in thrombocytes is extremely difficult because you have to control so many factors. I'll give you one example. We measured platelet monoamine oxidase activity and expressed it in terms of serum iron. We got a very good correlation between serum iron level and MAO activity. But there is something very odd about MAO-B. We don't know what are the endogeneous substrates for this enzyme. MAO-B metabolizes phenylethylamine and maybe dopamine, but also many other substrates like N-acetylputrescine.

Histamine is not metabolized by MAO-B but N-methylhistamine is a substrate. We have recently found that only MAO-A activity is regulated by steroids, while MAO-B activity has defied all attempts to be regulated by any sort of hormones. What actually regulates it? Looking at the developmental aspects of MAO-A and -B in the brain and other tissues of the rat, MAO-A activity appears well before MAO-B activity. So, there is something regulating these enzymes, which we don't know. Therefore, it may well be that some factor regulates the enzyme in Alzheimer's disease. This could possibly be some sort of marker.

Riederer: Coming back to Prof. Oreland's last statement, I wish to say that we cannot confirm the data on vitamin B_{12} deficiency in Alzheimer patients. None of our patients showed that. I would agree that MAO-B estimation would be a nice parameter in correlation with vitamin B_{12}, if there is a dose-dependent change in MAO-B activity with this vitamin. Is there any evidence of that?

Gottfries: I am surprised that you didn't find any vitamin B_{12} deficiency in your material of Alzheimer's dementia. Did that material include patients with an onset above the age of 75? As I said, it was in this group we found the decreased levels of vitamin B_{12}. As shown by Prof. Oreland, when we measured MAO-B in patients with pernicious anemia, with very severely reduced vitamin B_{12}, MAO-B activity was 3 or 4 times higher than in the controls and went steeply down to normal when compensating this deficiency. This really confirms that there is a correlation between megaloblastic anemia and MAO activity. In fact, we're not searching for a marker for Alzheimer's disease; what we are searching for is a marker for a subgroup of senile dementia, where we assume that there are disturbances in the form of atrophic gastritis, which cause a reduced absorption of vitamin B_{12}

and, eventually, a reduced absorption of other essential nutrients. So, I very much agree with the one who said: are these Alzheimer's dementia? I don't think they are. It would be better to call them secondary dementias, but they fulfill the criteria we today use for senile dementia of Alzheimer's type.

Oreland: I would just like to say that maybe we shouldn't over-emphasize the importance of MAO; to us, the importance of this MAO finding is that it has opened up the possibility to further investigate the role of B_{12} in this deficiency.

Kay: I was just going to say how interesting was this discussion about vitamin B_{12}. In the UK, the studies that have been done have tended to come to the conclusion that vitamin B_{12} deficiency in patients with senile dementia are usually regarded as secondary to the senile dementia and there has been, in most cases, no improvement with treatment. What I wanted to ask was: have these patients, or some of them, shown to have pathological changes typical of Alzheimer's disease in the brain? That seems to be the critical question. And I wonder whether I could ask another question: Prof. Nagatsu, unlike in other countries, in Japan multi-infarct dementia is said to be commoner than Alzheimer's disease. Can one say that this is due to an excess of multi-infarct dementia or is it due to a deficiency of Alzheimer's dementia?

Gottfries: Before Prof. Nagatsu answers your question, I would like to say that the total percent of dementias is the same in Japan and in other countries.

Nagatsu: In Japan it is said that 70% of dementias are of vascular type and 30% are of Alzheimer's type, which is different from what you are finding in USA and Europe. However, the percent of cases of Alzheimer's type dementia is really increasing in Japan, possibly due to the change in the diagnostic criteria used. It seems now that 50% of dementias is really of the Alzheimer's type in Japan.

Gottfries: I would like to come back to your question, Dr. Kay. Regarding malnutrition, as I said before, folic acid was not reduced in the demented patients and that's usually the best marker for mal-nutrition. These patients had no clinical signs of malnutrition. So, we don't think that is an explanation to the finding. Concerning the improvement of the patients with B_{12}, many have claimed that, by giving vitamin B_{12}, the patients do not improve. But there are case reports indicating that there are improvements. I think that if we have a disorder due to vitamin B_{12} deficiency, you have come so far as to have a degeneration, for instance of white matter, causing a dementia. Then you are so far away down the road that you cannot expect that this disorder is reversible; or you have to treat the patients during a very long time to estimate a possible improvement. Take for instance

Korsakoff's syndrome: as you know, this is due to thiamine deficiency; however, when you give thiamine, the patients are not improved at all. The question about Alzheimer's changes in crucial. We are following our patients and, hopefully, we will be able to answer that question.

Da Prada: I am just wondering whether MAO-B activity plays a role in Alzheimer's disease and I would like to ask Dr. Murphy if he has observed some improvement in Alzheimer patients treated with MAO-B inhibitors. I agree with Prof. Youdim when he said that we don't know the role of this enzyme and even its distribution in the brain. We have performed some autoradiographical studies with a very selective MAO-B inhibitor and the distribution is very puzzling. We found a nice localization in very specific regions, different from that of MAO-A; for instance, in the neurohypophysis there is only MAO-B.

Murphy: In some studies, which have now been published, based on the idea that deprenyl as an MAO-B inhibitor might be safer to use in depressed Alzheimer patients, we in fact did observe some behavioural improvement in those patients without hypotension or other problems. Probably because we were wondering about the importance of the increase in MAO-B with age and in Alzheimer's, and also to examine whether simply a treatment trial in the absence of any other effective drugs, essentially as a way to continue following our patients and having them on some form of treatment, we initiated a study, six months of deprenyl vs. six months of hydergine, and then we will cross over for a two-year period of time; and we won't know anything until this September. But we will be interested in whether there may have been any evidence of a plateau. In this case, we would use this model study for future studies, with what, I hope, might be more effective drugs in the future.

Bianchine: Dr. Herbert, who has worked with vitamin B_{12} for a long time, has spoken of the importance of the vitamin B_{12} transport system. He has defined that as vitamin B_{12} capacity. Could it be that there is a transport trouble with vitamin B_{12} in this setting? After all this is an inducible protein. Could there be an inhibition of this protein? My question then is: have you looked at this aspect of B_{12}?

Gottfries: We are going to look into that issue in a cooperation with a research group in Norway where they can measure the transcobalamins, the carrying proteins. They argue that it may be a question of reduced carrying proteins in these patients. I very much agree with you when you say there might be a reduced transport of vitamin B_{12}. The transport from the gut into the blood is decreased and, according to the pepsinogen findings, there is an artrophic gastritis, as is expected in these cases. But the transport into the brain over the blood-brain barrier, which also is an active transport, must also be investigated. Now we have a method to measure vitamin B_{12} in CSF. According to our preliminary findings, there is a group which is still lower in CSF, indicating that the serum vitamin B_{12} is not a very good marker. Therefore, we also have to investigate the CSF for its content.

Oreland: I would just like to reemphasize that just a few months ago a Japanese group reported on a study which was initiated by our results. They found a lower concentration of methylated B_{12} within the brain tissue in demented patients.

Gottfries: I think we'll stop the discussion about MAO and vitamin B_{12}. It's exciting but more work has to be done there. Now, I would like to close this session. Thank you to the panel members for their cooperation and the discussions and to the people of the auditorium.

Closing remarks

R. Guillemin

Distinguished Scientist

The Whittier Institute for Diabetes and Endocrinology,
La Jolla, California, USA

As you all know I am not directly working in the field of the studies of this Symposium. The closest I ever came to that was after we had isolated somatostatin some years ago, synthesized it, made antibodies to it and distributed those antibodies to quite a few colleagues, when it was found that somatostatin was co-localized with the dopaminergic neurons in the basal nuclei and when, later on, the group at Albert Einstein reported that somatostatin appeared to be dramatically decreased in its concentration in the cortex of patients with Alzheimer's, while other kinds of neuropeptides were not. So, I come here only as the neuroendocrinologist that I am and I am sure that you will have already thought yourselves about much of what I will say regarding this symposium.

My closing remarks will be very brief. First of all, I personally think that the purpose of such a meeting was highly meaningful. I personally think and agree that, establishing a consensus about early markers of Parkinson's and Alzheimer's diseases is very important, not only in terms of early diagnosis, but also in terms of the early application of therapies which, I also understand, are still to come, but will come. One should thus be prepared for such a time.

In my opinion, this is obviously a field in which clinicians and basic science people will have to work together in great harmony to ascertain and expand the current knowledge on these possible early markers. It seems to me also that, practically, for early markers to be useful to the clinicians, which is the whole idea, they will have to be relatively simple, as well as relatively inexpensive, in other words easily available and affordable. In this context, I think that the powerful physical methods, like positron emission tomography and the MRI's in their present and future forms will probably come as a second step, probably in confirmation and refinement as the powerful methodology that they are, being involved in the clinical definition or classification of what these early markers will have recognized, or observed.

It seems to me that the ideal methods for early markers will thus be mostly based on relatively simple measurements of one molecule or another in CSF or in plasma, or, more likely, rather than one molecule, several molecules. Before I say a few more words about these, let me say that I was impressed by so many statements made by several speakers about the heterogeneity, the multiple forms of both Parkinson's and the dementias. We heard about Parkinson's diseases and senile dementias and Alzheimer's type dementias and so on. Thus probably some of these early markers will be multiple molecules, I repeat, which in a second step, will lead either to physical methods or to more specific molecular markers for one type of disease or the other.

It seems to me that some major efforts should be directed to understand much better the control of the function of some of the genes on chromosome 21, for reasons known to all of you and as elegantly recalled during this meeting by Dr. Delabar.

I was surprised that so little was said about the constituents of the Alzheimer's plaque. The molecular biology of the amyloid-β-protein and its precursor as such, is now very well known if not totally understood. I'm referring particularly here to the overall contribution from the group at the NIH around Carleton Gajdusek. One marker I would look for myself, if I were busy at the laboratory in this field, is the calcitonin gene-related peptide, CGRP, or a closer small peptide (those can easily be synthesized with many variants, easily generating antibodies). I'm referring to the recent discovery, again by Gajdusek's group, of the constant presence of this small peptide or a related very close homologue of CGRP as a fragment of all precursors of the amyloid-β-protein. Moreover, there's even the more recent discovery, just a few months ago, of the amino acid sequence of this new peptide recently characterized in the pancreas, called amylin, apparently co-secreted with insulin by β-cells, and which is a constituent of the amyloid proteins of the pancreas and, at the end of the syndrome, of the spleen. With an 80% homology between CGRP, calcitonin and amylin, it is hard for me to believe that such a small peptide would not be involved in the constitution of the amyloid plaques and, to me, this would constitute a relatively easy marker to look into, both in plasma and, more likely, in CSF. These small peptides can be measured by highly quantitative radio-immunoassays, which are already available or can be available in the next few months. The same proposal could be made for the hybridization of the mRNA of these peptides. It may also well be that these small peptides are more important than I've just intimated in the etiology of the dementia, if the proposals by Gajdusek's group are confirmed, which would make these a new type of nucleic acid free viruses, to induce a sort of crystallization or triggering point for the polymerization of the proteins in the senile or Alzheimer plaques.

Let me also, in closing, drop a small pebble or plant a small seed in the minds of all of you. That of the existence of neurosteroids, this new class of steroids that are unquestionably made in the brain, specifically in the brain, probably by some neurons, certainly by glial cells. The role of these neurosteroids in the etiology or in the mechanism of either Parkinson's or Alzheimer's dementias is totally unknown at the moment. To my knowledge, no-one has even started studying the field or asking questions related to them.

Well, coming back to the first sentence of my concluding remarks, I think that a meeting like this was highly meaningful and very well worth all the efforts of organizers and participants. It was also well conducted, well organized. There are few meetings of this sort, where as much discussion and intercourse between the speakers and the audience actually take place as it did here. I personally enjoyed all parts of it. Thank you very much.

Subject index

A4 protein
  in AD and Down patients 168
A68 protein
  in AD brains 140
acetylcholine 246
  potassium-evoked release in
  SDAT brain tissue 251
  in SDAT 139
acetylcholine esterase (AChE)
  CSF activity in dementias 158
  in Down syndrome 166
  in PD 76
  in SDAT 139, 192, 248
acoustic reaction time test in PD 51
S-adenosylmethionine
  in CSF of AD patients 279
ADR 529 128
AF64A
  effect on MAO activity 270
aiming test in PD 42
alcoholism
  malnutrition and 210
aluminium and AD 152
Alzheimer's disease (AD)
  differentiation from SDAT 155,
  158
  environmental factors in 173
  familial AD 173
  genetic factors in 156, 172
  overexpression of chromosome
  21 genes in 173
  parieto-temporal syndrome in 157
D,L-α-aminoadipate (DL-AA) toxicity 213

γ-aminobutyric acid (GABA) 234
L-2-amino-4-phosphonoburyrate
  (L-AP4) 235
D,L-aminophosphonovalerate
  (APV) 214
amyloid-β-protein 300
anthracyclines
  cardiotoxicity and iron 128
anticholinergics
  in postencephalitic PD 15
APP gene
  in AD and Down syndrome 167,
  174
  expression in brain structures in
  AD 168
arecoline
  in SDAT 141
aspartate aminotransferase 238
aspartic acid 234
  in cortex of SDAT patients 238
auditory evoked potentials (AEP)
  comparison with PET and
  SPECT 207
  P300 component 198
  in SDAT 197

beta 2-microglobulin
  anti-human antibodies 72
  in PD 72
Binswanger's disease 19, 189, 292
biopterin and PD 71
bradyphrenia in PD 27
budipine 3
butyrylcholine esterase

in PD 76
in SDAT 249

calcitonin 299
calcitonin gene-related peptide 300
calcium homeostasis in SDAT 239
catechol-0-methyltransferase
    (COMT) inhibitors
    in PET scanning 62
    in PD 152
CGS 19755
    effect on experimental brain
    ischaemia 278
m-chlorophenylpiperazine
    in SDAT 141
chlorpromazine 117
choline
    plasma content in SDAT 247
choline acetyltransferase (CAT)
    in Down syndrome 166
    in relation to PDH 234
    in SDAT and PD 139, 225, 246
cholinesterase activity
    in SDAT and PD 248
choline uptake
    in SDAT 140
chromosome 21 300
    and AD/SDAT 151, 290
    Down syndrome and 167
    gene localization on in Down
    syndrome 169
citrate synthase 216
computer tomography (CT)
    in Binswanger's disease 19
    in the diagnosis of dementias 142,
    157, 185, 188
    in Fahr's disease 18
    in MSA 17
    in Pick's disease 18
    in Wilson's disease 18
copper
    in Wilson's disease 17
creatine kinase

isoenzymes 211
    serum levels in myocardial infarc-
    tion 211
    after exercise and head injury 212
Creutzfeldt-Jacob disease 189
cystathionine β-synthetase 278

1,2-dehydrosalsolinol
    urinary levels in healthy subjects
    96
dementia
    association with aging 159
    benign forms 159
    clinical characteristics 147
    causes of 148
    vascular dementias 148, 159
l-deprenyl 273
depression, in association with PD 27
desferrioxamine 117, 123, 126
    in rheumatoid arthritis 128
diadochokinesis test in PD 51
differential scanning calorimetry 225
3,4-dihydroxybenzoic acid 118
5,6-dihydroxytryptamine 127
dimercaprol
    in Wilson's disease 18
dipeptidyl peptidase II and IV acti-
    vities in PD 76
L-dopa (levodopa)
    in the diagnosis of PD 13
    in MSA 16
    in postencephalitic PD 15
    prediction of response to by MR
    scan 103
    in PSP 16, 66
    use in PD 134
dopamine (DA)
    brain tissue levels in PD and
    SDAT 223
    synthesis and tissue levels in PD
    patients 86
dopamine agonists
    use in PD 134

dopamine beta-hydroxylase (DBH)
 in PD 71
 anti-human antibodies 73
dopamine-neuron antibody
 in patients with adrenal medulla-
 to-brain transplantation 32
 in PD and other neurological
 disorders 32
dopamine receptors
 and PD 134
dopamine sulfoconjugates
 DA 3-0- and 4-0-sulfates in plasma
 80
 DA 3-0- and 4-0-sulfates in urine
 82
 effect on vascular α-adrenergic
 receptors 89
 origins 80, 87
 ratio in PD and depressed patients
 84
 stability 85
 storage functions 89
 synthesis in brain after MAO
 inhibition 87
Down syndrome
 absence of visible karyotype
 anomaly in 170
 brain lesions in 166
 chromosome 21 genes in 167
 clinical characteristics 166
 dementia and 166, 172
 expression of trisomy 21 in 166
 glutathione peroxidase activity in
 167
 partial trisomy 21 and 169

electro-encephalography (EEG)
 differential diagnosis of dementias
 with 157
enthalpy
 changes in SDAT and PD brain
 tissues 227

excitatory amino acids
 neurodegeneration and 277

Fahr's disease
 differential diagnosis from PD 18
ferritin 115, 126
2(^{18}F)-fluoro-2-deoxy-D-glucose
 (FDG) 181
L-^{18}F-fluoro-dopa
 use in PET studies in PD 61, 129,
 187
 brain uptake in PD 63
frontal lobe dementia 160

Giessen-Test personality inventory
 24
glial cells
 involvement in cytotoxicity 215
global field power (GFP)
 in P300 identification 199
glucose metabolism (brain)
 differential diagnosis of dementias
 with 181
 effect of hypoglycemia on 238
 effect of muscarinic cholinergic
 agonist on 192
 effect of piracetam on 192
 effect of phosphatidylserine on
 193
 factors influencing 182
 in Huntington's disease 185
 in mammalian brain 234
 in Parkinson's disease 187
 in Pick's disease 185
 regional differences 183, 191
 in SDAT 185, 192, 237
 in vascular dementias 189
glutamate dehydrogenase (GDH)
 association with OPCA 215
 forms of 215
 in human leukocytes 216
 in rat brain 216
glutamic acid 234
 in cortex of SDAT patients 238

plasma levels in OPCA 216
receptors in AD brains 277
role in brain damage 213
glutamine 234
glutaminergic receptors
brain localization 235
glutamine synthetase
effect of DL-AA and APV on 214
glycine 236
Guam Parkinson dementia complex
126
A68 protein in 140

haemosiderin 115
Haller-Varden-Spatz disease 113
herpes encephalitis 189
hexokinase 234
L-homocysteic acid
possible involvement in neurodegeneration 278
homovanillic acid (HVA)
brain tissue levels in PD and
SDAT 223
CSF levels in dementias 158
Huntington's chorea 19
differential diagnosis from AD/
SDAT 160
genetic factors in 210
6-hydroxydopamine 123
hydroxyl radical
role of iron in the formation of
113
5-hydroxyindolacetic acid (5-HIAA)
brain tissue levels in PD and
SDAT 225
CSF levels in dementias 158
8-hydroxyquinoline 117
hypoglycemia
effect on brain amino acid levels
238

immunocytochemical reactivity in
PD 32

immunoglobulins
presence and abnormalities in
various degenerative diseases 32
in SDAT CSF, correlation with
BBB function 263
insulin
effect on PDH and CAT activities
236
effect on phosphoenolpyruvate
carboxykinase 236
receptors 236
receptors in olfactory structures
240
iron
chelation of, effect on lipid peroxidation 123
chelators 117
consequences of iron overload
118
correlation with MR signal
attenuation 101
distribution in human brain structures 101, 113
in PD 108, 112
effect on the binding of MPTP
and MPP+ to melanin 124
effect of DA on iron binding to
melanin 116
ferric and ferrous ions in PD 113
serum levels and MAO activity
295

kainic acid 215
effect on MAO activity 270

lactate dehydrogenase 216
lazaroids 123
leucoaraiosis 158, 292
Lewy bodies
in MSA 17
in postencephalitic PD 15
in tissues of PD patients 86
line tracing test in PD 42

lipid peroxidation
    in the SN of PD patients 117
    in vivo measurement 125
lipoic acid 118
long-latencies reflexes in PD 50

Machado-Joseph disease 16
Madopar
    effect on the ratio of DA sulfo-
    conjugates in PD 83
manganese
    role in PD 118, 126
melanin
    as a radical scavenger or promoter
    114
    iron binding to 115
    MPTP binding to 116
memory impairment
    associated with aging and SDAT
    159
methionine
    effect of vitamin B-12 deficiency
    on 276
    role in myelin production in the
    white matter 275
methionine synthase 271
N-methyl-D-aspartate (NMDA) 213,
    215, 236
    receptors in SDAT and Hunting-
    ton's chorea 277
    receptor subtypes 277
1-methyl-4-phenyl pyridinium ion
    (MPP+)
    binding to melanin 116
^{11}C-methyl-spiperone
    brain uptake in PD 66
    use in PET studies 66
1-methyl-4-phenyl-1,2,3,6-tetrahy-
    dropyridine (MPTP) 64, 111
    binding to melanin 116, 124
    effect on the binding of iron to
    melanin 124
    and parkinsonism 123, 126, 229

MK-801
    effect on experimental brain
    ischaemia 278
monoamine oxidase (MAO)
    activity in AD/SDAT brain 272
    A- and B-activities in brain tissues
    269
    activity in Down syndrome 274
    A- and B-activities after experi-
    mental lesions 270
    activity in Huntington's chorea
    273
    activity in patients with megalo-
    blastic anemia 276
    activity in PD 274
    activity in thrombocytes of AD/
    SDAT patients 273
    changes in activity with aging 114,
    268
    extra- and intra-neuronal activi-
    ties 268
    platelet activity in dementias 141,
    158, 294
    subtypes 267
movement time test in PD 51
multi-infarct dementia (MID)
    differential diagnosis from AD/
    SDAT 159
    differentiation by CT and MRI
    189
multiple system atrophy (MSA)
    differential diagnosis from PD 16
muscarinic receptors
    in lymphocytes of SDAT and PD
    patients 249
    in SDAT 246

nasal epithelium abnormalities in
    SDAT patients 239
nerve growth factor (NGF) 1
    antibodies to in AD 141
nicotine
    in SDAT 141

uptake in SDAT brain 251
nicotinic receptors
    changes in SDAT 246, 249
    in lymphocytes of MID patients 249
nigral-subthalamic-pallidal atrophy 16
nitrous oxide
    effect on vitamin B_{12} 271, 275
[11]C-nomifensime
    brain uptake in PD 65
    striatal uptake after MPTP 64
    use in PET studies 63
noradrenaline (NA)
    brain tissue levels in PD and SDAT 223
nuclear magnetic resonance (NMR)
    in arteriopathic parkinsonism 19
    in Binswanger's disease 19
    in the detection of Parkinson Plus syndrome 103
    in MSA 17
    signal attenuation in PD 99
    in dementias 157, 188, 292
    in Wilson's disease 18

olfactory tests deficits in SDAT patients 239
olivopontocerebellar atrophy (OPCA) 16, 215
o-phenanthroline 117
oxidative stress
    disorders associated with 112
    in PD 113, 229
oxygen cerebral metabolic rate in SDAT 237

P300 recordings 198
    in elderly subjects 206
    in epilepsy 204
paired helical filaments (PHF)
    antibodies against in Down syndrome and AD 166

in SDAT 233
paraquat 229
Parkinson's disease (PD)
    arteriopathic parkinsonism 19
    brain levels of neurotransmitters in 223
    CAT activity in brain tissues 225
    classification of parkinsonisms 14
    clinical signs 6, 14
    cognitive slowness in 27
    differential diagnosis of 42, 101
    dopamine sulfoconjugates urinary excretion in 79
    drug-induced parkinsonism 15
    early diagnosis of 5, 46, 49, 129
    electrophysiological methods in 49
    genetic factors in 219
    immunological changes in 75
    instrumental methods for an early diagnosis of 7
    lipid peroxidation in 113
    motor disorders in 41, 49
    motor performance tests in 41, 50
    MR signal attenuation in putamen in 101
    "nigral" and "striatal" types 107
    noradrenergic neuron changes in 75
    obsessive compulsive behaviour and 28
    postencephalitic parkinsonism 14
    post-intoxication parkinsonism 19
    premorbid personality traits in 24
    role of environmental factors in 126
    smoking behaviour and 25
Parkinson Plus syndrome 99
penicillamine
    in Wilson's disease 18
phenolsulfotransferase (PST)
    forms 87
    in human platelets 88

role in the metabolism of brain
DA 87
phenylethanolamine-N-methyltransferase (PNMT) in PD 71
phosphatidylserine 193
3'-phosphoadenosine-5'-phospho-
sulfate 87
phosphofructokinase 234, 237
in SDAT and Down patients 239
physostigmine 192
effect on choline release 252
Pick's disease
A68 protein in 140
differential diagnosis from AD/
SDAT 160
differential diagnosis from PD 18
piracetam 192
platelet membrane fluidity (PMF)
in SDAT 140
plugging test in PD 42
positron emission tomography
(PET)
in dementias of various origins
189
in Huntington's chorea 186
measurement of human dopaminergic function with 59
with (+) and (−) ¹¹C-nicotine in
SDAT 251
in Pick's disease 185
in SDAT 185, 190, 247
posttraumatic encephalopathy 189
progressive supranuclear palsy (PSP)
A68 protein in 140
differential diagnosis from PD 16
proline endopeptidase activity in PD
76
pseudobulbar palsy 19
pyruvate dehydrogenase (PDH)
activation by Ca²⁺ 234
role in brain glucose metabolism
234
in SDAT 237

pyruvate kinase 234

quinolinic acid 215, 278
3-quinuclidinyl-4-iodobenzilate
(QNB) 142, 250

¹¹C-raclopride
striatal uptake after MPTP 67
radionuclides
use in PET 60
renin activity in PD 86
RS86 192

salsolinol
biosynthesis in humans 93
enantiomeric composition in
human urine 85
in PD patients 94
scopolamine
in SDAT 141
senile dementia of the Alzheimer type
(SDAT) 210
A68 protein in 140
as a subcortical dementia 157
brain atrophies in 157
CAT activity in brain tissues in
225
clinical characteristics 139, 157
CSF antibrain antibodies in 257
diagnostic criteria 149
differential diagnosis 137
differentiation from AD 155, 158,
287
differentiation from depression
160
epidemiology of 148
genetic factors in 151
genetic markers of 138
head trauma and 152
incidence rate 150
MAO activity in 271
neurotransmitters brain levels in
223, 247

parkinsonism associated with 160
PMF in 140
prevalence 149
regional brain degeneration in 222
smoking and 152
vitamin B_{12} serum levels in 276, 295
Shy-Drager syndrome 16
single photon emission computed to-
    mography (SPECT)
    differential diagnosis of dementias
    with 142, 157
somatostatin 299
    in SDAT 140
steadiness test in PD 42
Steele-Richardson-Olszewsky syn-
    drome 16
substantia nigra (SN)
    degeneration in PD 114, 221
superoxide dismutase Cu/Zn (SOD_1)
    Down syndrome, AD and 167
    gene mRNA expression in fibro-
    blasts of AD and Down patients
    175

tapping test in PD 42
tetrahydroaminoacridine (THA) 192
    effect on choline release 252
thermostability
    changes in SDAT and PD brain
    tissues 227
thiamine 210
tracking test in PD 51
tremor in PD 50
tremor recording 51
trisomy 21 mosaicism 168, 172
tyrosine hydroxylase (TH) in PD 71

ubiquinone 124
UK4006F 118, 123

visual reaction time test in PD 51
vitamin B_{12} 209
    correlation with MAO activity
    277, 294
    deficiency and experimental neu-
    ron degeneration 275
    deficiency in relation to white
    matter changes 275
    levels in brain of demented
    patients 275
    and SDAT 295
    serum concentration in dementias
    158
    syndromes caused by deficiency of
    276

Wernicke-Korsakoff syndrome 209
white matter
    MAO-B activity in, in AD/SDAT
    272
    changes in dementias 157, 275,
    291
Wilson's disease 128
    differential diagnosis from PD 17
Wisconsin Card Sorting Test 26

xenon regional cerebral blood flow
    (rCBF)
    differential diagnosis of dementias
    with 142, 157
    differentiation between SDAT
    and depression with 160

K. Maurer, P. Riederer,
and H. Beckmann (eds.)

Alzheimer's Disease.
Epidemiology, Neuropathology,
Neurochemistry, and Clinics

(Key Topics in Brain Research)

1990. 118 figs. (9 in colour). XIX, 581 pages.
Soft cover DM 176,–, öS 1230,–
ISBN 3-211-82197-X

Prices are subject to change without notice

The book "Alzheimer's Disease – Epidemiology, Neuropathology, Neurochemistry, and Clinics" is derived from an International Symposium on the occasion of the 125th Anniversary of Birth of Alois Alzheimer (14. 6. 1864–19. 12. 1915).
Over the past decade, as the elderly have become the fastest growing segment of the population in industrial countries, Alzheimer's disease has emerged as one of major mental health problems. The contributors to this book represent internationally recognized authorities in the field of dementia and present new information about epidemiology, neuropathology, neurochemistry, and clinics in Alzheimer's disease. This book comprises a rich and valuable up-to-date resource for psychiatrists, neurologists, scientists working in the field of neuropathology, neurochemistry and molecular genetics, behavioral scientists, family physicians and all who share an interest in understanding and treating the older individual with Alzheimer's disease/dementia.

Springer-Verlag Wien New York

Horst Przuntek, Peter Riederer (eds.)

Early Diagnosis and Preventive Therapy in Parkinson's Disease

(Key Topics in Brain Research)

1989. 59 figures (1 in color). XIV, 442 pages.
Soft cover DM 135,–, öS 950,–
ISBN 3-211-82080-9

Preisänderungen vorbehalten

At the time when "Parkinson's Disease" is diagnosed in a patient roughly two thirds of dopaminergic neurons of substantia nigra are already degenerated. Therefore, the onset of the disease must be much earlier. This book deals with early diagnosis and early preventive treatment which may sustain the process underlying the disease.

By use of psychometric, kinesiologic, physiological, histological, biochemical, endocrinological, pharmacological and imaging techniques including positronemission tomography and brain mapping specialists tried to focus new diagnostic criteria. New methods including psychometric evaluation, apparative measurement of movement, analysis of peripheral blood and urinary constituents have supplemented this approach. Early preventive therapy has been agreed to consist of low dosis L-DOPA plus benserazide, L-deprenyl and dopaminergic agonists.

Springer-Verlag Wien New York